Gerhard Fritz
Eberhard Matern

Carbosilanes

Syntheses and Reactions

With 40 Figures and 74 Tables

Springer-Verlag
Berlin Heidelberg NewYork Tokyo

Prof. Dr. rer. nat. Dr. h. c. Gerhard Fritz
Dr. rer. nat. Eberhard Matern

Institute of Inorganic Chemistry
University of Karlsruhe
P.O. Box 6380
D-7500 Karlsruhe 1
Federal Republic of Germany

ISBN 3-540-15929-0
Springer-Verlag Berlin Heidelberg New York Tokyo

ISBN 0-387-15929-0
Springer-Verlag New York Heidelberg Berlin Tokyo

Library of Congress Cataloging in Publication Data.

Fritz, Gerhard, 1919–. Carbosilanes: syntheses and reactions.
Bibliography: p. 1. Silane. I. Matern, Eberhard, 1945–. II. Title.
QD412.S6F75 1986   547!08   86-1808
ISBN 0-387-15929-0 (U.S.)

Offsetprinting: Ruksaldruck, Berlin
Bookbinding: Lüderitz & Bauer, Berlin
2152/3020-543210

Foreword

Carbosilanes are compounds in which the elements silicon and carbon alternate in the molecular skeleton [1]. Just as the alkanes are formally derived from the diamond lattice and the aromatics from the graphite lattice, the carbosilanes are structurally derived from silicon carbide. Because of the tetravalent nature of silicon and carbon we can expect stable linear, cyclic and polycyclic compounds to occur. However, carbosilanes do not exist in nature.

This book is an attempt to give a summarized presentation. Carbosilanes are, of course, part of organosilicon chemistry, but their behavior differentiates them distinctly from other organosilicon compounds. The differences result primarily from the alternating $Si-C-Si$ arrangements in the molecular skeleton, and especially the various methylene bridges (CH_2, CHX, CX_2; $X = $ halogen) cause changes in $Si-C$ bond polarization and hence influence the reaction possibilities. It is convenient to regard carbosilanes as similar to silicones except that the oxygen bridges of silicones are replaced by methylene units. However, this does not accurately account for all the chemical properties of these compounds. Carbosilanes are related more directly to silicon carbide, as shown occasionally by the reactive behavior of polycyclic carbosilanes. Therefore, in view of the present interest shown for thermally stable ceramics of unusual character, interesting possibilities arise for further development.

Most cyclic carbosilanes can be classified in two groups: the carborundanes and the Si-scaphanes. Compounds belonging to the carborundane class maintain $Si-C$ six-membered rings in the boat conformation.

It appeared useful in the text to designate compounds by arabic numerals or in a somewhat abbreviated fashion, for example as 1,3,5,7-tetrasilaadamantane, in referring to a particular basic structure, still in accordance with IUPAC nomenclature.

Initial access to the carbosilanes resulted essentially from the gas phase pyrolysis of methylsilanes. The wide spreading development of this research is presented along the following lines and employed as an organization scheme. After a description of the various means for obtaining these compounds, an account of their chemical reactivity and of the influence of substituents on the $Si-C-Si$ skeleton is given. This is followed by a summary of the results of X-ray crystal structure determinations of selected characteristic compounds.

The identification of compounds centers essentially around NMR spectroscopic and mass spectrometric techniques. Detailed and general conclusions resulting from these investigations are not discussed here, nor can a comprehensive description of the important separatory techniques be included.

It is not the intention of the authors, nor is it even possible in this text to summarize the present situation in the broad field of organosilicon chemistry. Rather, emphasis is placed on a comprehensive account of the chemistry of carbosilanes carried out during the last 35 years in order to hopefully provide a favorable basis for further experimental development, as well as for better theoretical understanding.

The results to be presented here could not have been achieved without the contributions of many co-workers and PhD students whose names are stated in the list of references. The senior author wants to thank all those who participated in the research described, especially Dr. H. Scheer (mass spectrometry) and Mr. H. Domnick (NMR spectroscopy). He gratefully acknowledges Prof. Dr. H. G. von Schnering and his research group for their fruitful and pleasant cooperation over many years.

Furthermore this research would not have been possible without the support of "Deutsche Forschungsgemeinschaft" and "Fonds der Chemischen Industrie".

We would also like to thank Dr. R. Payne for translating the German manuscript, and we are greatly indepted to Prof. Dr. E. G. Rochow who went over the full text improving the readability and the spelling to American standards.

Karlsruhe, December 1985   G. Fritz · E. Matern

Contents

I. Introduction

1. Characteristic Differences between the Elements Silicon and Carbon

Before considering compounds in which silicon and carbon in an alternating fashion are used to build the molecular skeleton [1], it is useful to give a short account of the characteristic differences between the properties of the elements. Diamond is a non-conductor whereas graphite is a semiconductor. The element silicon, which is not freely available in nature and is produced by reduction of SiO_2, crystallizes in the diamond lattice and is a semiconductor. Silicon carbide is also a semiconductor. Pure monocrystalline silicon has increasing technical importance because of these properties. It is produced appropriately from $HSiCl_3$ by reduction with H_2, or through thermal decomposition of SiH_4.

Elementary silicon is also of principal importance to the covalent chemistry of silicon, because all pathways to such silicon compounds involve the element itself. Of course, the silicate systems are distinctly separate. Their properties and uses arise from the high strength of the silicon—oxygen bond. To illustrate the differences between related compounds of carbon and silicon, some bond energies and bond distances are shown in Table 1. More recent compilations give considerably higher Si—X bond energies [66].

A comparison of the Si—Si and C—C bond energies and lengths exemplifies the greater stability of the C—C bond. Similar differences exist between C—H and Si—H bonds, in contrast to which the exceptional stability of the Si—O bond is noteworthy. It is understandable, then, that SiO_2, silicic acid or silicates are usually the compounds generated from reactions involving silicon compounds in the presence of air and water. Carbon is the primary building block element of living systems, and chemical reactions within living systems are based on the fact that C—O, C—C and C—H bond energies are close enough so as to allow enzymatically-controlled reaction pathways. It is then immediately obvious from the differences in the bond energies of Si—H, Si—Si and Si—O that analogous processes cannot occur in silicon chemistry. The high stability of the Si—O bond indicates why silicon is the primary building

Table 1. Bond energies (kJ/mol) [38] and bond lengths (pm)

C—H	414 kJ/mol	Si—H	314 kJ/mol	Si—C	314 kJ/mol	C—C	154 pm
C—O	355 kJ/mol	Si—O	460 kJ/mol			Si—C	187 pm
C—C	334 kJ/mol	Si—Si	·196 kJ/mol			Si—Si	234 pm

block in inanimate systems, as for instance the silicates of the earth's surface. However, the significance of silicon compounds in life processes has been affirmed [2].

A further appreciation of the difference between carbon and silicon compounds can be obtained on consideration of the respective elemental electronegativities: $Si = 1.8$, $C = 2.5$, $H = 2.1$. Accordingly, the H atom in the $Si-H$ bond will maintain a hydridic character, while the hydrogen in the $C-H$ bond has a positive partial charge. The Si atom in the $Si-C$ bond is somewhat more positive than the C atom. Hence it can be understood that $Si-H$ containing substances in alkaline solution produce H_2 on formation of $Si-OH$, while the $C-H$ bond under these conditions remains unaffected. Due to the different electronegativities, silicon compounds, for example silicon halides, are generally more polar than their corresponding carbon analogues.

The differing electronic configurations of silicon and carbon must also be taken into consideration. Silicon d orbitals are arranged in such a way as to make possible the formation of a pentacovalent or hexacovalent state. The hydrolytic cleavage of the $Si-Cl$ bond, for example, may proceed through a pentacovalent transition state. The existence of silicon compounds with coordination numbers 5 and 6 can be substantiated through the same reasoning [3].

The variety of reactions possible for any element arises mainly from the number and properties of its functional groups. To name just one important branch of the many possibilities in the chemistry of carbon, the

$$-C{\overset{O}{\diagup}}_{\diagdown H}\ ,\quad -C{\overset{O}{\diagup}}_{\diagdown R}\ ,\quad -C{\overset{O}{\diagup}}_{\diagdown OH}\ ,\quad -C{\overset{O}{\diagup}}_{\diagdown Cl}\quad\text{and}\quad -C{\overset{O}{\diagup}}_{\diagdown NH_2}$$

groups shall be mentioned here.

The number of functional groups associated with the element silicon is considerably smaller, and confined primarily to the Si-halogen, $Si-H$, $Si-OH$, $Si-OR$, $Si-N$ and $Si-P$ groups. The $Si-OH$ group tends to eliminate H_2O and form $Si-O-Si$ bonds provided no steric reasons prevent this condensation. Silicon is not capable of forming stable $Si=O$ double bonds.

The chemistry of carbon is determined to a considerable degree not only by the functional groups mentioned above, but also by its ability to form molecules with stable single or multiple $C-C$ bonds as well as aromatic systems. These classes of compounds enable a large number of reaction possibilities to be classified. The element silicon is neither in the position to build molecules with comparably stable $Si-Si$ bonds, nor under comparable conditions to form double or triple bonds as is possible for carbon. Reports claiming the synthesis of compounds with $Si=C$ [4, 75] and $Si=Si$ [5] bonds eventually set a conclusion to the discussion of the possible existence of such compounds:

The stability of the shown compounds depends upon the space requirement of the substituents. Proof of existence for the silaethylene $R_2C=SiMe_2$ in an argon matrix at $-180\ ^\circ C$ was obtained; under different conditions the silaethylene dimerizes [6]. Most of the numerous reaction mechanisms postulated to involve such double bonds at intermediate stages need a better experimental foundation. Substantial differences between multiply-bonded carbon and silicon compounds have been established, and it is obvious that the latter cannot reach a comparable importance.

2. Functional Groups in the Chemistry of Silicon

Functional groups on silicon are of great importance for the chemical behavior of carbosilanes. The properties and behavior of these groups will be summarized here.

The most widely used functional groups are the Si halogen groups (F, Cl, Br, I) [81].

Through these groups and especially through Si—Cl groups, reactions with organometallic reagents build important synthetic pathways in organosilicon chemistry, such as

$$\equiv SiX + RMgX \rightarrow\ \equiv Si-R + MgX_2$$

The hydrolytic cleavage of the Si—Cl group causes formation of Si—OH groups, giving rise to a class of compounds called silanols

$$\equiv Si-Cl + H_2O \rightarrow\ \equiv Si-OH + HCl$$

The Si—OH group of the silanols shows an extreme tendency to undergo condensation, enabling the formation of siloxanes and silicones

$$2\equiv SiOH \xrightarrow{-H_2O}\ \equiv Si-O-Si\equiv\ \rightarrow \left(-\overset{|}{\underset{|}{Si}}-O-\right)_n$$

Reaction of Si—Cl groups with amines yields silylamines

$$\equiv Si-Cl + HNR_2 \rightarrow\ \equiv Si-NR_2 + HCl$$

Comparable to the Si—OH group is the Si—OR group, which can be formed by reaction of Si—X with alcohols. This group reacts with organometallic reagents

$$\equiv Si-OR + R'MgX \rightarrow\ \equiv SiR' + ROMgX$$

and enables the formation of siloxanes through condensation

$$2\equiv Si-OR \rightarrow\ \equiv Si-O-Si\equiv\ + ROR$$

The SiH group containing a hydridic H-atom is most readily formed by the reaction of silicon halides with $LiAlH_4$. It oxidizes rapidly and undergoes reaction with water in alkaline medium to produce hydrogen. Alcohols react analogously

$$\overset{}{\underset{}{>}}Si-H + H_2O \rightarrow\ \overset{}{\underset{}{>}}SiOH + H_2$$

The Si—H group also readily undergoes nucleophilic substitution by reaction with LiR, is readily attacked by radicals, can be metallated to yield R_3SiLi and reacts with carbonyl hydrides

$$\text{$>$}SiH + HCo(CO)_4 \rightarrow \text{$>$}Si-Co(CO)_4 + H_2$$

Catalytic addition of the Si—H group to multiple C—C bonds is widely used [83]:

$$-C\equiv C- + HSiR_3 \xrightarrow{\;H_2PtCl_6\;} \begin{array}{c} \diagdown \quad \diagup \\ C=C \\ \diagup \quad \diagdown \\ H \qquad SiR_3 \end{array}$$

In this context also those groups have to be considered, which on reaction are converted into functional groups on silicon.

Disilanes are cleaved by bromine:

$$\text{$>$}Si-Si\text{$<$} + Br_2 \rightarrow 2\ \text{$>$}SiBr$$

The Si—P group in silylphosphanes reacts with HBr [84]:

$$\text{$>$}Si-P= + HBr \rightarrow \text{$>$}SiBr + HP=$$

Vinyl groups bound to silicon will act as shielding groups, because addition of HBr or Br_2 yields a Si—Br bond by β-elimination [85]

$$\text{$>$}Si-CH=CH_2 + HBr \rightarrow \text{$>$}Si-CH_2-CH_2-Br \rightarrow \text{$>$}SiBr + H_2C=CH_2$$

It is apparent that cleavage of the Si—Ph group with Br_2, I_2, HBr or HI is a prerequisite for the organometallic synthesis of carbosilanes and for the preparation of halosilanes containing Si—H groups, as in H_3SiBr [68]

$$H-\overset{|}{\underset{|}{Si}}-Ph + HBr \rightarrow H-\overset{|}{\underset{|}{Si}}-Br + C_6H_6$$

This often enables Si—Ph groups to be used as shielding groups in organometallic synthesis [60]. As more electronegative substituents are introduced onto silicon, cleavage is greatly retarded, and is eventually completely prevented.

Thus Ph_2SiH_2 reacts with HI (2 mols) to yield PhH and SiH_2I_2. However, Ph_2SiCl_2 does not react at all with HI [68].

Cleavage in such cases complies with the use of $HCl/AlCl_3$, as is evident from the formation of $(SiCl_2)_5$ derived from $(SiPh_2)_5$ [69].

The halogenation of Si—H groups determines the reaction of SiH-containing phenylsilanes with halogens. As long as Si—H groups are present in the molecule, halogenation will occur if the cleavage of the Si—Ph groups with the HX formed is not faster. The chlorination of SiH is extremely fast. For this reason, the influence of negative substituents on silicon becomes so large that cleavage is completely suppressed [68].

II. The Formation of Carbosilanes

The synthetic routes known to date to form carbosilanes are based on the gas phase pyrolysis of methylsilanes, on the reaction of CH_2Cl_2 and $CHCl_3$ with silicon (Cu-catalysed), on the reaction of Si-methylated carbosilanes with $AlBr_3$ or $AlCl_3$, and on organometallic syntheses.

1. The Formation of Carbosilanes by Thermal Decomposition of Methylsilanes

1.1 The Homolytic Cleavage of Silanes

From consideration of the bond energies Si—H: 314 kJ/mol and C—H: 414 kJ/mol, one would expect the thermal stability of SiH-containing compounds to be considerably lower than that of CH-containing compounds. Indeed, SiH_4 decomposes thermally to elemental silicon and hydrogen, but CH_4 does not at the same temperature. The course of this reaction is relatively complex, as yellow SiH-containing polymers are formed at about 400 °C together with hydrogen [7]. The formation of SiH_3 and hydrogen radicals in the first reaction step is proposed according to the following equation:

$$SiH_4 \rightarrow SiH_3 + H$$

Formation of Si_2H_6, H_2 and (to a lesser degree) higher silanes was achieved by passing SiH_4 through a reaction tube heated to 470–500 °C [7].

It is also known that higher silanes are produced by thermal decomposition of Si_2H_6 but will readily deteriorate at approximately 500 °C. Polymeric SiH-containing compounds, for example the hydrolysis products from CaSi of approximate composition $[SiH_2]_x$ [8] and the siloxen obtained from $CaSi_2$ of composition $Si_3O_{3/2}H_6$ [9], decompose at about 200 °C and 300 °C respectively, yielding H_2 and SiH_4 [10]. The formation of SiH_4 has been suggested to occur through radical reactions.

Consecutive reactions of silyl and hydrogen radicals with other molecules can only occur if reactivity and lifetime of these radicals are sufficient to enable these successive reactions to predominate over the further decomposition of the radicals to Si and H_2. Such a reaction occurs between SiH_4 and hydrocarbons above the decomposition temperature of SiH_4, and results in the formation of heterogeneous organosilicon compounds. Saturated and unsaturated hydrocarbons react with SiH_4

in this way. The compound types produced from the reaction of SiH$_4$ with ethylene at 450 °C are shown in Table 2 [10].

Table 2. Reaction products from SiH$_4$ and ethylene

1	2		3
MeSiH$_3$	Si$_2$C$_9$H$_{24}$:	Si$_2$Me$_3$(C$_2$H$_5$)$_3$	(SiMe)$_x$
C$_2$H$_5$SiH$_3$		Si$_2$HMe(C$_2$H$_5$)$_4$	
Me$_3$SiH		Si$_2$H$_2$(C$_2$H$_5$)$_3$(C$_3$H$_7$)	
Me$_4$Si	Si$_3$C$_{12}$H$_{32}$:	Si$_3$Me$_4$(C$_2$H$_5$)$_4$	
MeH$_2$SiC$_2$H$_5$	Si$_4$C$_{16}$H$_{42}$:	Si$_4$Me$_4$(C$_2$H$_5$)$_6$	
MeH$_2$SiC$_3$H$_7$	oil		
(C$_2$H$_5$)$_2$SiH$_2$			
(C$_2$H$_5$)$_3$SiH			

The reaction products are arranged into 3 distinct classes according to their structure. In the first class of compounds, one or more hydrogens from SiH$_4$ are replaced by an organic group. The compounds of the second group have high boiling points, are oily, viscous and tar-like in nature, and to some extend have not been clearly identified. These compounds generally contain Si—Si bonds; however, there may also a carbon atom be included in the molecular chain, as in Si—Si—C—Si. Isomers appear also which differ primarily in the number of Si—H bonds and in the position of the organic substituents. The third group contains (SiMe)$_x$, a lemon-yellow solid polymer in which each Si atom is directly bound to three other Si atoms while the remaining valence is occupied by a methyl group [11]. This substance resembles that of composition (SiCl)$_x$. Analogous compounds formed from reaction of SiH$_4$ and vinyl chloride [12], naturally contain SiCl groups, as in MeSiHCl$_2$ and (C$_2$H$_5$)$_2$SiCl$_2$. High molecular weight products also arise, namely, (Si$_2$ClR)$_x$, analogous to (SiMe)$_x$. The reaction time has a considerable influence on the resulting products. Their composition changes with increasing reaction time in such a way that higher molecular weight compounds are preferred.

The experimental data obtained allow the following reaction course to be drawn. Figure 1 shows the variation of pressure with time for the reaction of SiH$_4$ and C$_2$H$_4$.

Curve I (Fig. 1) shows the decay of SiH$_4$ which coincides with an increase in pressure. Under the same conditions for C$_2$H$_4$ alone, no comparable change in pressure is detected (Curve II). A mixture of SiH$_4$ and ethylene, under identical conditions, reacts with decreasing pressure (Curve III). No polymeric ethylene compounds were detected. The reactions begin with the decay of SiH$_4$. This is confirmed by the results presented in Fig. 2. Ethylene reacts not only with SiH$_4$ producing organosilicon compounds but also with polymeric (SiH$_2$)$_x$ and siloxen. The reaction course is identified through a pressure decrease at the temperature at which these SiH-containing compounds decompose, producing SiH$_4$ and H$_2$.

The intermediate presence of H atoms is evident from organosilicon products containing a carbon count in the organic moiety differing from that of the hydrocarbon starting material, for example MeSiH$_3$ or C$_3$H$_7$MeSiH$_2$. This means that a cleavage of the C—C bond must have occurred but which, on consideration of reaction con-

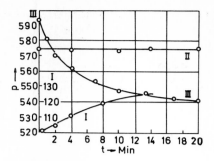

Fig. 1. $SiH_4 + C_2H_4$. Reaction Temperature 460 °C. $p(C_2H_4) = 60\%$; $p(SiH_4) = 40\%$. I SiH_4; II C_2H_4; III $SiH_4 + C_2H_4$

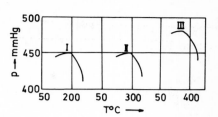

Fig. 2. Curve I $= (SiH_2)_x + C_2H_4$. Curve II $=$ Siloxen $+ C_2H_4$. Curve III $= SiH_4 + C_2H_4$

ditions, was not achieved thermally and must arise from well-known dehydrogenating and chain cleaving effects of the H atoms [13].

The distribution of compounds in the pyrolysis mixture depends on decomposition temperature and lingering time in the reaction tube, in such a way that the higher the temperature and the longer the lingering time, the higher is the tendency to form compounds of higher boiling point and greater insolubility (Table 3).

Table 3. Influence of reaction time on reaction of SiH_4 with ethylene

No.	Reaction time	Temperature	Percentage of products	
	min	°C	Gases	Liquids + solids
1	60	450	46	54
2	80	450	16	84

Reactions which occur through the formation of free atoms and radicals can also be initiated by other means. For instance, the reaction between SiH_4 and ethylene can also be initiated by irradiating at 254 nm using photochemical sensitization by means of mercury vapor [13, 86].

A further example, which demonstrates how thermal decomposition of SiH_4 can be utilized in reactions with other hydrides, is shown by the formation of H_3Si-PH_2 and other silicon-phosphorus compounds from SiH_4 and PH_3 [14].

It is recognized from the reaction of SiH_4 and polymeric SiH-containing compounds already described that subsequent reactions with volatile hydrides are indeed feasible, which in turn can lead to a new class of compounds. Because the use of SiH_4 on a preparative scale seemed too problematic, it was originally of interest to establish whether similar reactions could be generated through thermal decomposition of SiH-containing alkylsilanes in the temperature range of 400—500 °C. The thermal decomposition of $EtSiH_3$, Et_2SiH_2 and Et_3SiH can easily be measured in the gas phase above 450 °C. Gaseous, liquid and low volatile reaction products were observed when this thermal decomposition was carried out in the presence of C_2H_4 under static conditions [15]. Obviously, the homolytic cleavage of the $Si-H$ bond initiated the reactions in all cases. However, reaction products other than those arising from the decomposition of SiH_4 could be encountered from the subsequent reactions of silyl radicals with organic groups. Furthermore, these products are constructed in a more complicated manner than those derived from the reaction of SiH_4 with ethylene.

1.2 Continuous Flow Pyrolysis

Quite similar to the thermal $Si-H$ cleavage of alkylsilanes R_xSiH_y $(x + y = 4)$ yielding silylradicals and initiating a series of consecutive reactions, an equivalent synthesis should be possible starting from $SiMe_4$ or the methylchlorosilanes Me_xSiCl_y $(x + y = 4)$, provided the decomposition temperature in the gas phase is exceeded. It has been established that the thermal decomposition of methylchlorosilanes at 800 °C/100 mm Hg produces not only gaseous components but also a mixture of liquid and high-boiling-point compounds with Me and Cl substituents on silicon [16]. The decomposition products of $SiMe_4$ at 700 °C (100–200 mm Hg, 2–5 min. reaction time) contain gaseous, liquid and solid compounds of which the cyclic compounds $(Me_2Si-CH_2)_3$, $(Me_2Si-CH_2)_4$ and the crystalline 1,3,5,7-tetrasilaadamantane were readily identified [17].

It was ascertained from these observations that very complicated subsequent reactions result from the thermal decomposition of $SiMe_4$ and of the methylsilanes in the gas phase, in which compounds are formed containing a molecular skeleton of alternating Si and C atoms. Because a kinetic investigation into the complexities of these reactions would have little chance of success without previous knowledge of how these compounds are formed, and secondly, the influence of reaction conditions, the investigations were directed towards the preparation and identification of reaction products.

For the preparative formation of pyrolysis products it was necessary to provide an arrangement in which the gases involved flowed continually through the apparatus, the lingering time of these gases in the reaction zone lasting approximately one minute. The results are shown in Table 4 [18, 19, 20].

The apparatus used is schematically depicted in Fig. 3. The reaction tube (Rotosil opaque fused silica, diameter 12 cm, length 80 cm) is located inside a furnace heated to 720 ± 5 °C. Liquid $SiMe_4$ is stored in a flask K in front of the reaction tube. This flask is connected to the previously evacuated apparatus so that the pump P will direct vaporized $SiMe_4$ (bp. 26 °C) into the reaction tube in the direction of the arrows. Soon after reaction began, white fumes appeared below the reaction tube and condensed in the product traps F_1, F_2, F_3 at room temperature. Later yellow molten products dripped from the reaction tube, from which crystals separated on the walls of the glass tubing underneath the furnace. At the same time H_2, CH_4, C_2H_4 and C_2H_6 formed. Due to the vapor pressure of $SiMe_4$ (720 mm Hg at 20 °C) and those of the individual product gases, atmospheric pressure was reached after 2–3 hours. Any excess pressure was released by means of a mercury bubbler which also served as a manometer. The vapor pressure of $SiMe_4$ in the system remained relatively constant, provided a constant supply of liquid $SiMe_4$ was present in the storing flask. In this

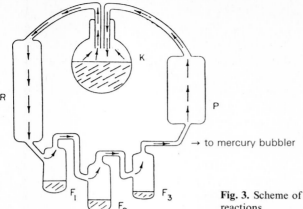

→ to mercury bubbler

Fig. 3. Scheme of apparatus used in decomposition reactions

way, it was possible to maintain constant conditions within the system. During the total reaction period, $SiMe_4$ and compounds of higher boiling point already produced were driven through the reaction zone in a ratio governed by their individual vapor pressures, together with small amounts of H_2, CH_4, C_2H_6, C_2H_4 and SiH-containing silanes. The increased pressure in the system caused volatile products to escape. These consisted of hydrogen, hydrocarbons of low molecular weight, and small amounts of $SiMe_4$. The latter was recovered. To achieve the results presented in row 3 of Table 4, the apparatus was in continual use for 3 weeks. During this time approximately 50% of the $SiMe_4$ was converted. In order to achieve reproducible results, the reaction conditions had to be rigidly maintained.

Table 4. Yields of liquid and solid silicon compounds from pyrolysis of $SiMe_4$ at 720 °C; various lingering times

	1	2a	2b	3a	3b	4a	4b	5a	5b
	ling. time (min)	ml. substance Bp. 90–180 °C 760 mm Hg	Vol. %	ml. substance Bp. 120–180 °C 100 mm Hg	Vol. %	ml. substance Bp. 90–300 °C 1 mm Hg	Vol. %	ml. substance Solid, Insoluble	Vol. %
1	3	8	20.5	10.7	27	6	15.4	14.5	37.2
2	2	12.2	23.3	7.7	14.8	21.9	42	10.5	20.2
3[a]	1	38.9	15.6	27.4	11	143.6	57	40	16

[a] Streaming

1.3 The Thermal Decomposition of $SiMe_4$

1.3.1 The Separation of Pyrolysis Products

During the investigation of reaction products, considerable difficulty was encountered in separating the reaction mixtures [21], so a detailed study was made. Several litres of pyrolysis product were available. From the reaction mixture, 18 fractions were separated in the boiling point range from 45 °C (760 mm Hg) to 280 °C (10^{-3} mm Hg). In these, some 45 different compounds were detected by preliminary gas chromatographic investigation [22]. Their respective contributions to the total product mixture were between 0.01 and 10.0%. Of these substances only 12 were present in higher than 1% contribution to the total product mixture, 9 between 0.5 and 1.0%, 11 between 0.1 and 0.5%, and 13 under 0.1%. These 45 compounds comprised approximately 53% of the total product mixture. The other 47% consisted of high-molecular-weight silicon compounds. Hydrocarbons up to hexane in the alkane homologous series were found in the liquid fraction, and also benzene, which comprised 0.9% of the total mixture and was the most common hydrocarbon component.

Apart from traces of SiH_4 and Me_2SiH_2, the gaseous mixtures formed in the pyrolysis consist of the following compounds in the given ratios [18]:

$$H_2 : CH_4 : C_2H_4 : C_2H_6 : C_3H_6 : C_3H_8 : Me_3SiH = 100 : 80 : 4 : 4 : 4 : 2 : 6 .$$

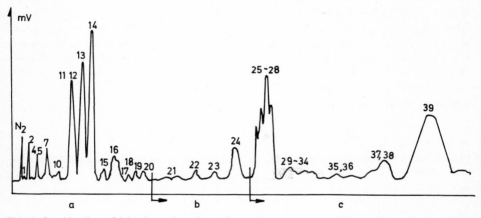

Fig. 4. Combination of 3 isothermal sections of gas chromatograms of the product mixture of the $SiMe_4$ pyrolysis in the boiling point range of 20 °C to 200 °C/760 mm Hg [18]

1.3.2 Compounds of Low Boiling Point Containing up to Four Si Atoms

Separation of the volatile compounds was achieved using gas chromatography (Fig. 4). The structural formulae of the numbered compounds are shown in Table 7, together with their percentage contribution to the total product mixture (Tables 5 and 6).

An inspection of the compounds in Table 7 shows that they contain only a few basic structures. The valencies of silicon which are not fixed in the carbosilane molecular skeleton are linked to methyl groups and (to a minor degree) to H atoms. A notable result was that only one compound (34) with a Si—Si bond occurred.

Table 5. Linear silanes produced by the pyrolysis of $SiMe_4$

No.	Formula	Percentage (Weight %)	No. in Fig. 4	Bp. °C 760 mm Hg
1	$H_2C=CH—SiMe_3$	0.014	4	56–54.4
2	$Me—CH_2—SiMe_3$	0.022	5	60.5–63
3	$HC\equiv C—SiMe_3$	0.003	6	52
4	$PhSiHMe_2$	0.14	23	157.5–158
5	$PhSiMe_3$	2.5	24	172–170.6
6	$Me_3Si—CH_2—SiMe_3$	6.7	14	132.3–133
7	$Me_3Si—CH_2—SiHMe_2$	3.52	12	120.5
8	$Me_2HSi—CH_2—SiHMe_2$	0.17	10	105–107
9	$Me_3Si—CH_2—SiH_2Me$	0.05	9	98
10	$Me_3Si—CH_2—SiH_3$	0.001	9	91
11	$Me_2HSi—CH_2—SiH_2Me$	0.003	8	88
12	$Me_2HSi—CH_2—SiH_3$	>0.001	6	70
13	$MeH_2Si—CH_2—SiH_3$	>0.001	3	—
14	$C_2H_5Me_2Si—CH_2—SiMe_3$	0.04	20	159.5–159
15	$C_2H_5Me_2Si—CH_2—SiHMe_2$	0.003	19	—
16	$H_2C=CH—SiMe_2—CH_2—SiMe_3$	0.14	19	155
17	$Me_3Si—CH_2—CH_2—SiMe_3$	0.54	16	149.5–150
18	$Me_3Si—CH=CH—SiMe_3$	0.009	15	145–145.5
19	$Me_3Si—C\equiv C—SiMe_3$	0.12	15	133–135
20	$Me_3Si—CH=CH—CH_2—SiMe_3$	0.07	21	169–171
21	$Me_3Si—CH_2—SiMe_2—CH_2—SiMe_3$	3.54	28	204.5–206

Table 6. Cyclic carbosilanes produced by the pyrolysis of $SiMe_4$

No. of compound (cf. Table 7)	Percentage (Weight %)	No. in Fig. 4	Bp. °C 760 mm Hg
22	5.33	13	122–123
23	0.47	12	~110
24	0.009	10	97–99
25	0.65	16	139–138.5
26	0.39	15	126
27	0.008	14	
28	0.11	20	152
29, 30	0.018	19	
31	0.069	18	148
32	0.005	11	105.5
33	0.074	21	161
34	0.23	22	165
35	4.57	27	201.5–202
36	3.30	26	198
37, 38	1.66	25, 25	193
39	12	39	mp. 117 °C

Table 7. Cyclic compounds produced by the pyrolysis of $SiMe_4$

22

23

24

25

26

27

28

26

30

31

32

33

34

35

36

37

38

39

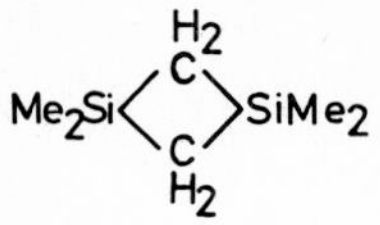

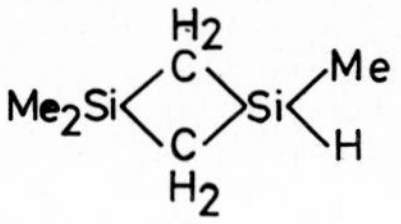

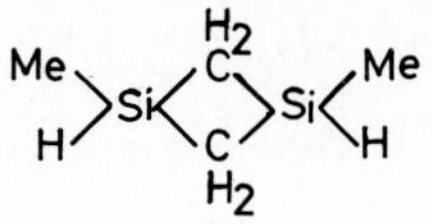

1.3.3 Compounds with Polycyclic Molecular Structures

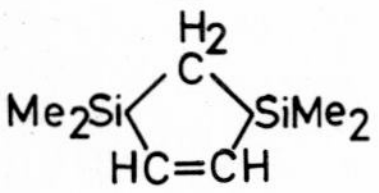

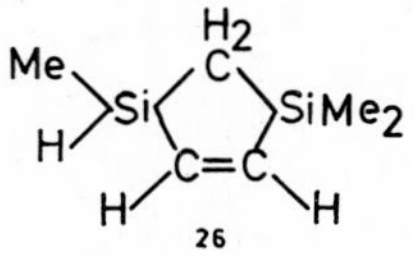

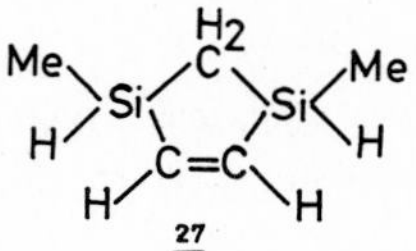

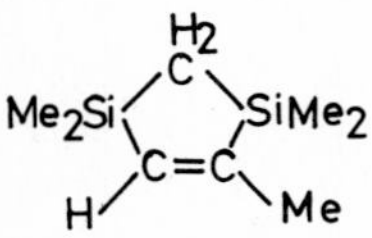

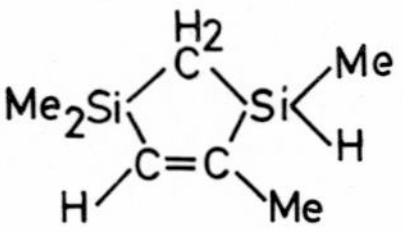

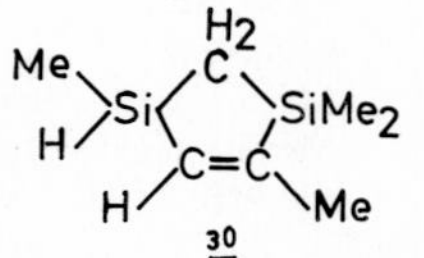

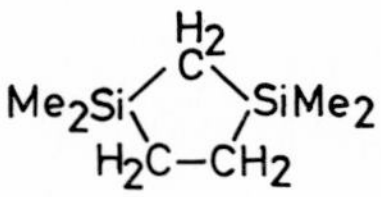

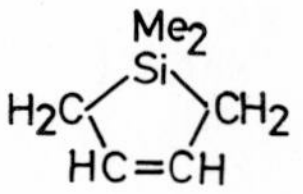

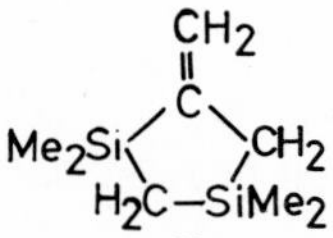

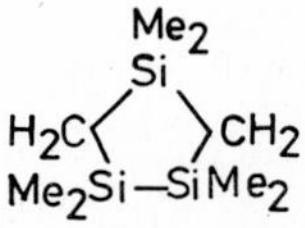

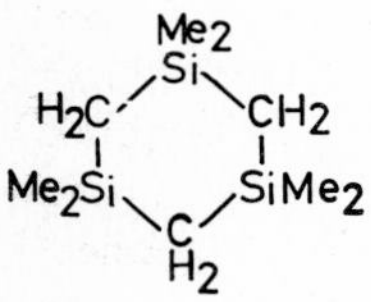

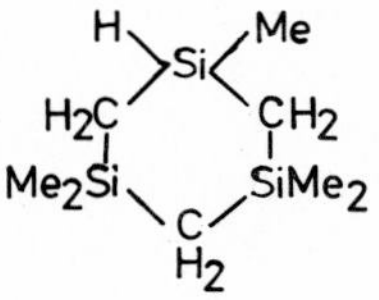

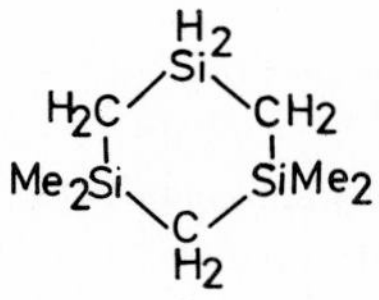

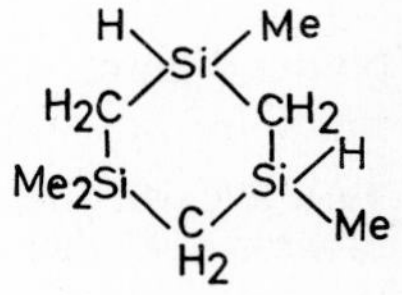

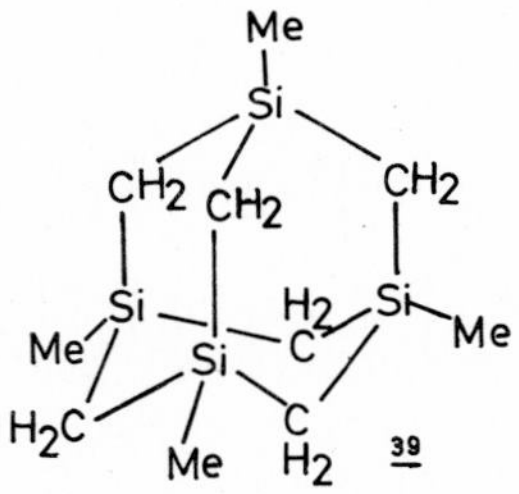

Considerably more difficulties were encountered in overcoming the problem of separation and identification of the compounds of high molecular weight. Eventually, such product mixtures could be separated at least into groups of compounds by column chromatography on Al_2O_3 according to their absorption ability. From these fractions, several readily crystallizing compounds were isolated and identified through NMR spectroscopy and mass spectrometry [23, 24].

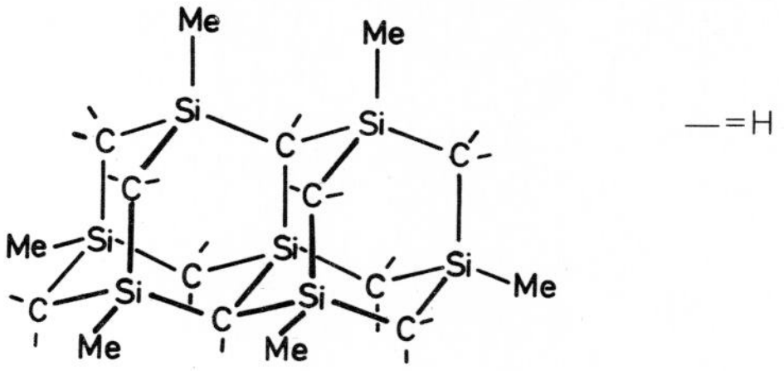

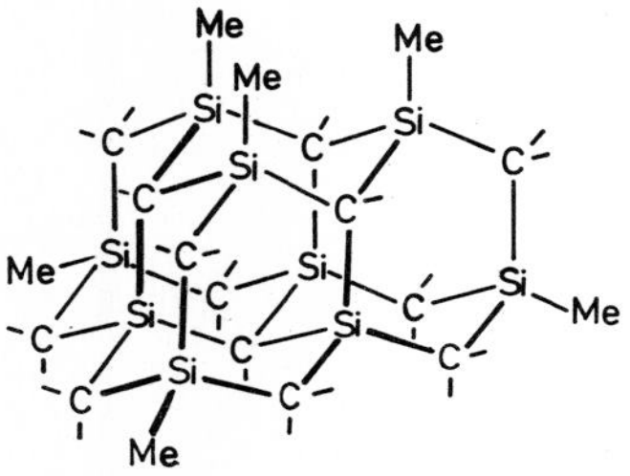

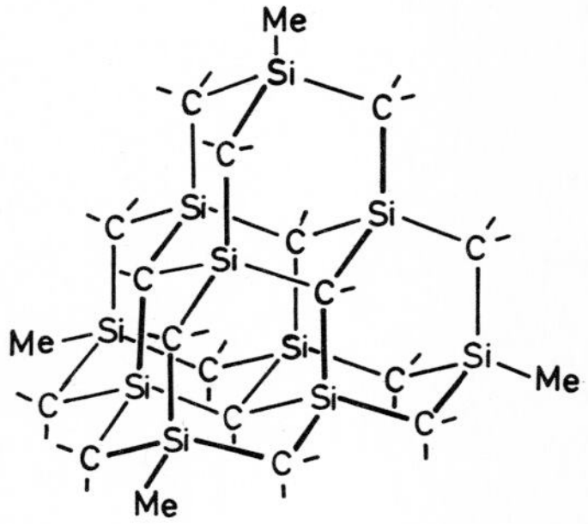

Because thermal after-treatment changed the pyrolysis mixture by forming more high-boiling products, isolation of pure compounds could only be achieved by combining different separatory methods. The first coarse separation by column chromatography was followed by molecular distillation. Those fractions, which contained only partially separated compounds, were resolved further by a Craig distribution process, and then by gel-chromatographic means. In this way, a reasonable concentration even of compounds with higher silicon counts (molecules with Si_4, Si_5, Si_6, Si_7 etc.) could be obtained in different fractions. Using gas chromatography, eventually the compounds shown in Table 8 were isolated [25, 26, 190, 27].

Although this separation led to further elucidation of the pyrolysis products, the separation of compounds by gas-chromatographic means did not proceed without difficulty. For instance, not all of the existing compounds could be adequately isolated by this method, and furthermore, the method cannot be applied to mixtures of low volatile, silicon-rich compounds of high molecular weight. Therefore in the subsequent stages of development of these separatory methods, high pressure liquid chromatography (HPLC) was employed in the separation of such compounds in the pyrolysis mixture [44]. After preliminary chromatographic separation, HPLC allowed for improved isolations and permitted systematic investigation of compounds containing from 5 to 8 Si atoms, as shown in Fig. 5 [37].

The compounds 60, 40c, 40d, 56, 40, 69, 61, 65, 67, 68 and 64 could be isolated in a pure form, and the compounds 59a, 59b, 63 and 62 were separated from their respective mixtures and clarified. A comparison of the compounds containing more than 5 silicon atoms from the pyrolysis of tetramethylsilane shows that of 22

Table 8. Compounds from the pyrolysis of $SiMe_4$

43

47

44

39g

39

39h

45

48

46

49

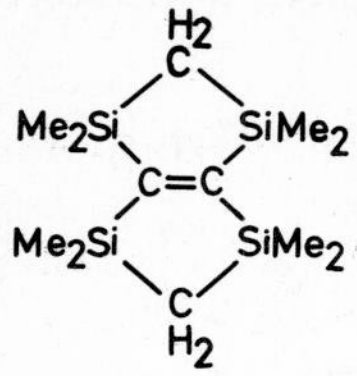

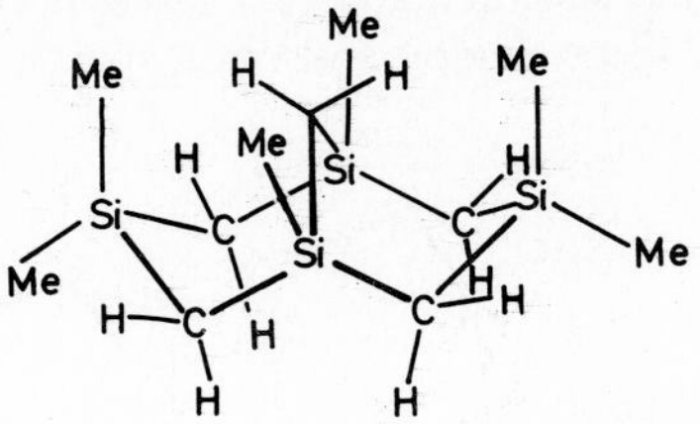

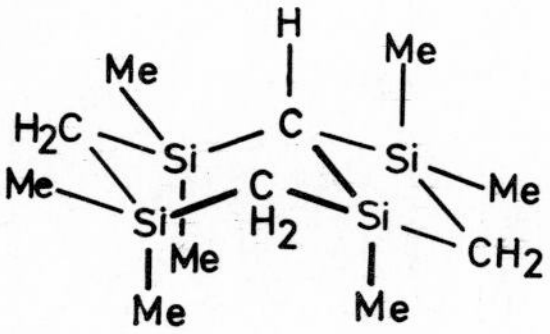

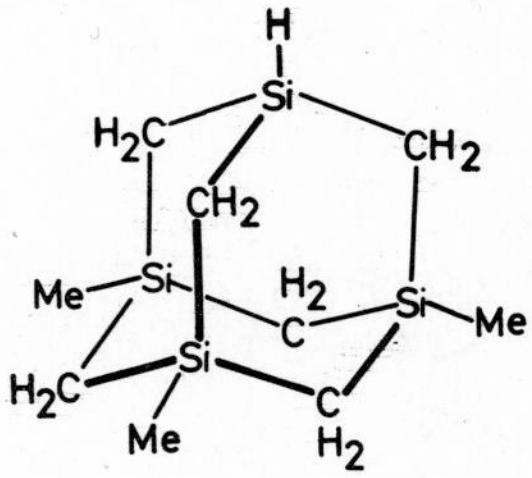

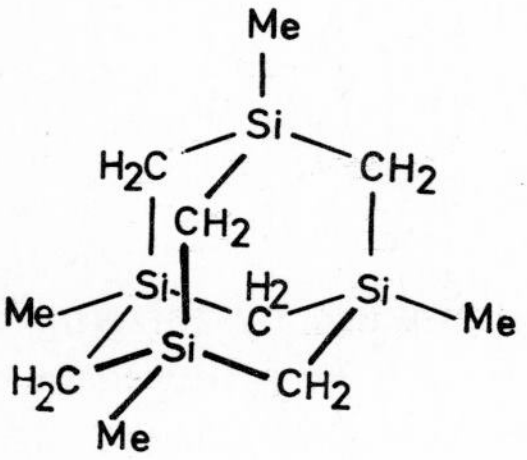

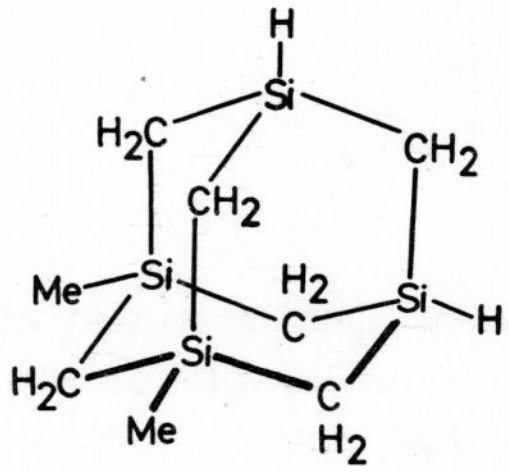

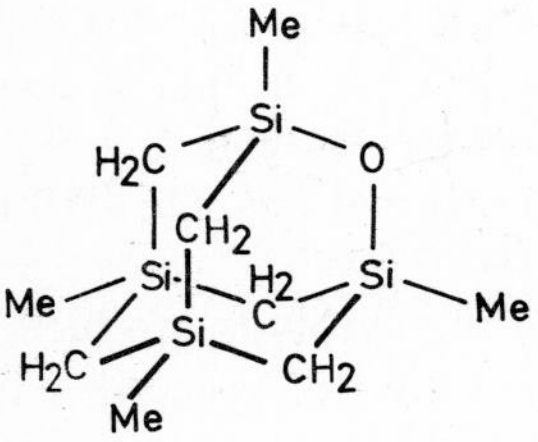

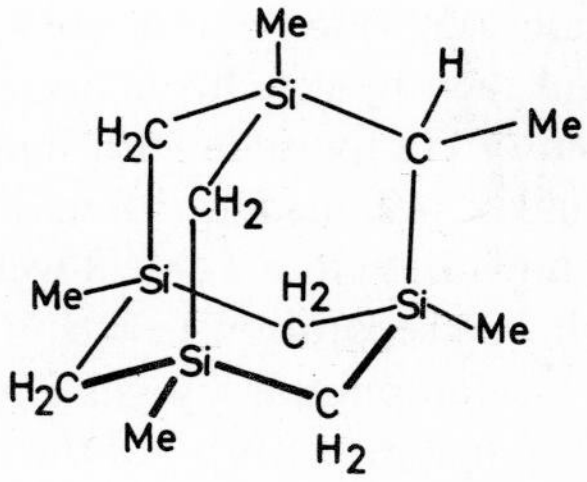

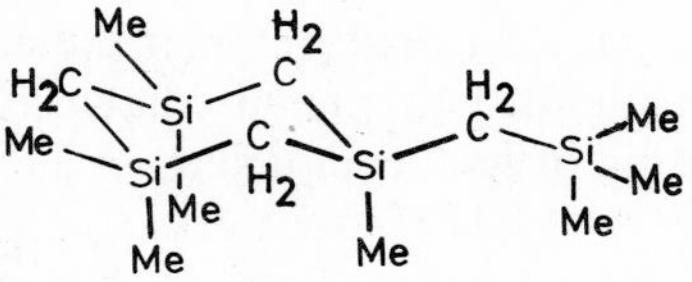

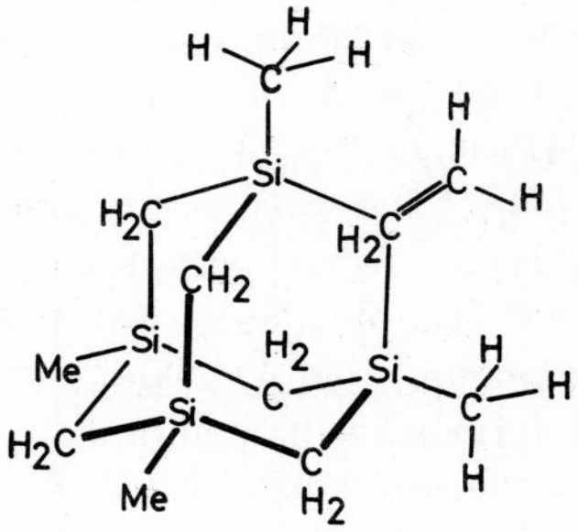

50

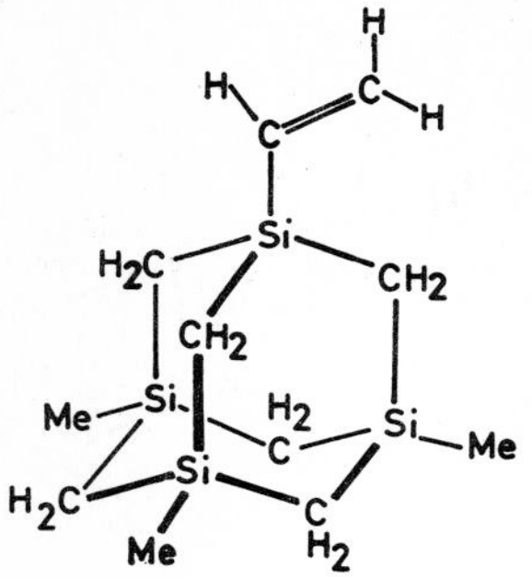

51

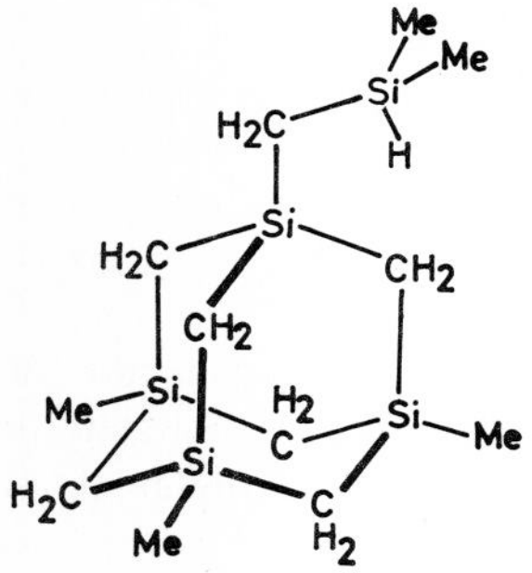

52

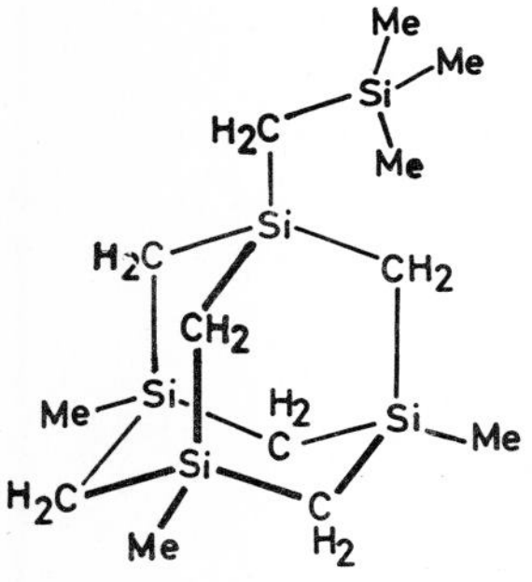

53

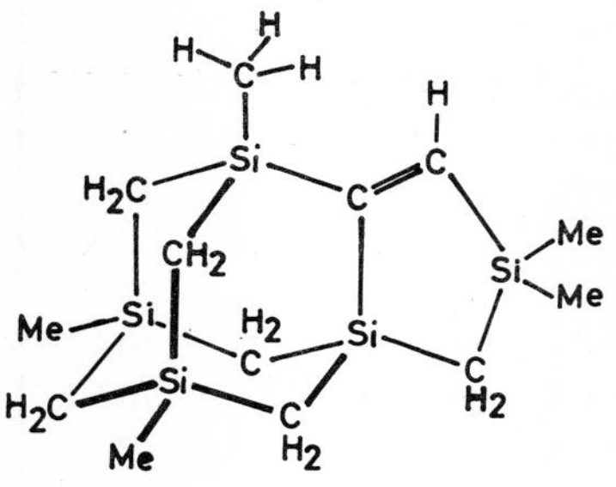

55

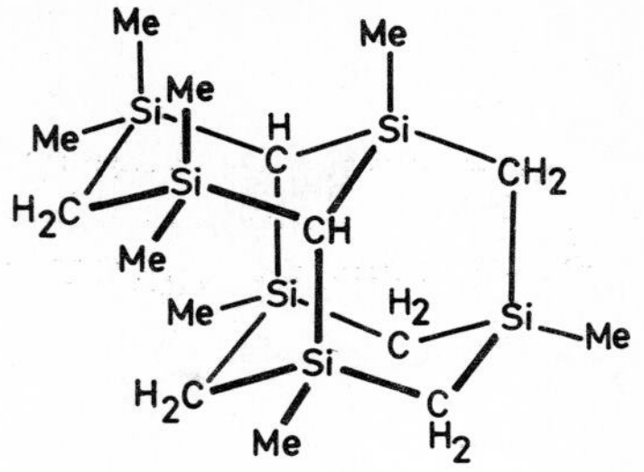

40

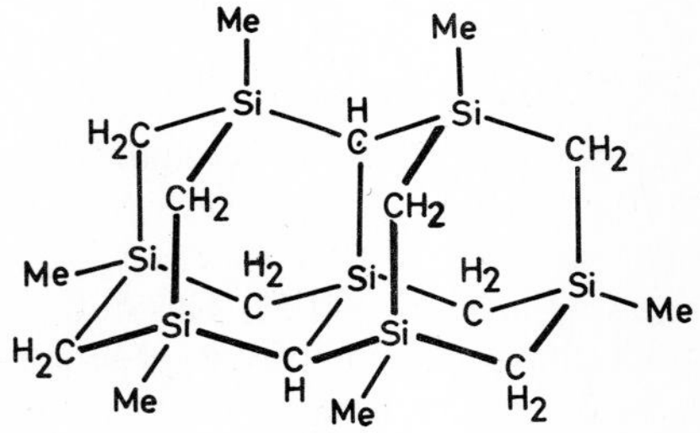

56

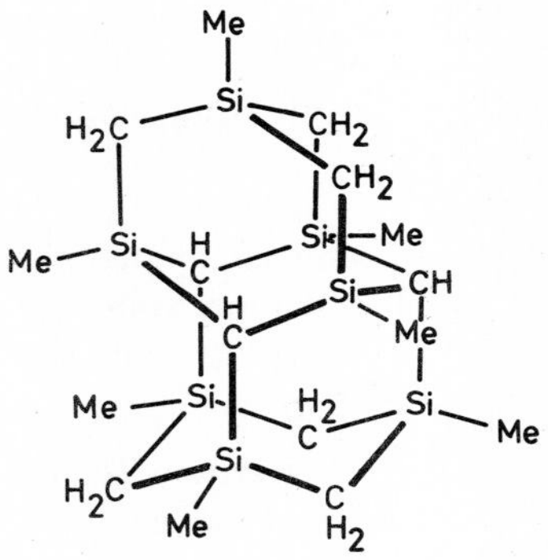

57

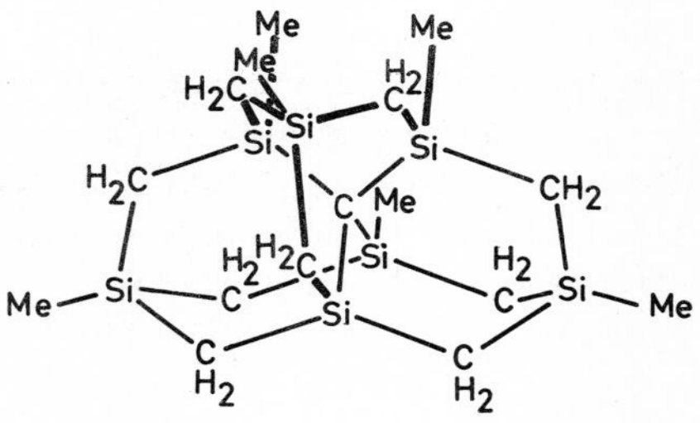

(Continued)

Table 8 (cont.)

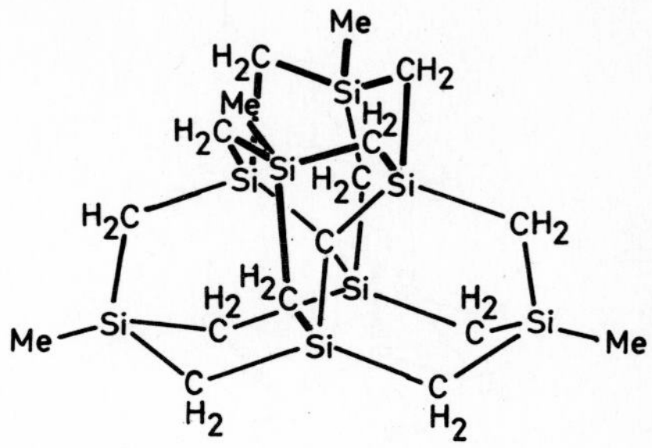

compounds present, 17 contain the adamantane molecular structure (Fig. 5). Two further compounds, 56 and 69, include the adamantane structure as a fragment of the basic structure. Only two compounds are based on a divergent structural skeleton, the scaphanes 57 and 58. This outcome complements the earlier investigation of low-molecular-weight pyrolysis products, and also points out the marked significance of the adamantane structure in the formation of carbosilanes. The transition of 60 to 40, 40 to 61 or 63, from 63 or 61 to 64, and from 64 to 66 outlines a system for the construction of carborundanes. It is not surprising, considering the preference shown for formation of 1,3,5,7-tetrasilaadamantane, that compounds 67 and 68 should appear, containing such structural units linked through CH_2 groups. The relatively high number of partly Si-hydrogenated compounds (40a, 40b, 40c, 40d, 63, 62, 65) also is noteworthy.

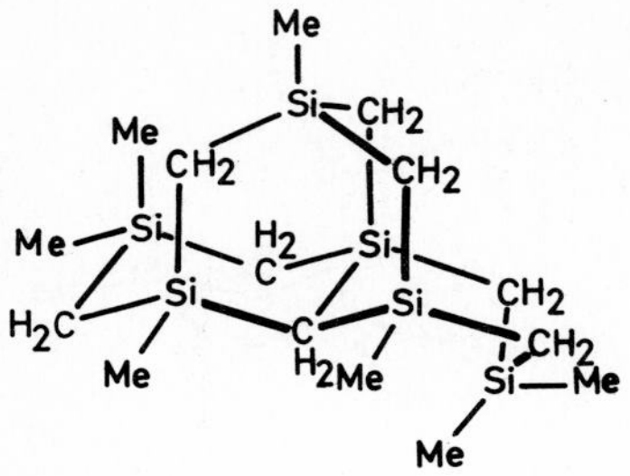

The compounds isolated as described here generally form colorless crystals which remain unharmed on sublimation. They are only slightly soluble in pentane, benzene or CCl_4, but slightly more soluble in toluene. The solubility decreases as the molecular weight of the compounds increases.

1.3.4 Influence of Distillation Temperature on the Composition of the Pyrolysis Products

The development of the chromatographic separation methods made it possible to investigate secondary changes observed on distillation of the pyrolysis mixture. The observed changes are shown in Table 9 [21]. The low boiling products formed in the gas-phase pyrolysis of $SiMe_4$ were separated from their more volatile components by distillation under reduced pressure, in the course of which the temperature did not exceed 20 °C (Column 1 of Table 9). From the second fraction of the original pyrolysis mixture, substances were distilled up to 110 °C at 1–2 mm Hg, which meant the pyrolysis mixture itself was heated to 200–260 °C (Column 2). A part of the remaining residue was distilled up to 260 °C at 1 mm Hg, which meant the pyrolysis mixture was heated to a temperature ranging between 340–400 °C. The products resulting from this further distillation are shown in column 3. A part of the remaining residue was heated for 4 hours to 430–450 °C (Column 4) and the rest to 600 °C for 5 minutes (Column 5).

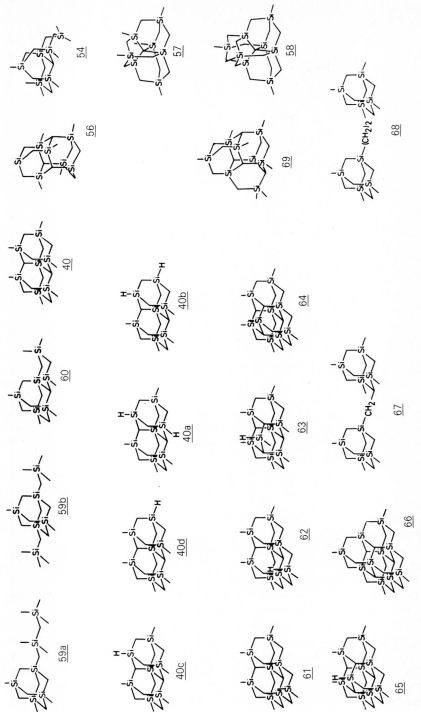

Fig. 5. Carbosilanes containing 5 to 8 silicon atoms, isolated by column chromatography and HPLC

Table 9. Composition of the pyrolysis products of SiMe$_4$ after thermal treatment at various temperatures and chromatographic separation

Group		Column 1 up to 20 °C	Column 2 10 hrs. at 200–260 °C	Column 3 5 hrs. at 340–400 °C	Column 4 4 hrs. at 430–450 °C	Column 5 5 min at 600 °C
A	Amount of pyrolysis product (Vol.-%)	60	48	11	6	1.4
	Molecular weight	515	550	800		
	Si:C:H	1:2.8:5.0	1:3.6:5.4			
B	Amount of pyrolysis product (Vol.-%)	Traces	0.56	0.2	—	—
	Molecular weight		550			
C	Amount of pyrolysis product (Vol.-%)	30	10	26.5	12	
	Molecular weight	770	660	1370		
	Si:C:H	1:2.8:4.0	1:3.3:6.5			
D[a]	Amount of pyrolysis product (Vol.-%)	10	40	62.3	28	not measured
E[b]	Amount of pyrolysis product (Vol.-%)	—	—	—	54	85

[a] High-molecular-weight substances,
[b] less soluble in benzene

In order to obtain a first impression of the influence of increasing thermal load on the five fractions, each one was chromatographically separated after the vacuum distillation into the substance groups A to E of increasing molecular weight. A progressively higher percentage of groups D and E was found. Beforehand these were practically non-existent. The higher molecular weight substances belonging to group C (no insoluble products present) also increased as the temperature in the distillation flask increased from 260 °C to 400 °C. The substance group B represented only a very small percentage of the product mixture. The higher molecular weight substances seemed to be produced only on thermal treatment and to disappear on further heating. As a result it was recognized that work-up via distillation causes complicated chemical changes in the pyrolysis mixture.

A qualitative perspective of the secondary processes occurring on distillation of the pyrolysis mixture was derived from an investigation into the stability of some selected pure compounds, as well as that of some mixtures. The temperature range in which changes would begin to occur had to be established by measuring the mean molecular weight. Moreover, the gases produced in these secondary processes had to be collected in order to discover more about the mechanism of the observed increase of molecular weight. The compounds chosen for this investigation were substance group A, compared to a full undistilled pyrolysis mixture, the adamantanes Si$_4$C$_{10}$H$_{24}$ (**39a**) and Si$_7$C$_{16}$H$_{36}$ (**40**) as well as the 1,3-disilapropane (Me$_3$Si)$_2$CH$_2$ (**6**).

The experiment showed that the previously untreated pyrolysis mixture was considerably changed in content on heating to 400 °C. There was an increase in the average molecular weight and some formation of benzene-insoluble components of group E.

Only a small amount of gaseous products was obtained. After keeping the pyrolysis mixture for 46 hours at 430–490 °C, 72% of the starting material was converted to give a residue insoluble in benzene, along with only small amounts of methane and hydrogen. A similar course of events occurred with the substance group A already at about 230 °C; however, no benzene-insoluble constituents appeared.

On the other hand, it was established that the Si-methylated 1,3,5,7-tetrasilaadamantane 39a is thermally exceptionally stable. After 37 hours at 520–540 °C no observable structural change in this substance was detected. Also, by heating to 550–560 °C for 70 hours no noticeable decomposition occurred. However, after 24 hours at 580 °C a distinct decomposition of this compound was detected, after which a benzene-insoluble fraction was isolated in approximately 33% yield. The benzene-soluble fraction consisted to a large extent of unchanged starting material.

The structurally similar crystalline diadamantane $Si_7C_{16}H_{36}$ (40) showed a thermal stability similar to that of the adamantane 39a. This compound showed almost no change on 54 hours of heating to 450 °C. Only on additional heating to 560 °C for 55 hours did an insoluble residue form corresponding to 42% of the starting substance. The benzene-soluble fraction consisted of unchanged starting material.

The decomposition temperature of $(Me_3Si)_2CH_2$ was established at about 500 °C, which is somewhat lower than that for the mentioned crystalline polycyclic compounds.

1.4 Formation of Si-Chlorinated Carbosilanes by Pyrolysis of Methylchlorosilanes

A large number of carbosilanes of different molecular weights can be obtained from the pyrolysis of tetramethylsilane. These compounds, to a large degree, are unreactive because they do not have functional groups. For many investigations this presented difficulties, as compounds containing the same molecular skeleton but containing reactive substituents bound to silicon were required. Such compounds were obtained from the pyrolysis of the three methylchlorosilanes [20, 28]. The thermal decomposition of methylchlorosilanes in the gas phase proceeds relatively quickly at about 700 °C, whereby pyrolysis products were obtained in a similar fashion as for tetramethylsilane.

Regarding the pyrolysis products of Me_3SiCl, approximately 60% of these products were separable by distillation at atmospheric pressure between 163–200 °C, and 40% existed as oils and solids which were soluble in nonpolar solvents. Approximately 85% of the pyrolysis products of $MeSiCl_3$ are colorless liquids, while the other 15% are oils or meltable solids. Using gas chromatography, it was possible to detect all compounds with boiling points up to 250 °C formed by pyrolysis of the three methylchlorosilanes, as well as to establish the ratios of amounts of each compound present.

All possible chloro- and methyl-substituted compounds with the basic structure $R_3SiCH_2SiR_3$ could be isolated from the pyrolysis mixture of Me_3SiCl. Table 10 shows the distribution of 1,3-disilapropanes in the pyrolysis mixtures from the three

Table 10. Carbosilanes produced by pyrolysis of MeSiCl₃, Me₂SiCl₂, and Me₃SiCl. The SiCl-containing compounds from the pyrolysis, as well as the products which are obtained by hydrogenation with LiAlH₄, are listed

	SiCl containing compounds in the pyrolysis mixture	Hydrogenated derivatives	Bp. (°C/mm Hg) of SiH compounds	Contribution to the hydrogenated pyrolysis mixture (volume ratio) from		
				MeSiCl₃	Me₂SiCl₂	Me₃SiCl
				referring to (Cl₃Si)₂CH₂ = 100		
70	Cl₃Si—CH₂—SiCl₃	H₃Si—CH₂—SiH₃	17/757	100	100	100
71	Me₂SiCl—CH₂—SiCl₃	Me₂SiH—CH₂—SiH₃	70.5–71/768	2.5	—	0.8
73	Me₂ClSi—CH₂—SiCl₃ } MeCl₂Si—CH₂—SiCl₂Me } (2:3)	Me₂HSi—CH₂—SiH₃ } MeH₂Si—CH₂—SiH₂Me } (2:3)	71/768	6.8	33	44.7
74	Me₃Si—CH₂—SiCl₃	Me₃Si—CH₂—SiH₃	91–92/768	0.5	23.7	141
75	Me₂ClSi—CH₂—SiCl₂Me	Me₂HSi—CH₂—SiH₂Me	88.5/768	—	10.7	71.4
76	Me₂ClSi—CH₂—SiClMe₂	Me₂HSi—CH₂—SiHMe₂	107/768	—	—	41.9
78	Me₂ClSi—CH₂—SiClMe₂ } Me₃Si—CH₂—SiCl₂Me } (1:3)	Me₂HSi—CH₂—SiHMe₂ } Me₃Si—CH₂—SiH₂Me } (1:3)	103/768	—	0.66	13.9
79	Me₃Si—CH₂—SiClMe₂	Me₃Si—CH₂—SiHMe₂	120/768	—	—	13.6
				referring to (SiH₂—CH₂)₃ = 100		
80	Cl₃Si—CH₂—SiCl₂—CH₂—SiCl₃	H₃Si—CH₂—SiH₂—CH₂—SiH₃	100/760	109	28	14
81	Cl₃Si—CH₂—SiCl₂—CH₂—SiCl₂Me	H₃Si—CH₂—SiH₂—CH₂—SiH₂Me	123/758	45	—	69
82	Cl₃Si—CH₂—SiCl₂—CH₂—SiClMe₂, a)	H₃Si—CH₂—SiH₂—CH₂—SiHMe₂, a)	133/762	—	—	155
		Me₃Si—CH₂—SiMe₃	135/768	—	—	64
83	(cyclic) Cl₂Si–CH₂–SiCl₂–CH₂–Si(Cl₂)–CH₂ ring: H₂C, Cl₂Si, SiCl₂, CH₂, H₂C, Si Cl₂	(cyclic) H₂C, H₂Si, SiH₂, CH₂, H₂C, Si H₂	142/760	100	100	100
84	(cyclic) H₂C, Cl₂Si, SiCl₂, CH₂, H₂C, Si Cl Me	(cyclic) H₂C, H₂Si, SiH₂, CH₂, H₂C, Si H Me	159/766	25	50	960

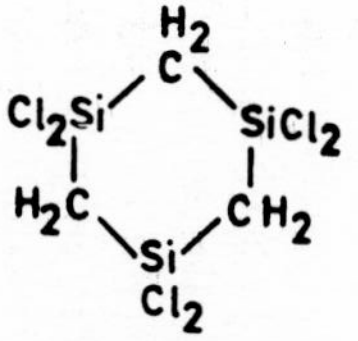

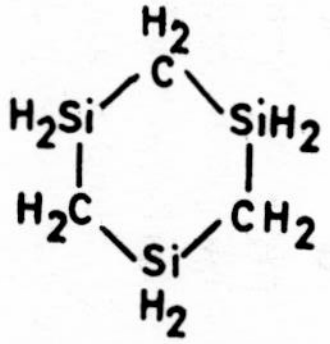

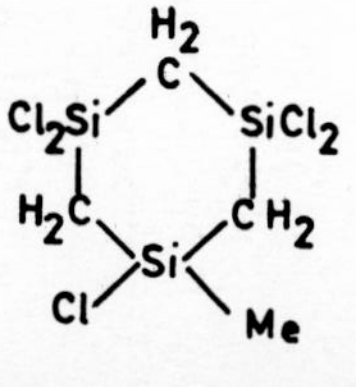

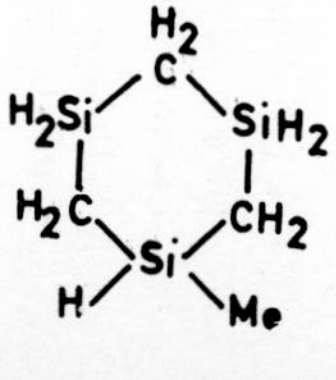

(Continued)

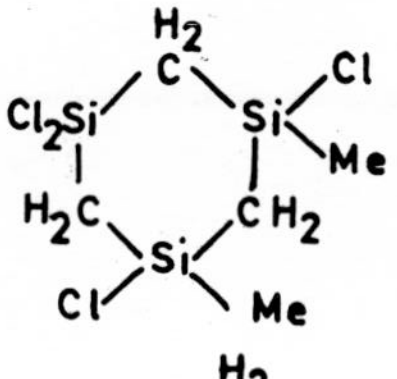

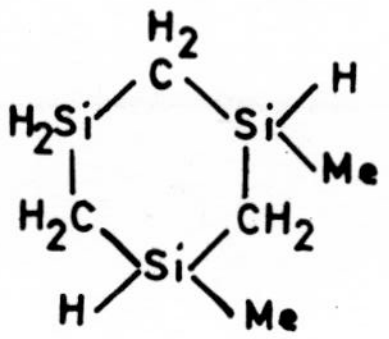

1690	1000	168	65
—	—	—	—
—	—	—	—
166/764	180/764	190/764	201/767

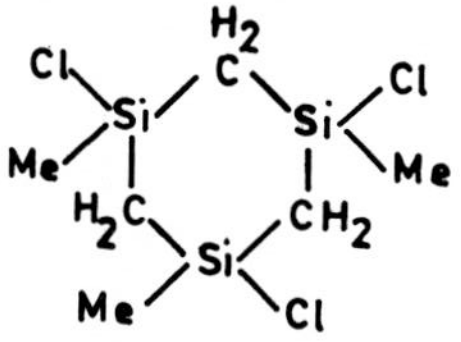

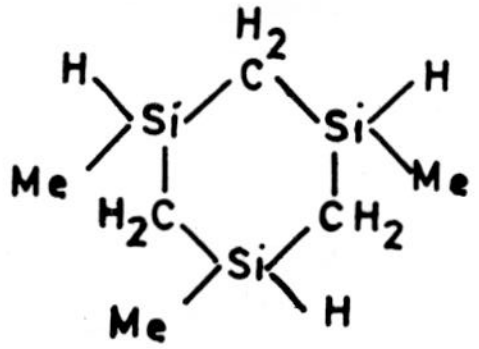

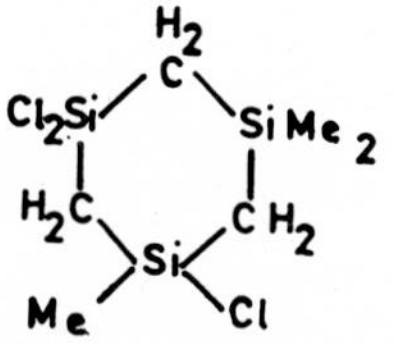

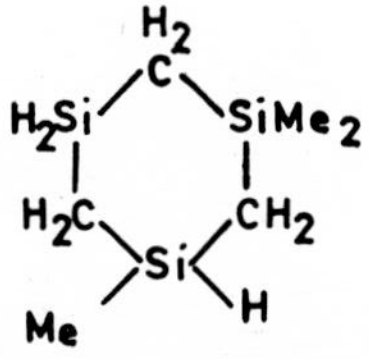

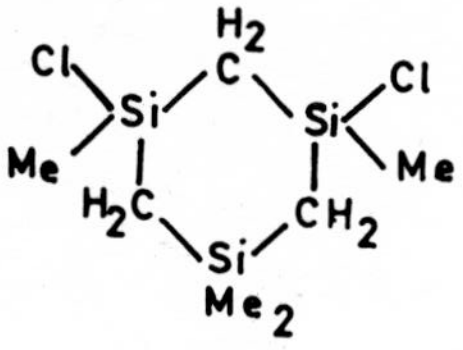

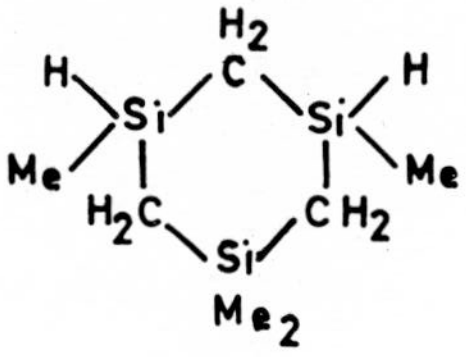

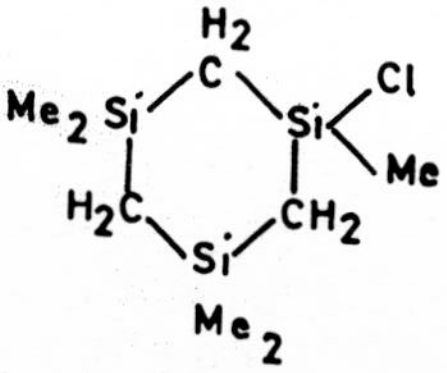

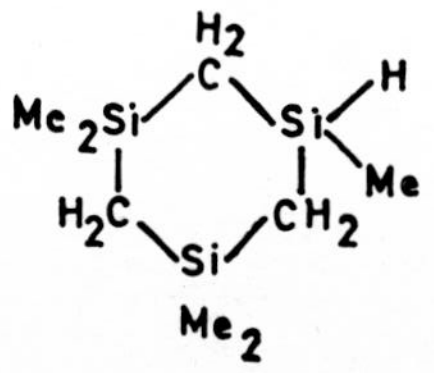

85

86

87

88

89

Table 10 (cont.)

SiCl containing compounds in the pyrolysis mixture	Hydrogenated derivatives	Bp. (°C/mm Hg) of SiH compounds	Contribution to the hydrogenated pyrolysis mixture (volume ratio) from		
			MeSiCl$_3$	Me$_2$SiCl$_2$	Me$_3$SiCl
90		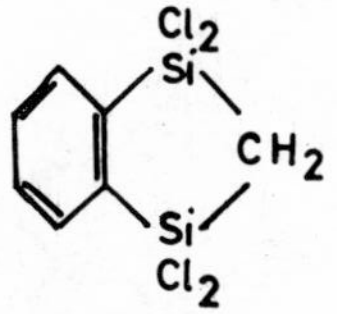 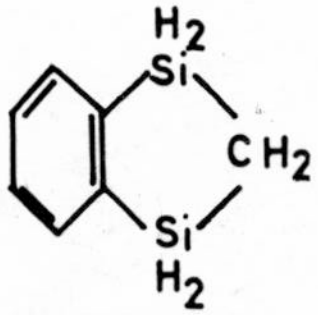 204/746	—	—	65
91		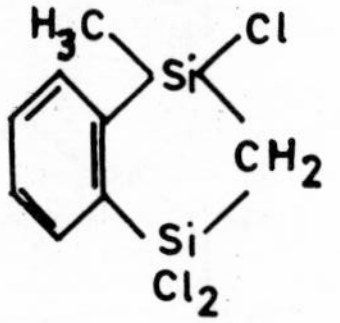 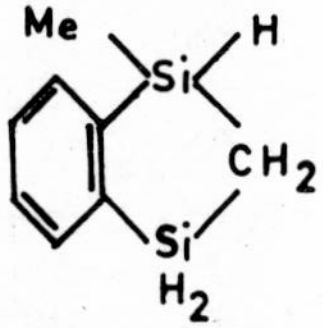 210/767	—	—	59
92			—	—	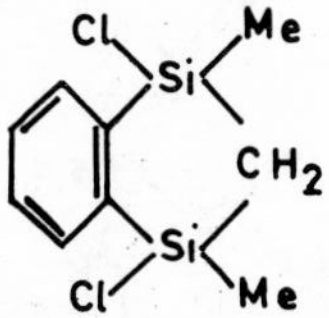 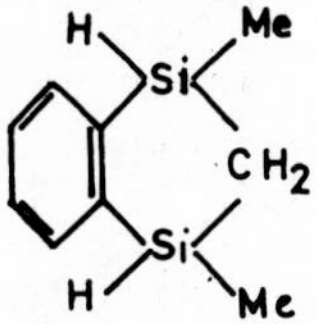 14

Table 11. Carbosilanes containing 4 or more Si atoms from the pyrolysis mixture of methylchlorosilanes

Formed from Me_3SiCl:

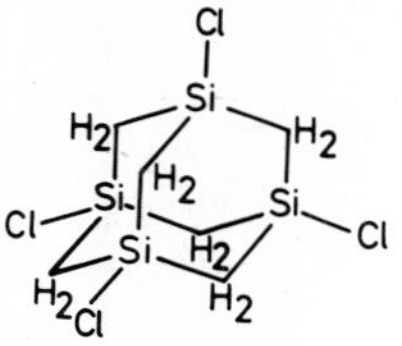

a

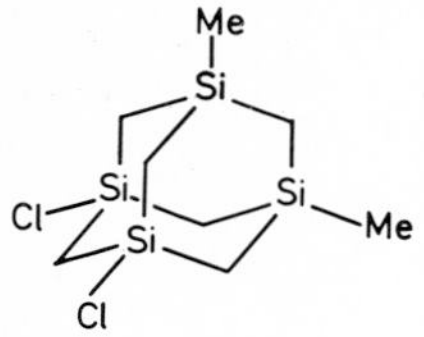

c

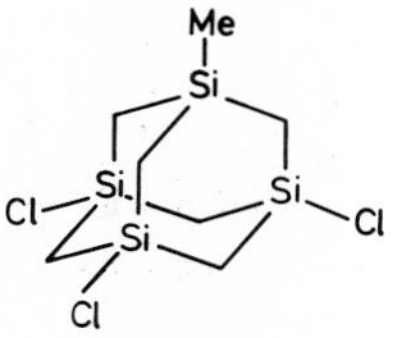

b

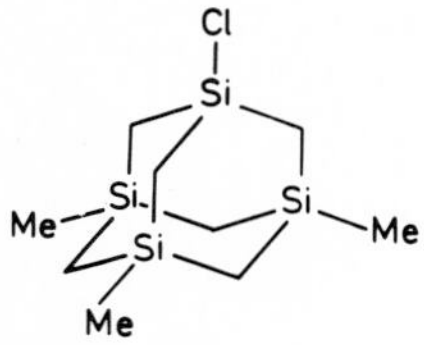

d

$a:b:c:d = 170:36:3:1$

Formed from Me_2SiCl_2:

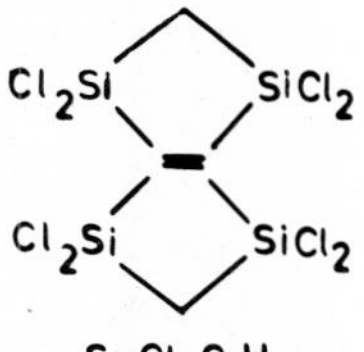

$Si_4Cl_8C_4H_4$

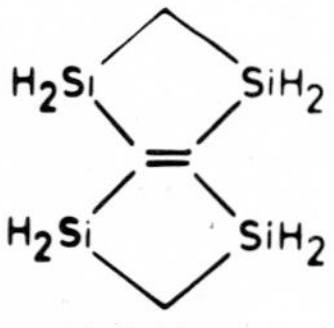

$Si_4Cl_5C_4H_5$

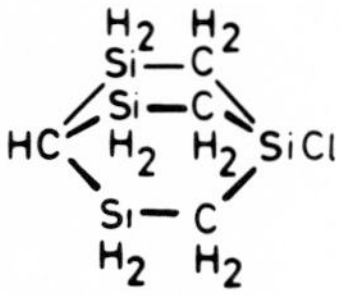

$Si_4C_5H_{10}Cl_6$

Converted by $LiAlH_4$ to

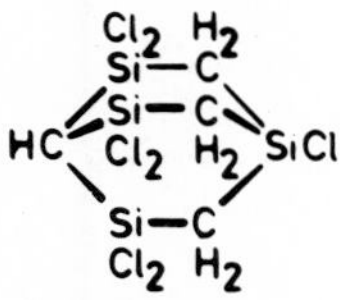

$Si_4C_4H_6$

$Si_4ClC_4H_{13}$

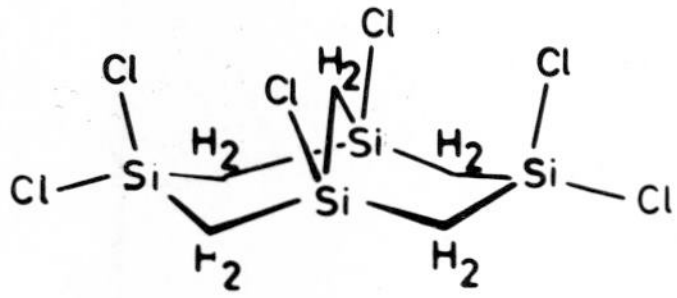

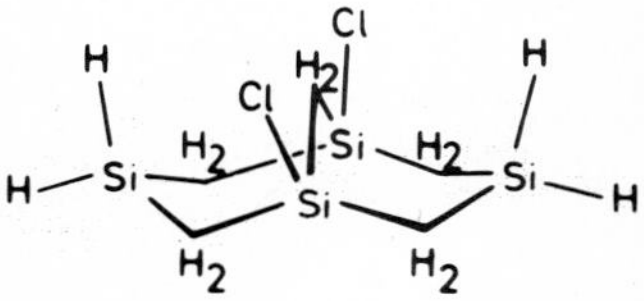

$Si_4C_5H_{14}Cl_2$

(Continued)

Table 11 (cont.)

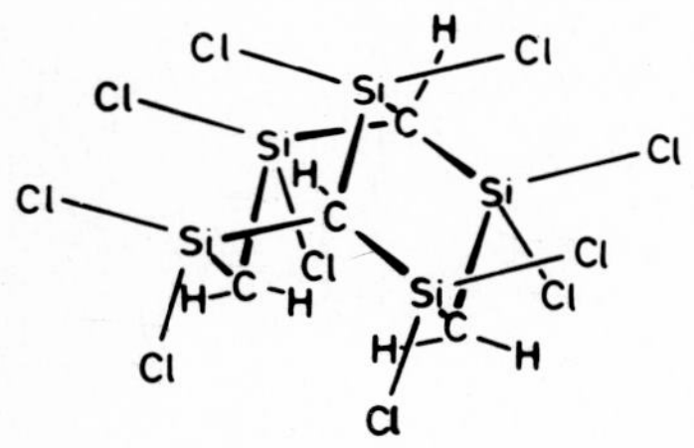

Si$_5$Cl$_{10}$C$_4$H$_6$

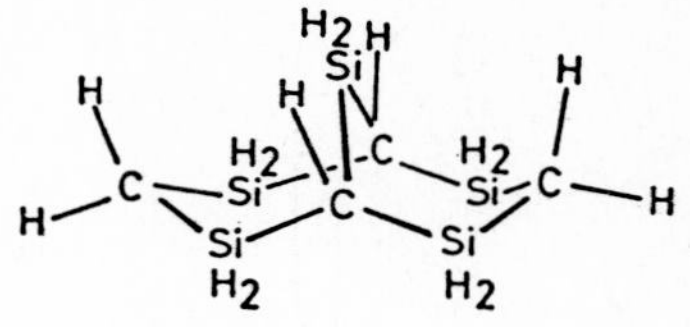

Si$_5$C$_4$H$_{16}$

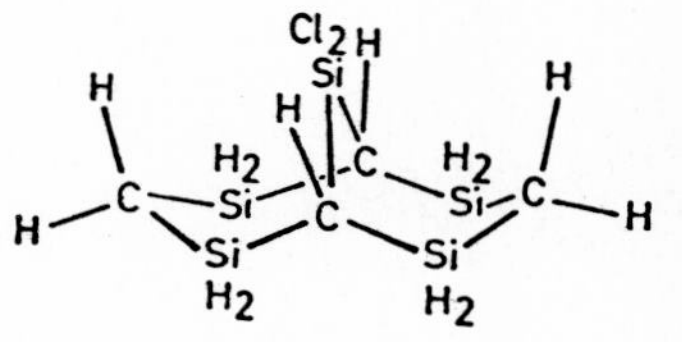

Si$_5$C$_4$H$_{14}$Cl$_2$

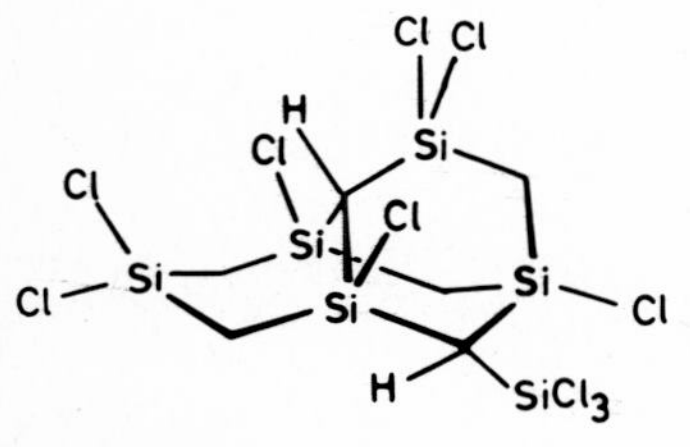

Si$_6$Cl$_{10}$C$_6$H$_{10}$

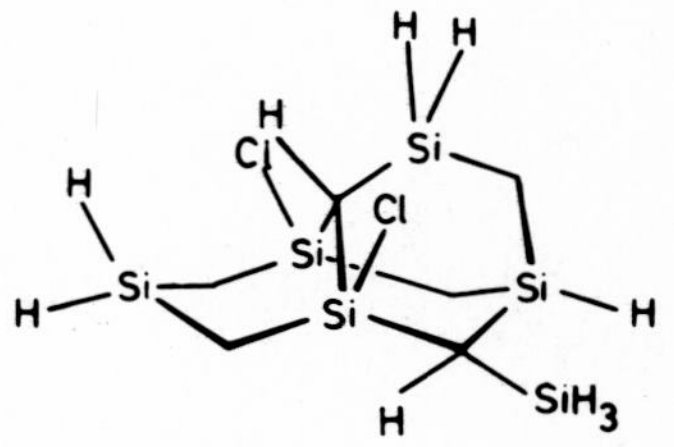

Si$_4$Cl$_2$C$_6$H$_{18}$ [29]

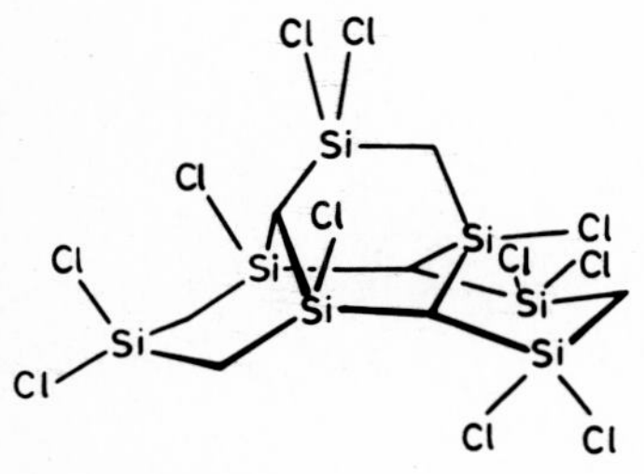

Si$_7$Cl$_{11}$C$_7$H$_{11}$

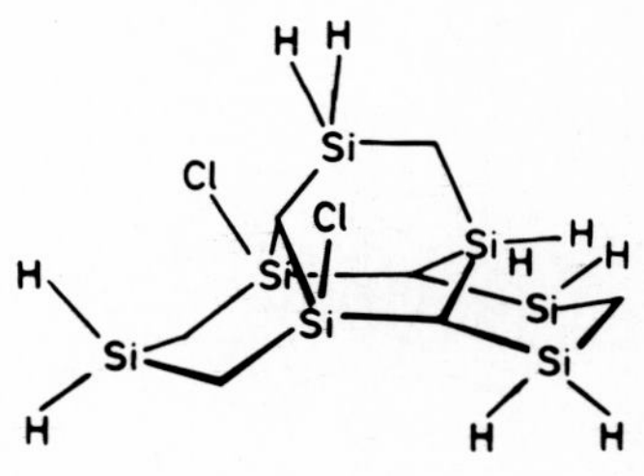

Si$_7$Cl$_2$C$_7$H$_{20}$ [29]

Formed from $MeSiCl_3$:

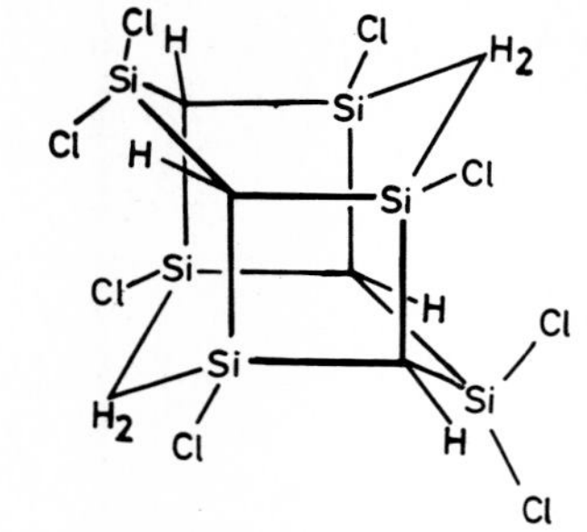

Converted by $LiAlH_4$ to

$Si_6Cl_8C_6H_8$ 335

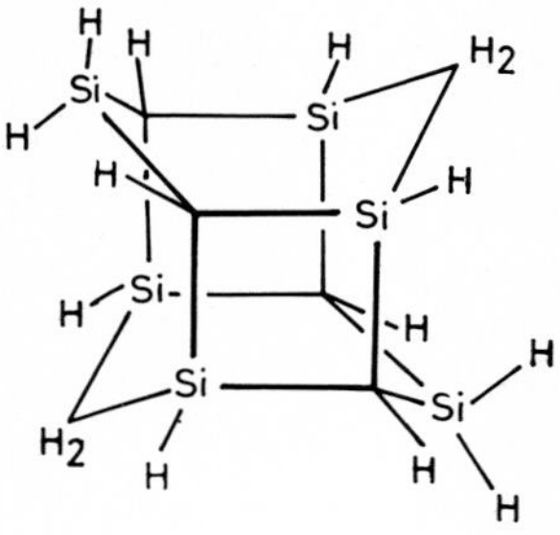

$Si_6C_6H_{16}$ 336

Formed from $Cl_3Si-CH_2-SiCl_3 + 17\% MeCl_2Si-CH_2-SiCl_3$ [29]:

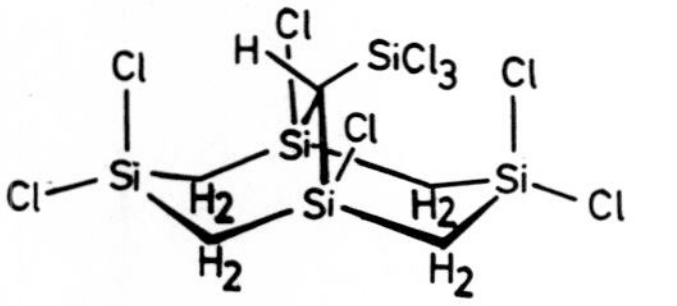

$Si_5Cl_9C_5H_9$

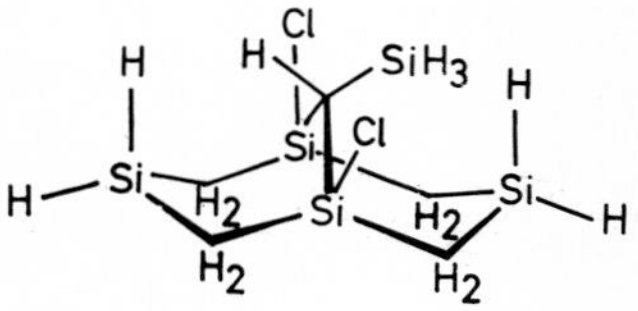

$Si_5Cl_2C_5H_{16}$

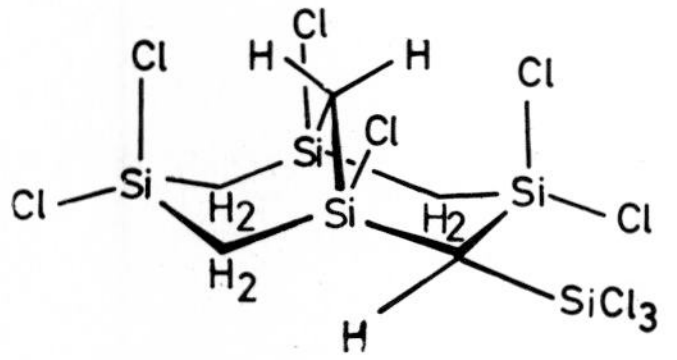

$Si_5Cl_9C_5H_9$

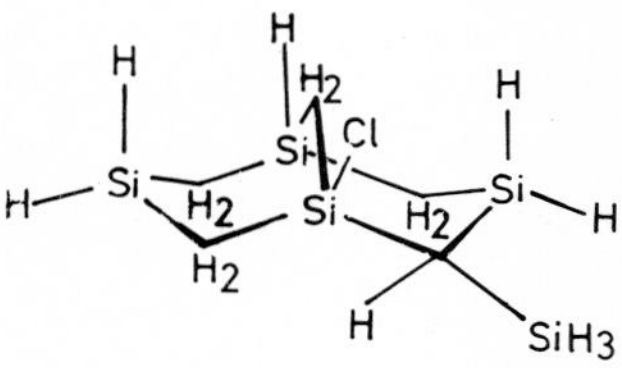

$Si_5ClC_5H_{17}$

methyl chlorosilanes. The percentage contribution of the 1,3-disilapropanes to the total product mixture was 32% from the Me_3SiCl pyrolysis, 43% from the Me_2SiCl_2 pyrolysis and 60% from the $MeSiCl_3$ pyrolysis.

The investigation of substances of high molecular weight from the pyrolysis of the methylchlorosilanes required the separation of individual compounds from the mixture. This involved difficulties, as the fractionating methods (distillation and gas chromatography) were not suitable for the high-boiling Si-chlorinated carbosilanes containing 3 or more silicon atoms in the molecular skeleton. However, it could be shown that the Si—Cl bond in these compounds could be substituted by an Si—H bond by using $LiAlH_4$. This hydrogenation does not affect the Si—C—Si skeleton, as was shown by hydrogenation of some Si-chlorinated 1,3-disilapropanes with known structures and of 1,1,3,3,5,5-hexachloro-1,3,5-trisilacyclohexane. The respective SiH-containing compounds have lower boiling points, and can therefore be separated through distillation and gas chromatography. This SiH-containing group of substances is included in Table 10.

Considering to what extent each compound is present in the hydrogenated pyrolysis mixture of the respective methylchlorosilanes, it is obvious that the more SiMe groups are present in the starting material, the more carbosilanes of the 1,3,5-trisilacyclohexane type will be formed.

The investigation into more silicon-rich compounds is not yet complete, but an understanding of this area can be gained by inspection of Table 11.

The substances formed by pyrolysis of methylchlorosilanes behave on thermal treatment in a manner similar to those from the pyrolysis of $SiMe_4$: on increasing the temperature of distillation, changes in product composition to oily as well as solid glassy substances were observed, and fractions of varying colors (yellow to red) appeared. From $MeSiCl_3$, a brillant red, meltable, at 20 °C glassy solid formed which was soluble in nonpolar solvents and distillable at 210 °C/10^{-3} mm Hg. This substance is very stable toward conc. H_2SO_4 and HNO_3, and contains no Si—Si or Si—H groups. The elemental analysis produced the values Si 30%; C 12,5%; H 1,5%; Cl 55,2% and a molecular weight of 790. The analytical data suggest that this compound is still not pure; however, it approximates the qualities of polycyclic compounds with 8 Si atoms in the molecular skeleton.

Similar observations existed in the separation of pyrolysis products of $MeSiCl_3$ by molecular distillation. A solid yellow-brown substance was obtained which produced a honeylike melt at 260 °C, an average molecular weight of 900, and the following elemental composition: Si 29.7%; C 16.3%; H 2.2%; Cl 48.7%. It is evident that here again a mixture of carbosilanes containing Si—Cl groups exists.

Very similar results were obtained from the pyrolysis products of Me_3SiCl. Distillation at 234 °C/10^{-2} mm Hg produced a yellow glassy substance of composition Si 33.4%; C 24.3%; H 4.0%; Cl 37.0%; mol. wt. 610.

1.5 Pyrolysis of the Methylsilanes Me_3SiH, Me_2SiH_2 and $MeSiH_3$

The pyrolysis of Me_3SiH was carried out at 620 ± 5 °C by a procedure analogous to that for $SiMe_4$. The identified compounds and their percentage yields (Fig. 6) were divided into their respective structural groups [30, 31]. It was observed that

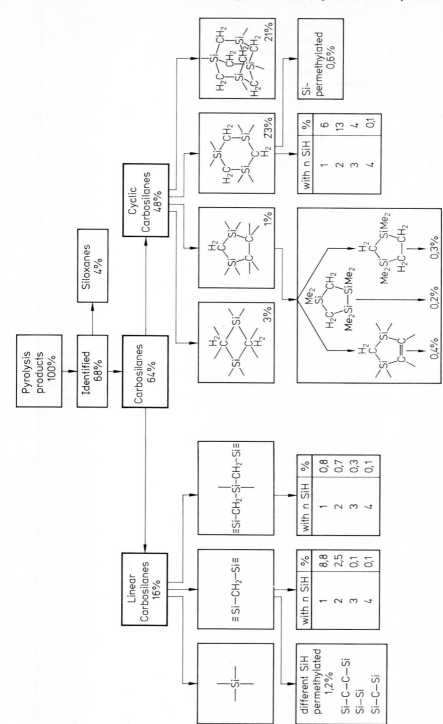

Fig. 6. Pyrolysis products of Me$_3$SiH

apart from simple linear carbosilanes, cyclic compounds with the molecular skeletons of 1,3,5-trisilacyclohexane and 1,3,5,7-tetrasilaadamantane were predominantly formed. Approximately 19% of the pyrolysis products are carbosilanes of high molecular weight which remain usually as distillation or condensation residues. These residues appear as either light, translucent, gelatinous products or yellowish viscous oils. The gelatinous residues are insoluble in organic solvents; on the other hand, the oily residue is soluble. The average cryoscopic molecular weight lies at around 600. To approximately 12%, the pyrolysis mixture consists of substances that are present only in very small amounts. The isolation of these compounds, which include 1,3-disilacyclobutanes with side chains, partly methylated five-membered rings with side chains, and 1,3,5-trisilacyclohexanes with side chains, would require a larger than the available amount of pyrolysis products. Definitely, no aromatic compounds nor any vinyl-substituted products were detected.

For the separation presented in Fig. 6, 2000 g of Me_3SiH were pyrolysed, from which 1240 g of liquid pyrolysis products were obtained. Gaseous and volatile compounds were also obtained from the pyrolysis of Me_3SiH, in volume %: H_2 41.0; CH_4 57.0; C_6H_6 0.5; $MeSiH_3$ 0.03; Me_2SiH_2 1.2 and $SiMe_4$ 0.2%.

The pyrolysis of Me_2SiH_2 was performed in a similar manner at 600 °C in a circulation apparatus. The separation and the distribution of pyrolysis products are shown in Fig. 7. It is apparent that the same basic carbosilane molecular skeletons were produced and that the compounds obtained, as expected, were richer in Si—H bonds. Comparing these products with those from the pyrolysis products of Me_3SiH, it is seen that linear compounds were preferred, especially the 1,3-disilapropanes. On the other hand, the percentage of cyclic compounds decreased generally from 48 to 22%. For example, the formation of 1,3,5,7-tetrasilaadamantanes was reduced markedly from 21% to 6%. Gaseous and volatile compounds also were present as pyrolysis products of Me_2SiH_2; in vol. % they were H_2 86; CH_4 11.4; C_2H_4 0.01; C_2H_6 0.3; C_3H_8 0.01; $MeSiH_3$ 0.7; Me_3SiH 1.5 and $SiMe_4$ 0.03%. The unclarified high-boiling compounds obtained were similar in properties to those from the pyrolysis of Me_3SiH. This group contained the silicon-rich compounds, for example $Si_6C_{11}H_{28}$. The compounds of this group, as well as the insoluble white powder obtained with an Si content of about 52%, which obviously contains a carbosilane molecular skeleton, all need to be investigated further.

The main product produced from the gas-phase pyrolysis of $MeSiH_3$ was a pale yellow substance. The pyrolysis of $MeSiH_3$ was carried out in the same manner as for the previously mentioned methylsilanes at 550 ± 5 °C; 87% of the starting material was converted. The pyrolysis product differed from those of the other methylsilanes in a characteristic way. From the 400 g of $MeSiH_3$, only 2 ml of liquid product was obtained, which consisted of $Me_2HSi—CH_2—SiH_3$ as major product, as well as Me_3SiH, $Me_2HSi—CH_2—SiH_2Me$, $Me_2HSi—CH_2—SiHMe_2$ and $Me_3Si—CH_2—SiHMe_2$. Compounds of the 1,3,5-trisilacyclohexane and 1,3,5,7-tetrasilaadamantane type were not detected. Inspection of this powdery main product through a microscope at a magnification of 500 × showed that it consisted of yellow transparent spherical balls of varying sizes. Elemental analysis showed the following composition: Si 72.5%, C 23.1%, H 4.1%, corresponding to the atomic ratio Si:C:H = 1:0.75:1.6. This substance does not melt or sublime and is amorphous on X-ray investigation. Letting this substance stand in air for several months, oxygen exposure reduces the

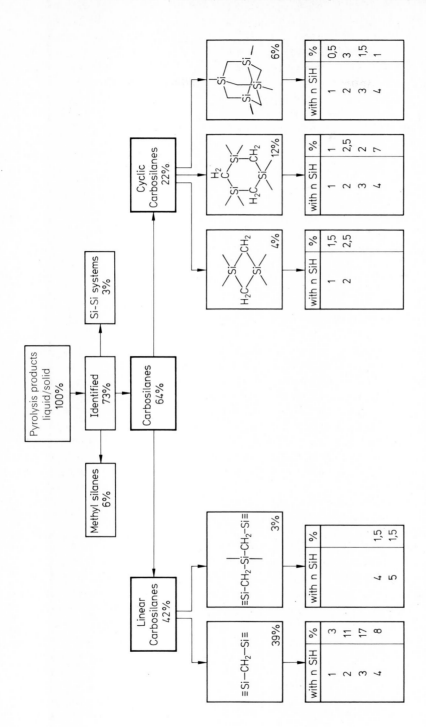

Fig. 7. Pyrolysis products of Me_2SiH_2

silicon content to 47 %. The IR spectrum of the unoxidized substance (KBr pressed disc) reveals a weak C—H band at 2900 cm^{-1} as well as a Si—H band at 2065 cm^{-1}. This value is low in comparison with that for normal Si—H bands but in agreement with certain literature values that constitute a generally accepted proof for the direct bonding of silicon to two or three further silicon atoms in such substances [32]. At 1400 cm^{-1}, a band appeared which was assigned to a CH$_2$- or CH$_3$-deformation. The band at 1350 cm^{-1} related to a Si—CH$_2$—Si group, while that at 1250 cm^{-1} was assigned to a Si—CH$_3$-deformation. A broad absorption band at 1000 cm^{-1} corresponds to the CH$_2$-wagging mode of the Si—CH$_2$—Si group. From comparable spectra, obtained for Si-asteranes and Si-adamantanes, it was seen that the position of this CH$_2$-wagging mode of Si—CH$_2$—Si was shifted to a lower wave number (1000–1020 cm^{-1}) compared to that for linear carbosilanes (1050 cm^{-1}), which means that this substance more likely resembles a polymeric compound with a complicated molecular structure. The yellow substance was oxygen free. Further information was obtained from the alkaline hydrolysis of this substance. Because Si—Si and Si—H groups produce H$_2$ on reaction with alcoholic KOH solutions according o the equations

$$\equiv SiH + H_2O \rightarrow \ \equiv SiOH + H_2$$

$$\equiv Si—Si \equiv \ + 2\,H_2O \rightarrow 2 \equiv SiOH + H_2$$

this enabled a quantitative analysis to be carried out on the groups in this yellow substance. The determination of hydrogen from this alkaline hydrolysis produced the ratio Si:H = 1:1.2. The oxidation of SiH groups by HgCl$_2$ and the subsequent oxidation of Hg$_2$Cl$_2$ by I$_2$ provided a means for determining SiH groups without affecting Si—Si bonds [33]. Using this method on the yellow substance the ratio Si:H = 1:0.06 resulted. Despite the insolubility of the substance giving rise to some uncertainty in the determination, there was a remarkably small number of SiH groups present, which was also reflected by the very weak Si—H stretch in the IR spectrum. This means that hydrogen produced by alkaline hydrolysis is derived primarily from the Si—Si groups.

It can be concluded Halogenating this substance with chlorine or bromine provides extra information. Halogenation is expected to convert Si—H groups to Si—X groups (where X = Cl, Br) without changing the structure of the molecular skeleton. From previous experience Si—C—Si groups are not attacked, while Si—Si bonds are cleaved. Proof for this came from the reaction of the yellow substance with chlorine and bromine, yielding MeSiCl$_3$ and MeSiBr$_3$, respectively. This meant that the Me—Si groups in the polymeric product must have been bound to three further silicon atoms. The composition of the unchlorinated substance was Si$_1$C$_{0.75}$H$_{1.6}$, and through chlorination a new composition of Si$_1$C$_{0.6}$H$_{1.15}$Cl$_{1.26}$ is achieved. This considerable uptake of chlorine is explained by the chlorination of Si—Si groups. The reaction carried out under stronger conditions (180 °C, suspension in silicone oil) produced SiCl$_4$ and a larger chlorine contribution to produce a Si:Cl ratio of 1:1.66 in the polymeric substance [34].

It can be concluded that the solid amorphous yellow substance is composed of Si—C—Si and Si—Si groups, and that the free valencies of silicon not contained within the molecular skeleton are bound to methyl groups, and to a small degree to H atoms, as substitutents.

1.6 Reactions of Methylsilanes by Gas Discharge

1.6.1 Reactions of $SiMe_4$, Me_2SiH_2 and $MeSiH_3$

Results obtained for the pyrolysis of $SiMe_4$ and other methylsilanes show that synthesis of carbosilanes can be achieved through any appropriate energy input. Photochemical reactions, for instance, have already been mentioned [13, 86]. In the high-frequency induced cold plasma state, transfer of energy to individual molecules proceeds by electron impact. Therefore, subsequent reactions are expected to be restricted particularly in comparison to those playing a major role in the course of the thermal decomposition of methylsilanes. In order to obtain a comparison between compounds produced from either pyrolysis or gas discharge of methylsilanes, gas discharge was carried out using gaseous $SiMe_4$ and other methylsilanes. Previously reported gas discharge reactions in $SiMe_4$ always produced compounds containing Si—H and Si—Si bonds [35]. The main products from these reactions are $Me_3SiC_2H_5$, $Me_3Si—SiMe_3$, $Me_3Si—CH_2—CH_2—SiMe_2H$ and $Me_3Si—CH_2—CH_2—SiMe_3$.

From corresponding reactions with SiH-containing methylsilanes, the main products synthesized were disilanes [36]: $MeH_2Si—SiH_2Me$ from Me_2SiH_2, and $Me_3Si—SiMe_3$ as the main disilane from Me_3SiH. In all cases CH_4 and H_2 were reported as byproducts in addition to some unidentified substances. All these investigations involved the use of high voltages and internal electrodes.

Compared to those just described, our investigations utilizing high frequency gas discharge [30] had the advantage that no contact is made between reactant and electrodes, and that reactions induced by this kind of energy transfer proceed at temperatures below 200 °C, whereas temperatures of the order of 600–700 °C are required for pyrolysis [31]. In order to sustain this plasma state, the pressure in the system was not permitted to exceed 10 mm Hg, while our pyrolysis reactions were carried out just below atmospheric pressure.

The compounds produced from $SiMe_4$ as a result of gas discharge are shown in Table 12. In a first run, the storage flask was cooled to −70 °C during the whole reaction. In a second run, the storage flask was allowed to warm gradually to 20 °C. By this means, further liquid compounds were obtained, due to the more volatile products participating in subsequent reactions. These additional compounds obtained are also shown in Table 12.

The gas chromatographic separation of the products from the reaction of Me_2SiH_2 in the cold plasma yielded the compounds Me_3SiH, $Me_2HSi—SiHMe_2$ and $Me_2HSi—CH_2—SiHMe_2$. By allowing this product mixture to react further in a cold plasma, carbosilanes containing up to three silicon atoms (in addition to yellow solids) were obtained.

Performing this reaction with $MeSiH_3$ under similar conditions, a pale yellow powder of composition $Si:C:H = 1.0:1.1:2.1$ was produced. The IR spectrum (KBr pressed disc) showed that CH, SiH, $SiCH_3$ and $SiCH_2Si$ groups were present in this substance. The small amount of liquid reaction product consisted essentially of $MeH_2Si—SiH_2Me$.

Table 12. Compounds produced from the decomposition of $SiMe_4$ in a plasma state

Compound	Weight percent of the liquid reaction products[a]	Weight percent of the liquid reaction products[b]
$C_2H_5SiMe_3$	15	12
$Me_2HSi-CH_2-SiHMe_2$	4	3
$Me_3Si-SiMe_3$	7	5
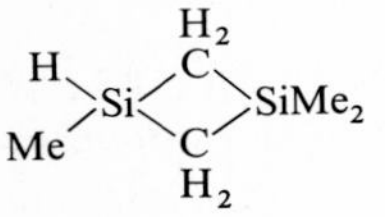 (cyclic structure: Me, H on Si, bridged by two CH_2 groups to $SiMe_2$)	2	1
$Me_3Si-CH_2-SiHMe_2$	21	17
$(Me_2Si-CH_2)_2$	6	4
$Me_3Si-CH_2-SiMe_3$	19	17
$Me_3Si-CH_2-CH_2-SiHMe_2$	2	1
$Me_3Si-CH_2-CH_2-SiMe_3$	5	3
$(C_2H_5)Me_2Si-CH_2-SiMe_3$	3	2
$(Me_2HSi-CH_2)_2SiMe_2$	—	2
$(Me_2Si-CH_2)_3$	—	3
$(Me_3Si-CH_2)_2SiHMe$	—	3
$Me_3Si-CH_2-SiMe_2-CH_2-SiHMe_2$	—	5
$(Me_3Si-CH_2)_2SiMe_2$	—	11

[a] TMS storage flask at $-70\ ^\circ C$ for the whole reaction
[b] TMS storage flask allowed to warm to $20\ ^\circ C$ during reaction

1.6.2 Comparison of the Products Obtained from Reactions in Cold Plasma with those from Pyrolysis

The compounds with 1 to 4 Si atoms within the molecule, formed by pyrolysis of Me_3SiH and Me_2SiH_2, possess the same molecular skeletons as those from the pyrolysis of $SiMe_4$. The percentage of SiH-rich compounds in these individual groups increases as one goes to Me_2SiH_2 as reactant. In order to compare compounds produced from the pyrolysis of $SiMe_4$ with those from the pyrolysis of Me_3SiH and Me_2SiH_2, as well as with those compounds produced from $SiMe_4$ in a cold plasma, Table 13 shows the ratios of the amount of certain basic molecular structures found in each product mixture. These ratios are derived by making the structural type $\geq Si-CH_2-Si\leq$ equivalent to 1. The amount of 1,3-disilapropanes produced as a result of pyrolysis increases on going from $SiMe_4$ to Me_2SiH_2. A corresponding decrease in the amount of cyclic compounds produced is especially noteworthy in the products from Me_2SiH_2, because all structural types up to 1,3,5,7-tetrasilaadamantane still exist in this mixture. On comparing the cyclic compounds from pyrolysis of $SiMe_4$ and Me_3SiH, one has to take into account that the investigation of carbosilanes of higher molecular weight from Me_3SiH is not yet brought to a similar stage.

The increase in the amount of 1,3-disilapropanes in the pyrolysis products of Me_2SiH_2 can be satisfactorily explained. Because the storage flask in the circulation apparatus had to be cooled to $-45\ °C$ to maintain an appropriate partial pressure of Me_2SiH_2, the 1,3-disilapropanes were trapped and could not participate in further

Table 13. Comparison of the amounts of carbosilane types present in the pyrolysis mixtures of $SiMe_4$, Me_3SiH, Me_2SiH_2 and from the reactions of $SiMe_4$ in the plasma state[a]

	$SiMe_4$	Me_3SiH	Me_2SiH_2	Compounds formed from decomp. of $SiMe_4$ in a plasma state
Methylsilanes	0.23	0.1	0.15	0.31
$\equiv Si-CH_2-Si\equiv$	1	1	1	1
$\equiv Si-CH_2-Si-CH_2-Si\equiv$	0.31	0.14	0.07	0.44
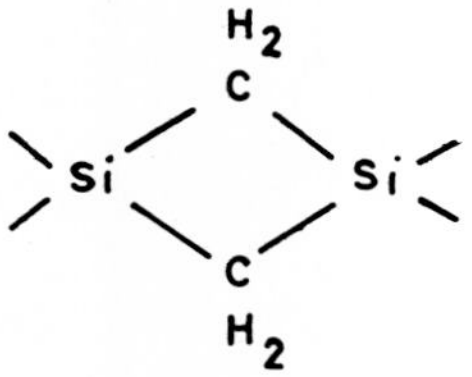	0.52	0.26	0.10	0.17
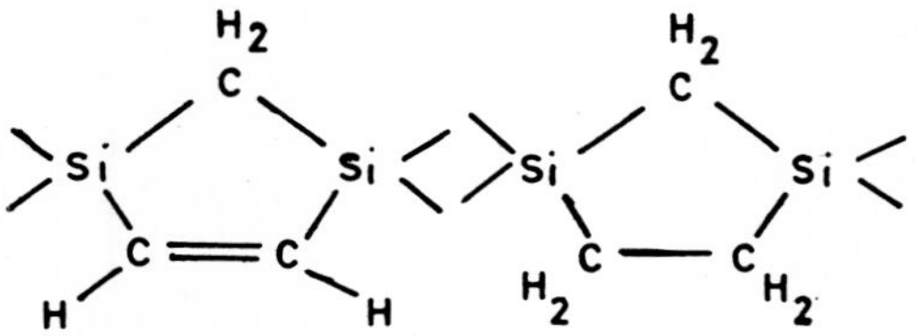	0.09	0.06	—	—
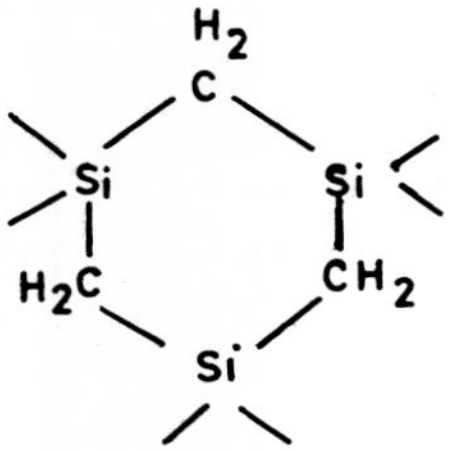	0.87	1.86	0.3	0.06
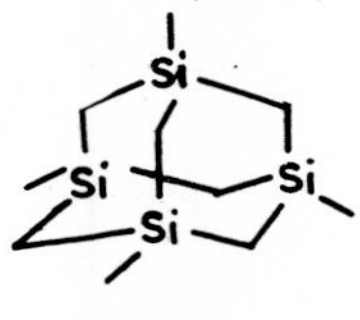	1.09	1.67	0.15	—

[a] Calculations of ratios were made relative to the amount of 1,3-disilapropanes ($= 1$; fixed) present.

reactions. From the results obtained for the decomposition of $SiMe_4$ in plasma, it was noted that the amount of 1,3-disilapropanes and 1,3,5-trisilapentanes produced was larger as opposed to the amount in the pyrolysis mixture of $SiMe_4$. The amount of cyclic compounds present was practically nonexistent. Because consecutive reactions in the case of plasma experiments are definitely limited in comparison to normal pyrolysis, this means that linear carbosilanes play an essential role in building cyclic compounds in the pyrolytic decompositions.

1.7 Mechanism of the Formation of Carbosilanes by Pyrolysis in the Gas Phase

Before a discussion on the reactions forming carbosilanes can be undertaken, the following comments should be made: carbosilanes from $SiMe_4$ are compounds that contain Si atoms with Me groups and to a lesser degree H atoms as substituents. The Si—Si moiety in the molecular skeleton is only observed in exceptional cases. From the methylchlorosilanes, compounds are formed in which Me or Cl groups will occupy those valencies of silicon not bound within the molecular skeleton. Carbon-chlorinated carbosilanes are not formed. The amount of higher-molecular-weight pyrolysis products increases remarkably by thermal work up of the pyrolysis mixture, but hydrocarbon polymers are not formed.

The observed pyrolysis products are formed through very complicated and still not sufficiently elucidated reaction pathways. This uncertainty is caused mostly by the fact that all initially formed compounds of low molecular weight in the continually-working gas circulation apparatus will participate according to their partial pressures in the overall reaction, in addition to the starting methylsilane.

Besides the initial reaction and subsequent reactions involving reactive intermediates, the thermal behavior of simple carbosilanes plays a considerable role in determining the formation of compounds of higher molecular weight. Taking into account the experimentally-proved facts, a mechanism of formation can be established by considering the following aspects:

1) The primary reaction involving the thermal decomposition of methylsilanes in the gas phase.
2) The significance of the insertion of the CH_2 group into the Si—Si bond.
3) The possibility that simple carbosilanes formed initially will react further to build larger compounds.

1.7.1 The Mechanism of Formation via Radical Reactions

Different workers have carried out kinetic investigations on the thermal decomposition of $SiMe_4$ in the gas phase. The first stage [39] is the reaction

$$SiMe_4 \rightarrow Me_3Si\cdot + Me\cdot$$

To explain how a compound forms with a Si—C skeleton it was necessary to include a second reaction

$$Me_3Si—Me + \cdot Me \rightarrow Me_3Si—CH_2\cdot + CH_4$$

so as to have the necessary radicals present to build such Si—C—Si skeletons, thus enabling the subsequent formation of the basic molecular units according to

$$Me_3Si—CH_2· + ·SiMe_3 → Me_3Si—CH_2—SiMe_3$$

The initial step of these reactions was unequivocally proved to proceed via cleavage of an Si—C bond producing methyl and silyl radicals [19, 20].

Of special significance is the isolation of the cyclic compounds 1,3-disilacyclobutane and 1,3,5-trisilacyclohexane. Their formation is connected to the previously mentioned reactions by assuming that a further reaction occurs,

$$Me_3Si—ĊH_2 → Me_2\dot{S}i—ĊH_2 + ·CH_3$$

followed by dimerization and trimerization. On the other hand, Davidson and co-workers [198] postulated the appearance of reactive intermediates with Si=C double bonds in the primary step of the decomposition. Such intermediates have not yet been proved to exist, and neither is any evidence available to suggest that Si=C double bonds could exist at 700 °C in those intermediates in the gas phase. Formulations of this kind have to be understood in a more general sense in describing highly reactive short-lived intermediates of yet uncertain structure.

1.7.2 Insertion of CH_2 Groups into Si—Si Bonds

It is noticeable, that compounds containing Si—H substituents appear in the pyrolysis products of $SiMe_4$ and to a degree in those of methylchlorosilanes. Because silyl radicals form in the pyrolysis of methylsilanes, it is logical to expect a recombination yielding Si—Si bonds. The experimental observation (no Si—Si bonds, but appearance of Si—H bonds) can be explained in terms of a rearrangement, namely, that a neighboring methyl group by insertion into the Si—Si bond will form an Si—CH_2—SiH group [40]. An example of this rearrangement is shown

$$Me_3Si—SiMe_3 → Me_3Si—CH_2—SiMe_2H$$

A further example is the rearrangement of 1,1,2,2,4,4-hexamethyl-1,2,4-trisilacyclopentane at 500 °C, which proceeds with 96% yield according to the following equation [41]:

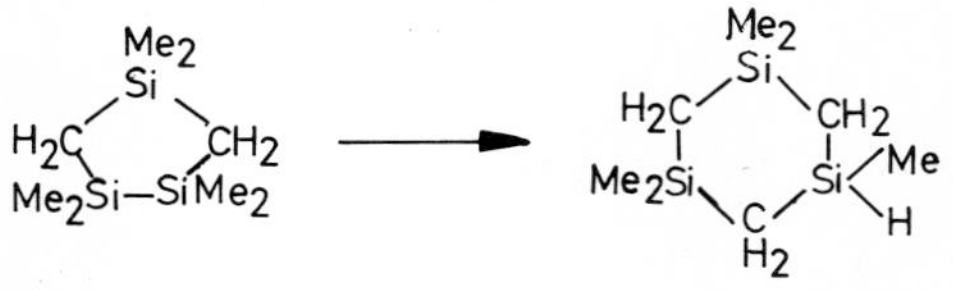

Similarly, the compounds $Me_3Si—SiMe_2—CH_2—SiMe_2H$, $Me_2HSi—CH_2—SiMe_2—CH_2—SiHMe_2$ and $Me_3Si—CH_2—SiMeH—CH_2—SiMe_2H$ are produced on rearrangement of $Me_2Si(SiMe_3)_2$ by methylene group insertion [42]. Methylated and halogenated disilanes, such as $BrMe_2Si—SiMe_2Br$, react by elimination of $SiMe_2$ followed by its insertion into the silicon-halogen bond to form trisilanes.

1.7.3 Participation of Volatile Cyclic Compounds in the Pyrolysis Reaction

In order to discuss the possible participation of simple carbosilanes in subsequent pyrolysis reactions, it was first of all necessary to investigate the thermal behavior of the linear compounds $Me_3Si-CH_2-SiMe_3$, $(Me_3Si-CH_2)_2SiMe_2$ and $(Me_3Si-CH_2-SiMe_2)_2CH_2$ as well as the cyclic compounds shown:

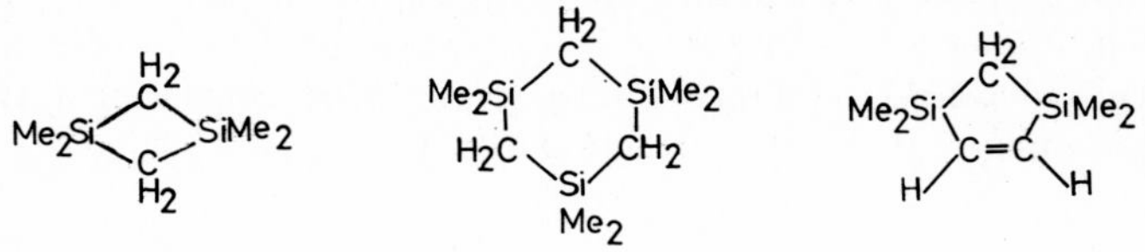

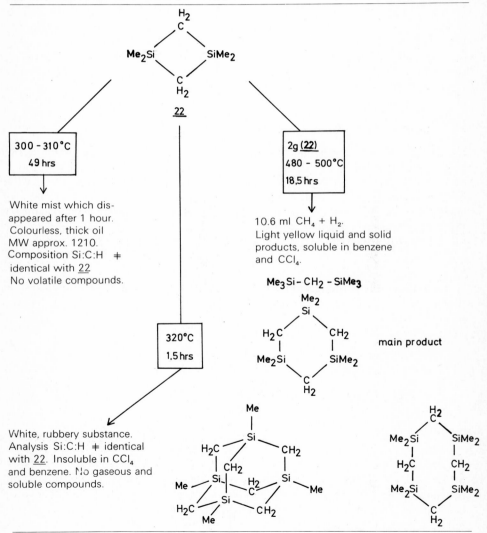

Table 14. Thermal behavior of $(Me_2Si-CH_2)_2$

300 – 310°C
49 hrs

White mist which disappeared after 1 hour. Colourless, thick oil MW approx. 1210. Composition Si:C:H ≠ identical with 22 No volatile compounds.

320°C
1,5 hrs

White, rubbery substance. Analysis Si:C:H ≠ identical with 22. Insoluble in CCl_4 and benzene. No gaseous and soluble compounds.

2g (22)
480 – 500°C
18,5 hrs

10.6 ml $CH_4 + H_2$. Light yellow liquid and solid products, soluble in benzene and CCl_4.

$Me_3Si-CH_2-SiMe_3$

main product

Table 15. Thermal behavior of 1,1,3,3-tetramethyl-1,3-disilapentene

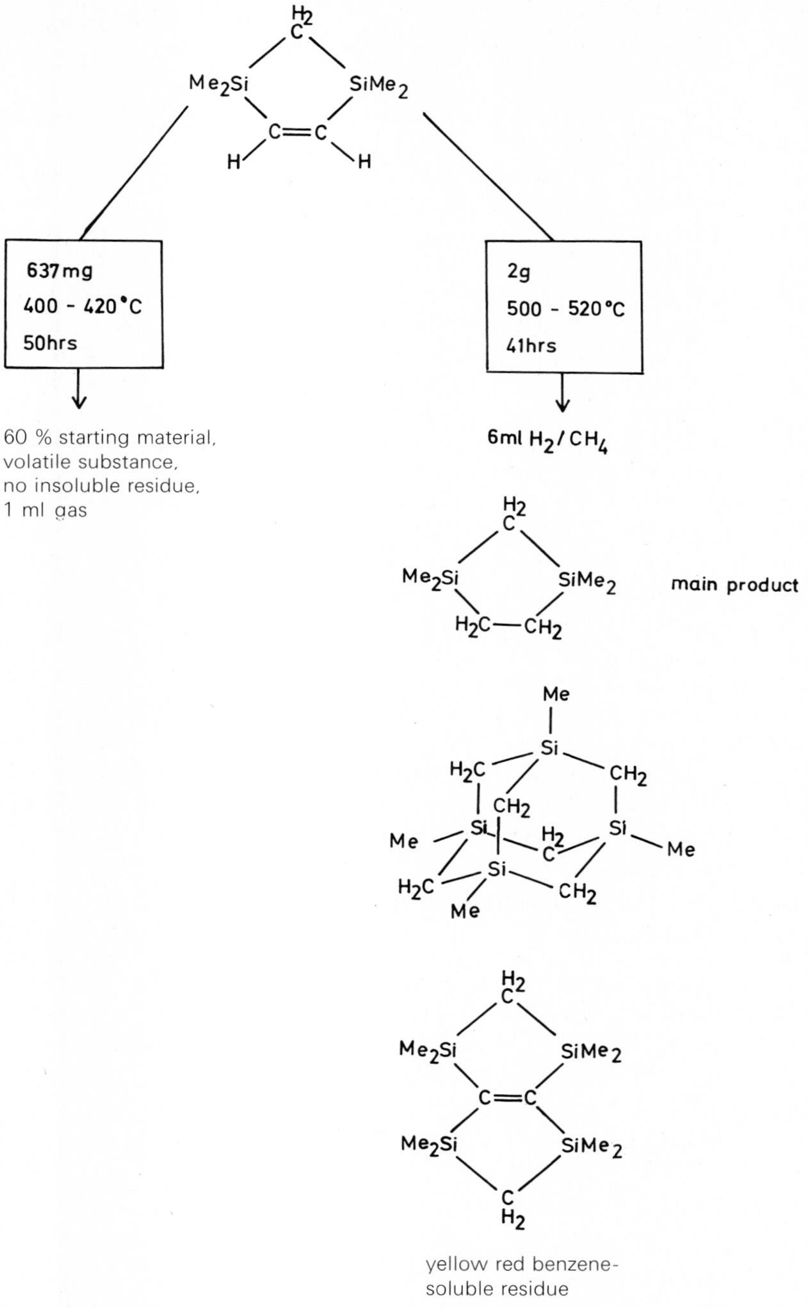

637 mg	**2 g**
400 – 420 °C	500 – 520 °C
50 hrs	41 hrs

60 % starting material, volatile substance, no insoluble residue, 1 ml gas

6ml H_2/CH_4

main product

yellow red benzene-soluble residue

Thermal rearrangement is generally the first thing to expect in compounds with a strained molecular skeleton. The behavior of $(Me_2Si—CH_2)_2$ is shown in Table 14. The reaction was performed in a sealed ampoule under the vapor pressure of the compound. The result showed distinctly the influence of temperature. While polymeric Si—C—Si chains occurred at approximately 300 or 320 °C, structural units were formed at 480–500 °C which correspond to those derived from the pyrolysis of tetramethylsilane whereby some compounds of higher molecular weight were not yet identified.

The corresponding six-membered ring $(Me_2Si—CH_2)_3$, on the other hand, is considerably more stable. A white rubber-like product formed within 3.5 hours at 380–400 °C, while 80% of the starting material remained unreacted along with no significant gas development. At 580 °C (18.5 hours) H_2, CH_4, and volatile silicon compounds of higher molecular weight were observed, along with still-soluble (benzene, CCl_4) silicon-containing compounds and also insoluble components.

Of special interest is the behavior of 1,1,3,3-tetramethyl-1,3-disilapentene as shown in Table 15. The formation of 1,1,3,3-tetramethyl-1,3-disilacyclopentane is attributed to a simple hydrogenation. Apart from unidentified compounds of higher molecular weight, formation of 1,3,5,7-tetrasilaadamantane as well as 2,2,4,4,6,6,8,8-octamethyl-2,4,6,8-tetrasilabicyclo[3.3.0]oct-1(5)ene should be emphasized.

The linear compounds previously mentioned appear more thermally stable than the cyclic compounds given. The decomposition temperature decreases, as expected, with corresponding increase in chain length. For example, 80% of $(Me_3Si)_2CH_2$ at 530 °C after 15 hours is left undecomposed, 75% of $(Me_3Si—CH_2)_2SiMe_2$ at 500 °C after 4 hours and finally 60% of $(Me_3Si—CH_2—SiMe_2)_2CH_2$ is undecomposed at 480 °C after 16 hours.

It follows that all inspected pyrolysis products of $SiMe_4$, under the conditions of pyrolysis, will react with different speeds to form other reaction products. Further, simple carbosilanes formed during the course of pyrolysis act as intermediates in the construction of larger carbosilanes. Mechanisms have been discussed to explain these reactions, but to date no appropriate experimental confirmation has been forthcoming [22].

1.7.4 Comments on the Formation of SiCl-Containing Carbosilanes from Methylchlorosilanes

Similarly to $SiMe_4$, the primary step in the pyrolysis of the methylchlorosilanes in the gas phase at about 700 °C is initiated through the cleavage of an Si—Me bond. In this manner, compounds can be formulated simply by dissociation and subsequent combination steps:

$$Me_3SiCl \rightarrow \cdot CH_3 + \cdot SiMe_2Cl$$

$$Me_3SiCl + \cdot CH_3 \rightarrow CH_4 + \cdot CH_2—SiMe_2Cl$$

$$Me_2ClSi\cdot + \cdot CH_2—SiMe_2Cl \rightarrow Me_2ClSi—CH_2—SiMe_2Cl$$

From the isolated compounds (Table 10) it is immediately apparent that the Si—Cl bond must have been attacked because more chlorine atoms are bound to silicon than in the starting compounds. This particularly applies to compounds formed from

Me_3SiCl and Me_2SiCl_2. A given methylchlorosilane does not give rise to all conceivable 1,3-disilapropanes in the same abundance, and the originally expected product is not necessarily the most abundant. Instead, group arrangements are preferred, the formation of which requires considerable bond breaking and bond making. For example, $Cl_3Si-CH_2-SiCl_3$ and $Me_2ClSi-CH_2-SiMe_3$ are produced from Me_3SiCl as well as from Me_2SiCl_2, indicating a considerable exchange of Cl atoms.

If the Si:C:H:Cl ratio in $MeSiCl_3$ (1:1:3:3) is compared with that in the total mixture of liquid and high molecular weight products, then one obtains the ratio Si:C:H:Cl = 1:0.75:1.82:3.06 for the reaction products (excluding compounds which boil at room temperature and contain no silicon). All the clorine present is bound to silicon and the Si:Cl ratio in the reaction products compared to that in the starting material remains unchanged, while the carbon and hydrogen content decreases.

In order to appreciate how individual elements are distributed in the pyrolysis mixture, an analytical investigation was undertaken into the pyrolysis products of $MeSiCl_3$ by dividing the overall product mixture into three classes:
1) compounds with one Si atom,
2) compounds with two silicon atoms,
3) products of higher molecular weight.

The compounds belonging to group two comprised 71% chlorine, 20% silicon and 6.2% carbon. In group three, the chlorine value decreased to 59.6%, while the silicon (26.4%) and carbon contents (12%) increased. From the results here and in Table 10, it is recognized that a large amount of the chlorine present is contained in the highly chlorinated 1,3-disilapropanes. This is entirely logical if one considers that these compounds possess the least ability to build carbosilanes of higher molecular weight. It is evident from Table 10, that this ability is also decreasing as the number of chlorine atoms in the starting methylchlorosilane of the pyrolysis is increased from Me_3SiCl to $MeSiCl_3$.

The Si:C ratio in the methylsilanes used will obviously influence the structural types produced. This applies in the case of 1,3,5,7-tetrasilaadamantanes which are formed preferentially from pyrolysis of $SiMe_4$ and Me_3SiCl. The Si:C ratio in this skeleton is 4:6, whereas in diadamantanes it is 7:10.

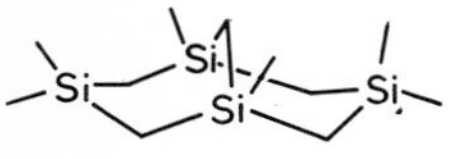

4 : 5 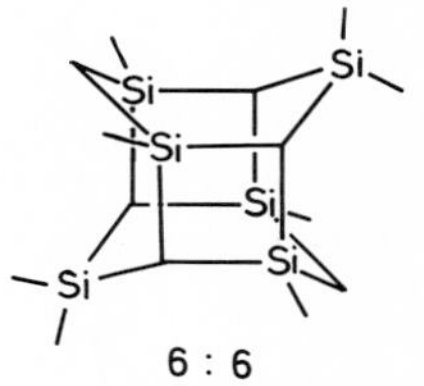   6 : 6 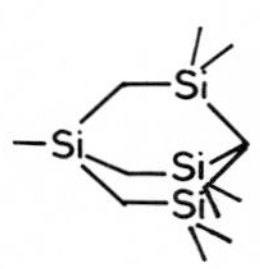   4 : 4

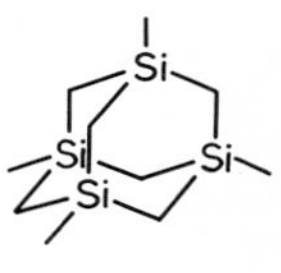

4 : 6 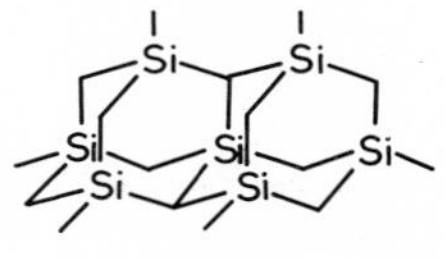   7 : 10

The Si-asterane formed from MeSiCl$_3$ (not observed in the pyrolysis products of SiMe$_4$ or Me$_3$SiCl) has an Si:C ratio of 6:6. In the series from SiMe$_4$ to MeSiCl$_3$, including MeCl$_2$Si$-$CH$-$SiCl$_3$, it is seen that as the number of methyl groups in these starting materials decreases, the Si:C ratio in the molecular structures of the products will tend towards 1:1.

1.7.5 Pyrolysis Products Obtained from EtSiCl$_3$ in Comparison to those from Me$_2$SiCl$_2$

The same carbosilanes produced by the pyrolysis of methylsilanes are obtained also from the decomposition of small carbosilane molecules [22]. One then poses the question of whether or not the pyrolysis of higher organosilanes will proceed to carbosilanes, or to what extent the C$-$C bonds in these silanes will influence the reaction pathway. To investigate this, EtSiCl$_3$ was pyrolysed at 600 °C in the gas phase under conditions given in Chapt. II.1.2 [43]. This pyrolysis does not yield the same carbosilanes as in the pyrolysis products of Me$_2$SiCl$_2$ or MeSiCl$_3$. Instead, linear and cyclic compounds are favored containing two or even more directly-linked carbon atoms, for example HC$\equiv$C$-$SiCl$_2$$-CH_2$$-CH_2$$-$SiCl$_3$, Cl$_3Si-C\equiv$CH, H$_2C=CH-$SiCl$_3$ and the cyclic compounds 1,1-dichloro-1-silacyclopentene(3), 1,1-dichloro-1-silacyclopentene(2) and 1,1,3,3-tetrachloro-1,3-disilacyclohexene(4). Other compounds detected were CH$_4$, C$_2$H$_6$, C$_2$H$_4$, pentane, cyclopentane and benzene, as well as an imperfectly-characterized highly viscous distillation residue. Characteristic carbosilanes such as 1,3,5-trisilacyclohexanes and 1,3,5,7-tetrasila-adamantanes did not form. This means that the C$-$C grouping in EtSiCl$_3$ causes a change in the principal mechanism of formation of the pyrolysis products. This is understandable because the Si$-$C bond is expected to be cleaved in the primary reaction step

$$C_2H_5SiCl_3 \rightarrow \cdot C_2H_5 + \cdot SiCl_3$$

In subsequent reactions various compounds will form in which the C$-$C grouping remains. The stability of C$-$C bonds and the possibility of producing multiple bonds insures that no carbosilanes result from the pyrolysis of EtSiCl$_3$.

The formation of carbosilanes in the gas phase pyrolysis of methylsilanes can be understood from an appraisal of the bond energies of possible reaction products. It is worth considering also the formation of molecules with Si$-$Si groups, as well as compounds with larger C$-$C units, as a result of the tetravalent nature of silicon and carbon. However, these were not observed; only molecular structures with an Si$-$C$-$Si skeleton were found.

1.8 Comparative Summary of the Carbosilanes Synthesized

The numerous carbosilanes produced by pyrolysis can be classified without referring to their substituents. Among the compounds with two or three silicon atoms, all the linear and cyclic molecules possess one of the following molecular skeletons:

$$\equiv Si-CH_2-Si\equiv \quad , \quad \equiv Si-CH_2-\overset{|}{\underset{|}{Si}}-CH_2-Si\equiv$$

Table 16 presents the molecular skeletons of the compounds containing more than 3 silicon atoms. The major structural types are derived from the structure of silicon carbide. Several modifications of silicon carbide exist which are differentiated by their stacking arrangement: type AAA corresponds to the zinc sulphide structure, and type ABA corresponds to the wurtzite structure. Two further modifications contain the same elements but a different stacking arrangement: AABBA and AAABBBA. The smallest structural unit of the zinc sulphide structure is the adamantane skeleton. To date a whole series of molecular compounds has been isolated which consist of 1,2,3 or 4 of these structural units and which are known as carborundanes (column 1). The structural unit of the wurtzite type is represented by the compound shown at the top of column 2. To date, this compound could not be isolated, but the compound underneath has been definitely identified. It consists of the structural units of adamantane and wurtzitane placed together. The compound underneath contains yet a further structural unit, namely, that of the barrelanes. This class of compounds represents the smallest molecular unit which incorporates all essential components of the 4H-SiC structure, namely, the adamantane as well as the complete wurtzitane skeleton [185]. While the first and second vertical columns of Table 16 contain carbosilanes which are derived from the structural elements of silicon carbides, columns 3 and 4 illustrate compounds not derived from silicon carbide, but composed of Si—C—Si six-membered rings in the boat form. The high symmetry of the compound at the bottom of column 3 is noteworthy. Compounds of this type are known as scaphanes [25]. Column 4 shows carbosilanes with barrelane or asterane structural units or a combination of both. In the horizontal rows, compounds maintain the same number of Si atoms in different structural units.

A comparison of the compounds now known containing more than 5 Si atoms in the molecule shows that 17 compounds contain the adamantane skeleton as the basic molecular unit (Fig. 5). Two other compounds contain the adamantane skeleton as a part of their overall skeleton. Only three other types are built based on a different structural principle. This outcome shows distinctly the significance of the adamantane structure in the formation of carbosilanes by pyrolysis reactions.

1.9 A Possible Chemical Pathway to the Synthesis of
of Polycyclic Molecular Skeletons

A number of carborundanes, each of which contains six or more Si atoms, has been isolated through systematic separation of the pyrolysis products of SiMe₄ by HPLC [37, 44]. These compounds in particular can be understood in terms of stepwise addition of carbon bridged silicon atoms into the molecular skeleton. To date, the

Table 16. Molecular skeletons of carbosilanes obtained from the pyrolysis of $SiMe_4$ and of the methylchlorosilanes

Columns					Number of Si atoms
1	2	3	4	5	
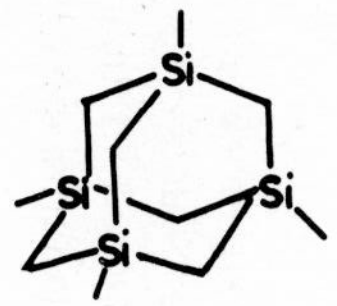		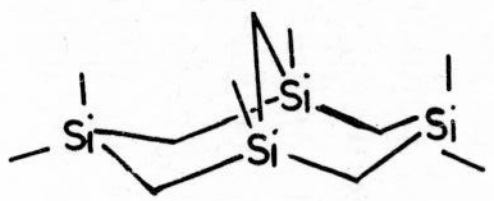	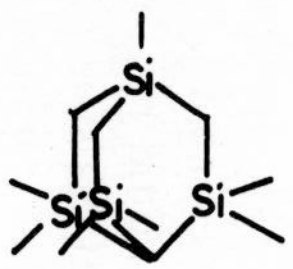	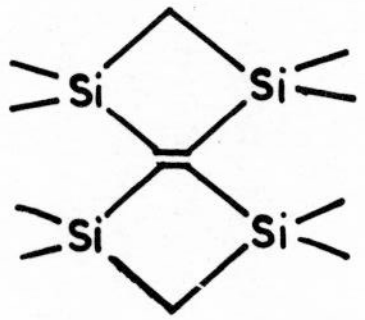	4
		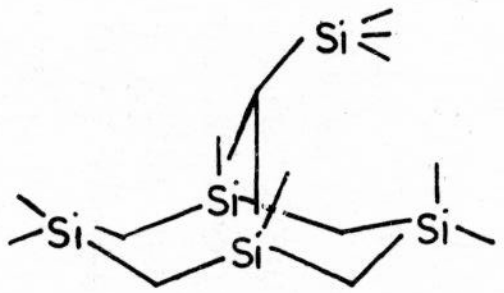			5

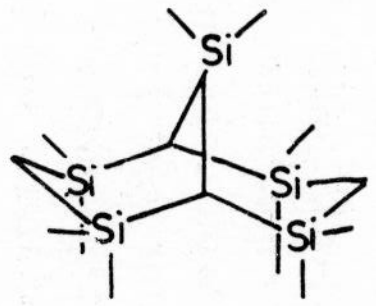

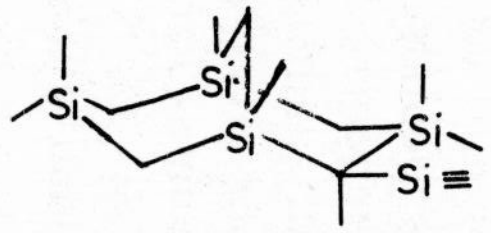

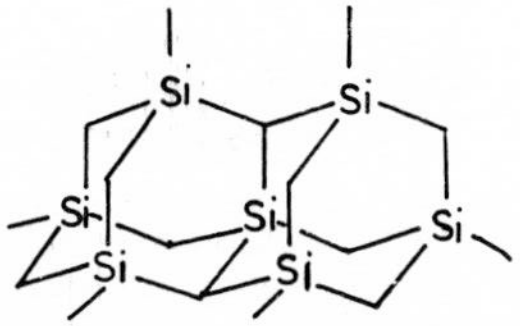

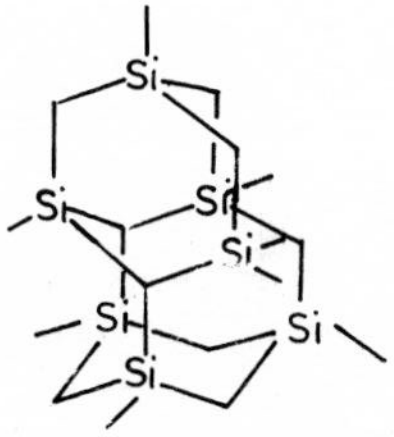

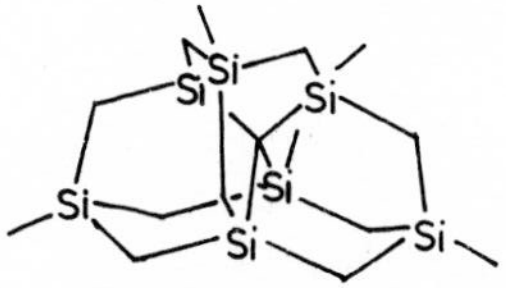

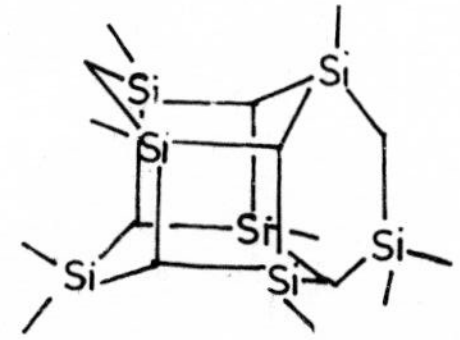

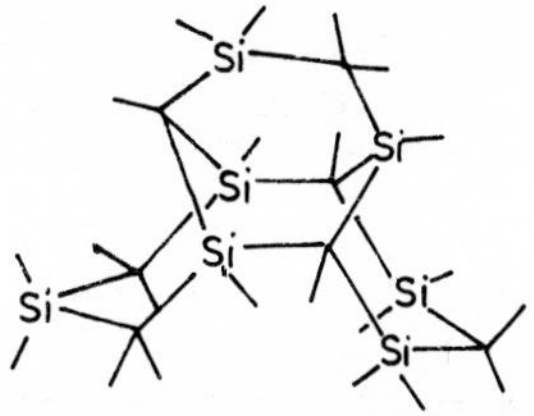

(Continued)

6

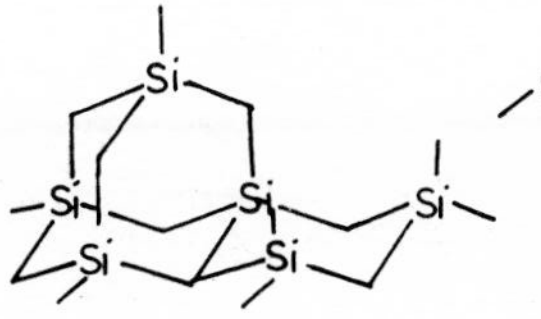

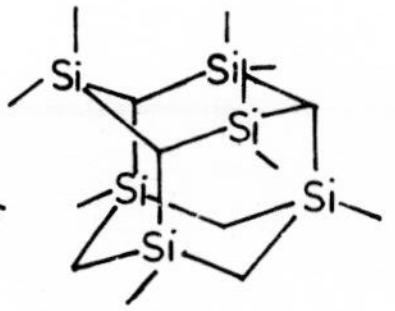

7

(not isolated)

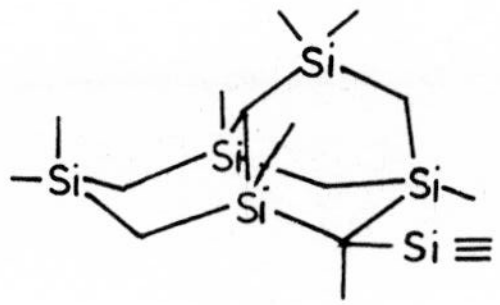

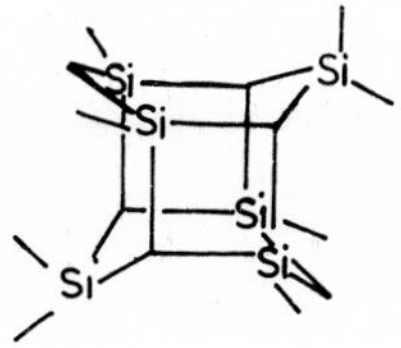

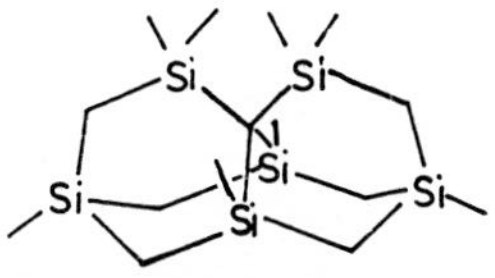

Table 16 (cont.)

Columns					Number of Si atoms
1	2	3	4	5	

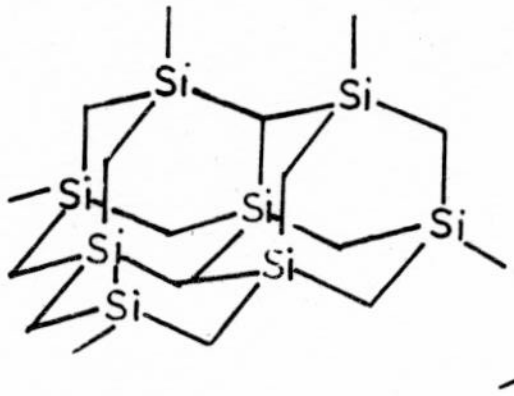

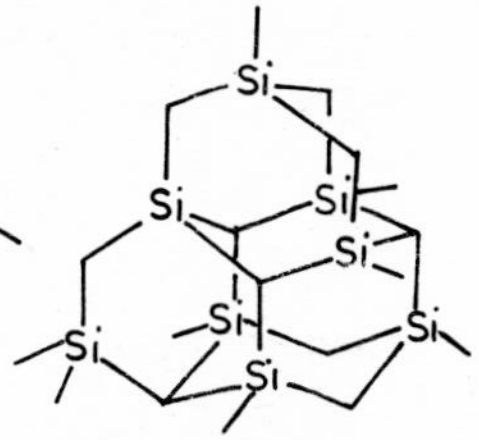

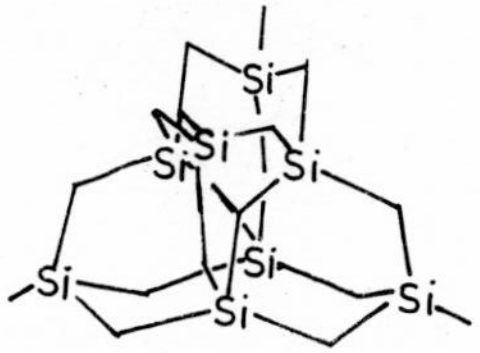

8

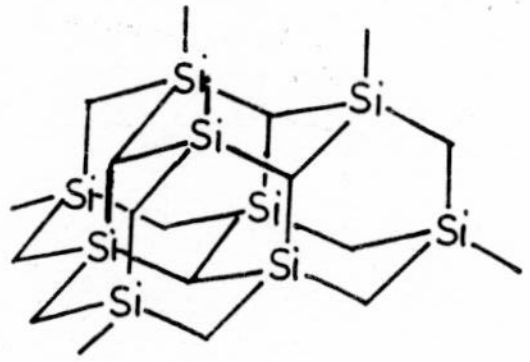

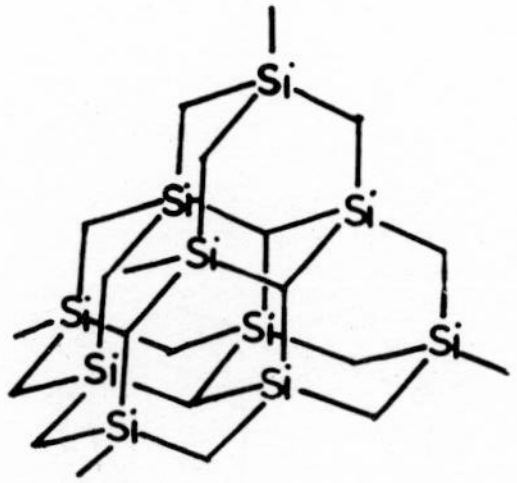

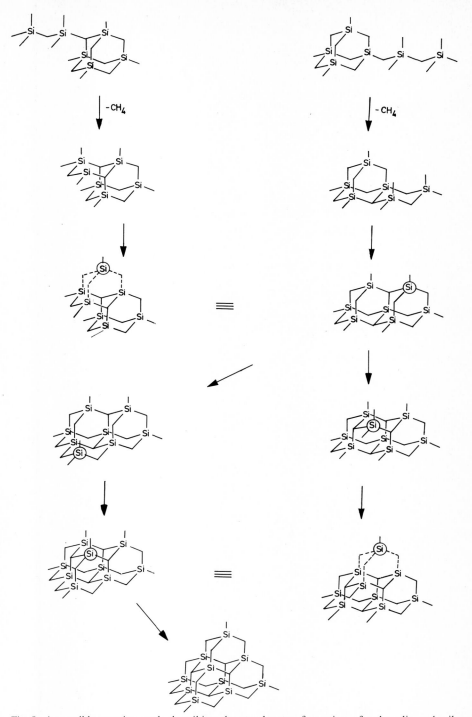

Fig. 8. A possible reaction path describing the step-by-step formation of polycyclic carbosilane skeletons

largest compound to be isolated and analysed is $Si_{10}C_{20}H_{40}$, the structure of which is shown in Fig. 8. Although the formation of this compound is satisfactorily explained by the sequence depicted in Fig. 8, it still leaves unsettled the question concerning the structure of the compounds of still higher molecular weight, namely those compounds existing in the pyrolysis mixture as solid, glassy, meltable substances. As a result of this investigation, two different synthetic routes can at least be proposed: either an extension of the molecules by insertion of SiMe groups into the skeleton, thus producing higher-molecular-weight carborundane units, or binding smaller skeletal units together (e.g. 1,3,5,7-tetrasilaadamantane) by CH_2-bridges, as illustrated by the structure of compounds 67 and 68.

2. Formation of Carbosilanes by Direct Reaction of Halogenomethanes with Silicon

2.1 Reactions of Silicon with Chloromethanes

2.1.1 The Reaction with CH_2Cl_2

Subsequent to the papers by E. G. Rochow [45a] and R. Müller [45b] on the direct synthesis of methylchlorosilanes, Patnode and Schiessler in an U.S.-Patent as early

Table 17. Distillable compounds from the reaction of silicon with CH_2Cl_2

Compounds	No.	Bp. (°C)	p (mm Hg)	%
CH_2Cl_2, Me_2SiCl_2, $MeSiCl_3$		40–60	760	10
$(HCl_2Si)_2CH_2$	113	150–170	760	20–30
$HCl_2Si-CH_2-SiCl_3$	114	30	10^{-2}	
$(Cl_3Si)_2CH_2$	70			
$(Cl_2Si-CH_2)_2$	115			
$HCl_2Si-CH_2-SiCl_3$	114	180	760	20–30
$(Cl_3Si)_2CH_2$	70	40	10^{-2}	
$HCl_2Si-CH_2-SiCl_2-CH_2-SiCl_3$	116	60–70	10^{-2}	20–30
$(Cl_3Si-CH_2)_2SiCl_2$	80			
$(Cl_2Si-CH_2)_3$	83			
$(Cl_3Si-CH_2SiCl_2)_2CH_2$	117	90–110	10^{-2}	10–20
$(Cl_2Si-CH_2)_4$	118			
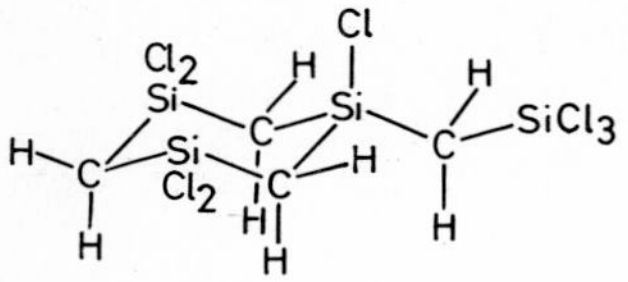 (structure below)	119			
			Residue	10–20

as 1947 reported on the reaction of silicon with CH_2Cl_2 in the presence of copper [45c]. The products given were linear compounds of formula $X_3Si(CH_2)_nSiX_2Y(X=Cl,$ $Y = H$ or Cl, $n = 1$–4) as well as cyclic organosilicon compounds of formula $(SiCl_2CH_2)_x$ ($x \geq 3$). Any further identification of these compounds was not given. Later, our research group isolated from a similar reaction mixture the crystalline compound 1,1,3,3,5,5-hexachloro-1,3,5-trisilacyclohexane $(Cl_2Si-CH_2)_3$ as well as the eight-membered ring $(Cl_2Si-CH_2)_4$ [46, 47]. In addition to these Si-compounds the corresponding linear analogues such as $Cl_3Si-CH_2-SiCl_2-CH_2-SiCl_3$, viscous and undistillable compounds resulted. Table 17 shows the fractions obtained by distillation.

On carrying out the catalysed reaction between Si and CH_2Cl_2 in a fluid bed at about 320 °C, an increase in the amount of high-molecular reaction products occured [48]. Considerable difficulty was encountered in separating this viscous mixture, since Si—Cl groups hinder chromatographic separation. Therefore, the Si—Cl groups were substituted by reaction with $LiAlH_4$, which by experience does not cause any change to the Si—C molecular skeleton. The corresponding SiH-containing mixture could then be separated by High Pressure Liquid Chromatography (HPLC) [44], enabling the isolation of individual compounds.

Table 18 shows the isolated and identified compounds which can be classified in the following way:
1) linear carbosilanes with two terminal SiH_3 groups, $Si_nC_{n-1}H_{4n}$ ($n = 4$–12)
2) linear carbosilanes with one terminal SiH_3 and one terminal Me group, $Si_nC_nH_{4n+2}$
3) Cyclic carbosilanes with a side chain on a ring Si atom,

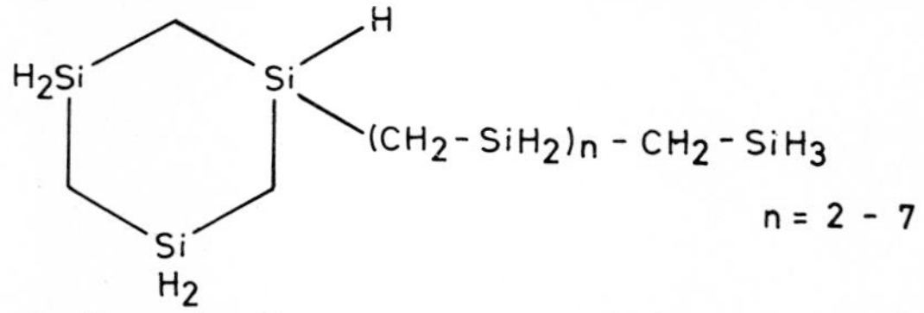

$n = 2 - 7$

4) Cyclic carbosilanes with a side chain on a ring C atom,

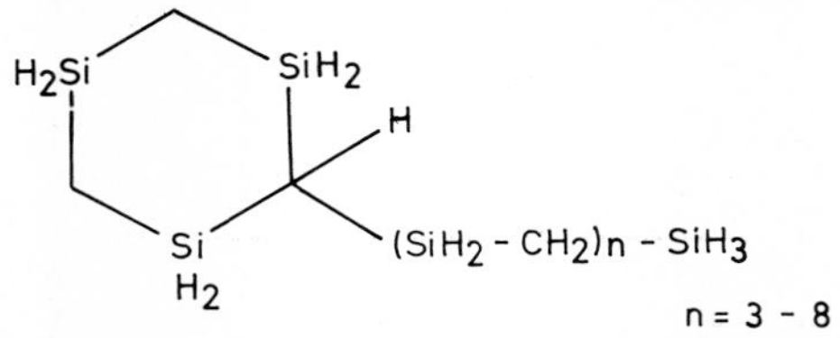

$n = 3 - 8$

The reaction product of Si with CH_2Cl_2 consists of the corresponding Si-chlorinated compounds. Compounds of higher molecular weight range from honey-like and tar-like to solid glassy substances. The viscosity of the SiH-containing compounds is distinctly smaller in comparison. The yield of compounds of higher molecular weight was around 20%, and their formation was favored by the lowest possible reaction temperatures in the fluid bed [48].

2.1.2 The Reaction with $CHCl_3$

Previously, Müller and coworkers [49] studied the reaction of Si and $CHCl_3$. They were able to isolate and identify a number of low molecular weight products which

Table 18. Compounds containing 4 or more Si atoms from the reaction of CH_2Cl_2 with Si, after hydrogenation

$H_3Si(CH_2-SiH_2)_2CH_2-SiH_3$ 120

$H_3Si(CH_2-SiH_2)_3Me$ 121

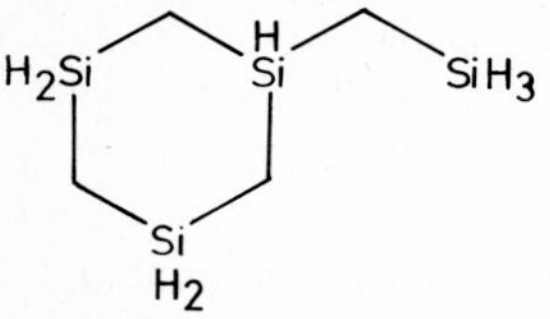

122

$H_3Si(CH_2-SiH_2)_3CH_2-SiH_3$ 123

$H_3Si(CH_2-SiH_2)_4CH_2-SiH_3$ 124

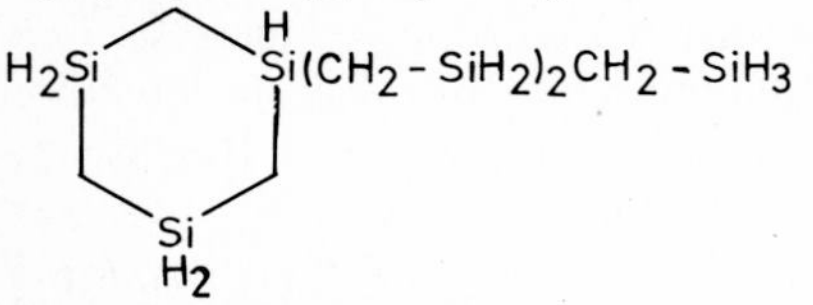

125

$H_3Si(CH_2-SiH_2)_5Me$ 126

$H_3Si(CH_2-SiH_2)_5CH_2-SiH_3$ 127

$H_3Si(CH_2-SiH_2)_6CH_2-SiH_3$ 128

$H_3Si(CH_2-SiH_2)_7CH_2-SiH_3$ 129

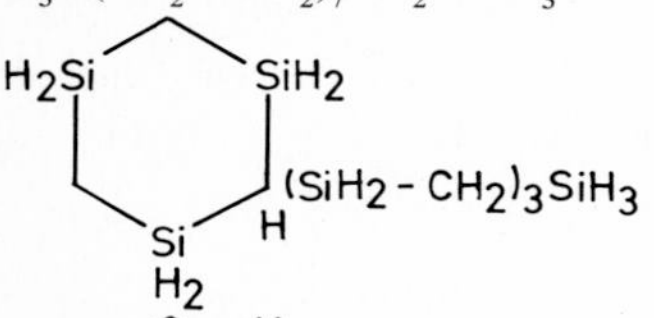

130

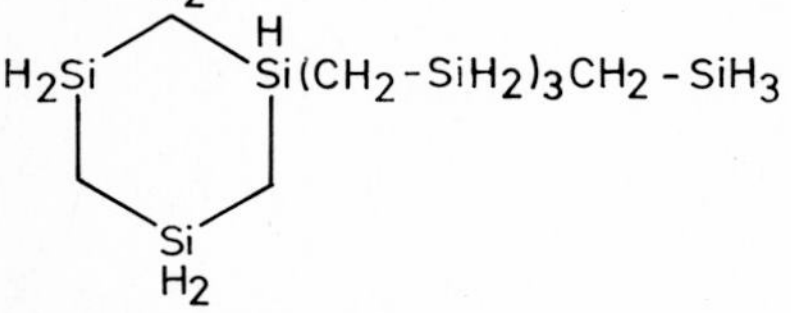

131

$H_3Si(CH_2-SiH_2)_6Me$ 132

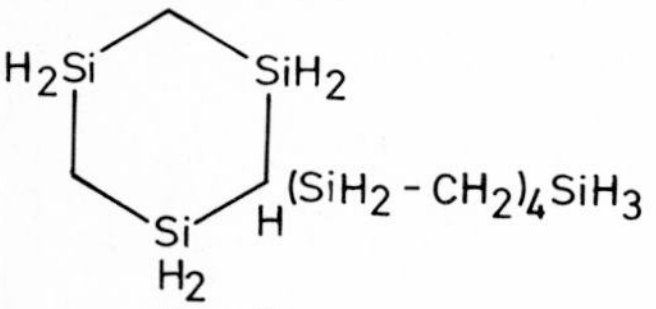

133

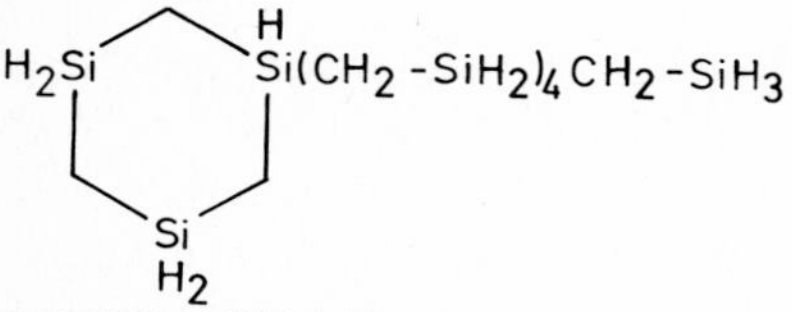

134

$H_3Si(CH_2-SiH_2)_7Me$ 135

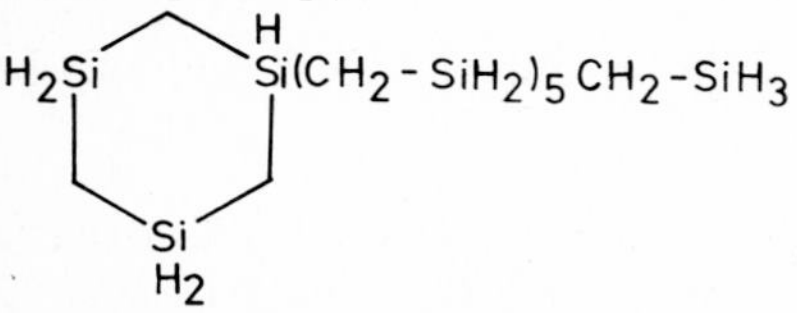

137

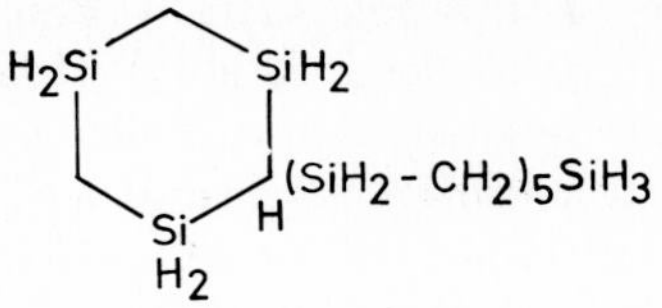

136

$H_3Si(CH_2-SiH_2)_8Me$ 138

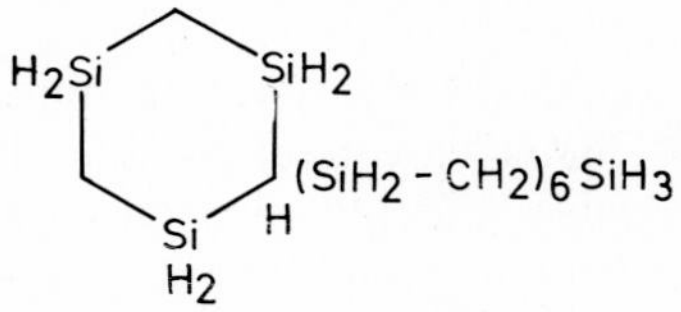

139

$H_3Si(CH_2-SiH_2)_8CH_2-SiH_3$ 140

$H_3Si(CH_2-SiH_2)_9CH_2-SiH_3$ 141

$H_3Si(CH_2-SiH_2)_{10}CH_2-SiH_3$ 142

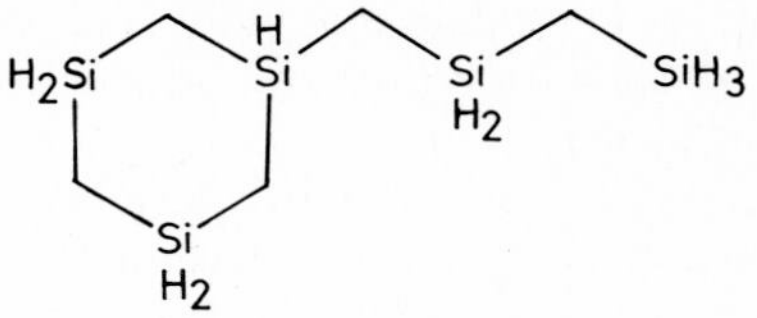

143

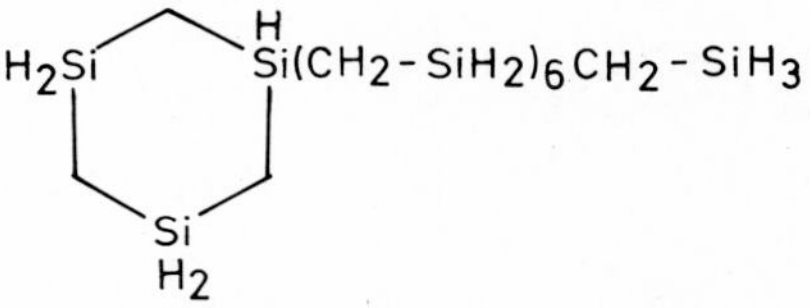

144

$H_3Si(CH_2-SiH_2)_9Me$ 145

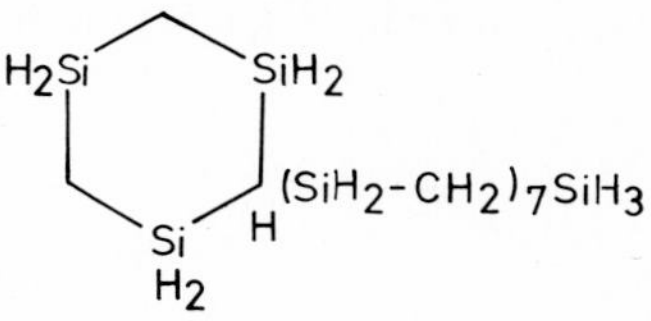

146

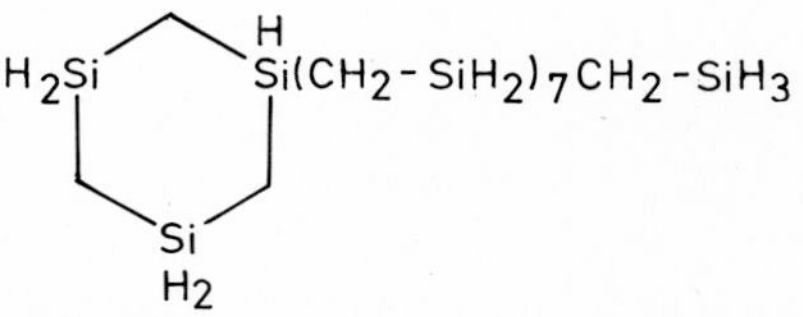

147

$H_3Si(CH_2-SiH_2)_{10}Me$ 148

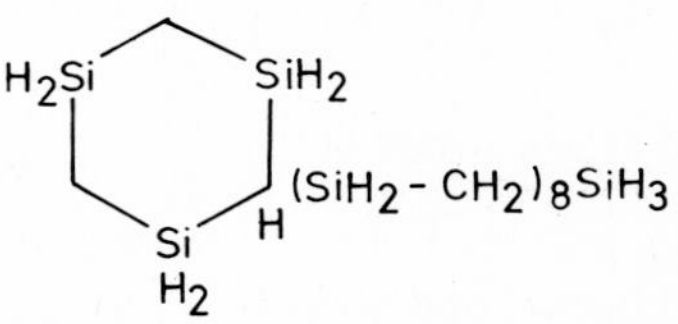

149

Table 19. Distillable compounds from the reaction of Si with $CHCl_3$

Compounds	No.	Bp. (°C)	p (mm Hg)	%
$HSiCl_3$, CH_2Cl_2, $CHCl_3$		40–65	760	10
$HCl_2Si-CH_2-SiCl_3$	114	180	760	20–30
$(Cl_3Si)_2CH_2$	70	40	10^{-2}	
$(Cl_3Si)_2CH_2$	70	50	10^{-2}	10–20
$Cl_3Si-CHCl-SiCl_3$	150			
$HCl_2Si-CH(SiCl_3)_2$	151	60–75	10^{-2}	30–40
$(Cl_3Si)_3CH$	152			
$(Cl_3Si-CH_2)_2SiCl_2$	80			
$Cl_3Si-CHCl-SiCl_2-CH_2-SiCl_3$	153	90–110	10^{-2}	10–15
$(Cl_3Si)_2CH-SiCl_2-CH_2-SiCl_3$	154			
$(Cl_3Si-CH_2-SiCl_2)_2CH_2$	117			

(Structure 155: a six-membered ring)
Cl_2Si — $C(H_2)$ — $SiCl_2$ — $CH(SiCl_3)$ — $Si Cl_2$ — $C(H_2)$ (ring)
No. **155**

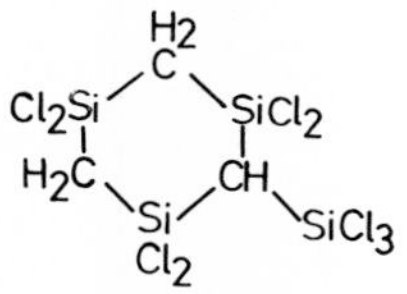

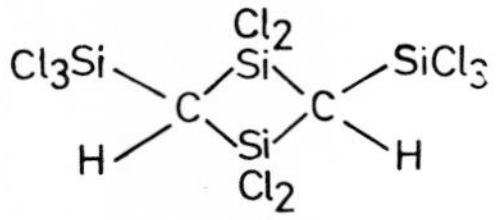

(Structure 156, cis)
Cl_3Si — $C(H)$ — $Si Cl_2$ — $C(H)$ — $SiCl_3$ with $Si Cl_2$ bridge
cis   No. **156**

(Structure, trans)
Cl_3Si — $C(H)$ — $Si Cl_2$ — $C(H)$ — $SiCl_3$ with $Si Cl_2$ bridge
trans

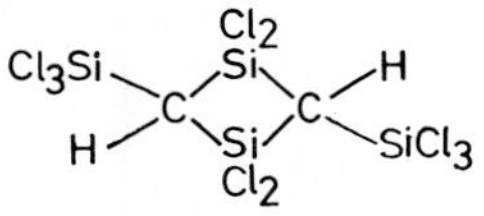

| | | | Residue | 10–15 |

are illustrated in Table 19. This same reaction carried out at about 320 °C in a fluid bed favored the formation of products of high molecular weight. After conversion of the SiCl groups to SiH groups, these compounds could be chromatographically separated, which in turn allowed for structural determinations to be carried out. The results obtained are shown in Table 20.

The compounds may be divided into four groups:
1. Unbranched carbosilane chains with terminal SiH_3 groups.
2. Carbosilane chains with one C-branch.
3. Carbosilane chains with two C-branches.
4. 1,3,5-trisilacyclohexanes with one, two and three Si-substituents on the carbon atoms of the carbosilane ring.

Table 20. Compounds obtained from the reaction of Si with $CHCl_3$ after conversion of SiCl to SiH groups. Only compounds containing more than 4 Si atoms are shown

157

158

120

159

160

161

162

163

164

123

124

165

166

167

168

169

170

171

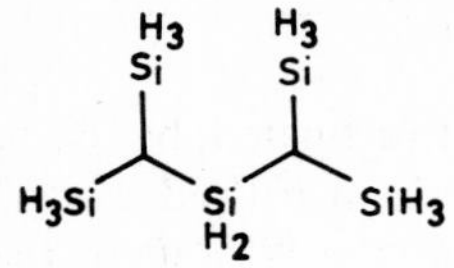

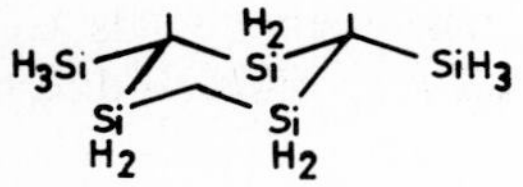

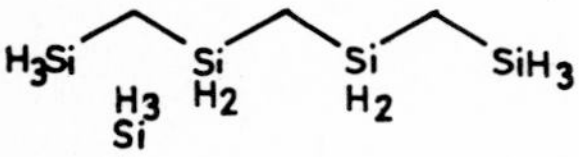

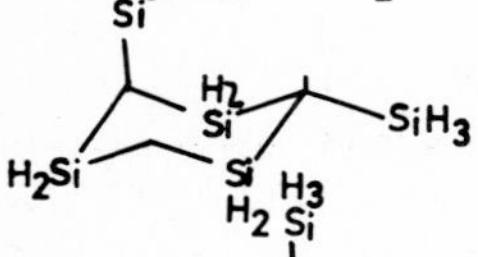

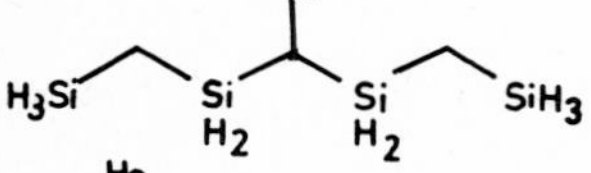

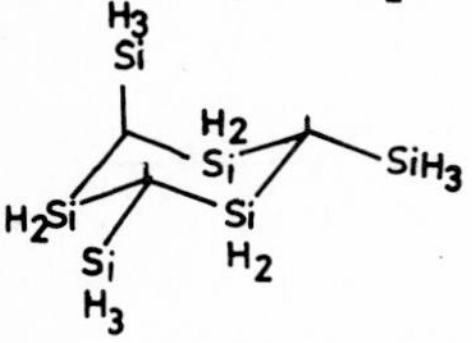

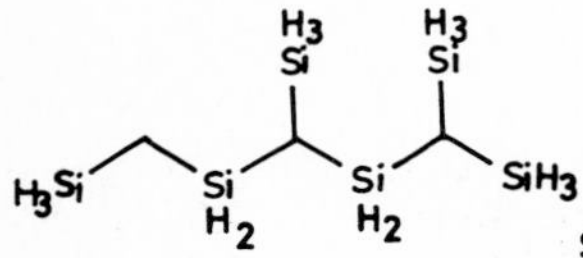

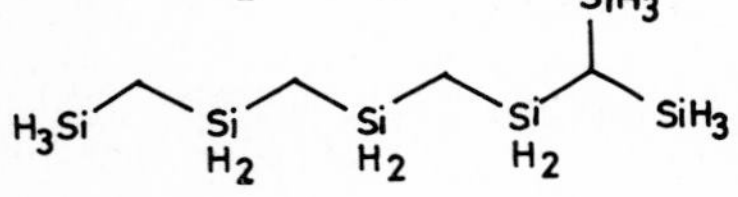

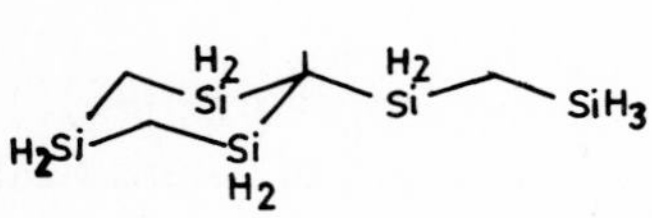

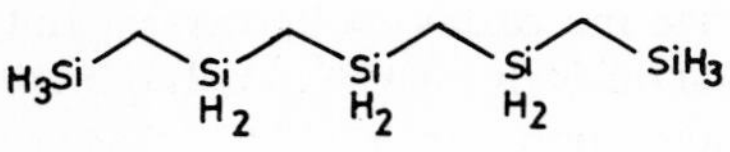

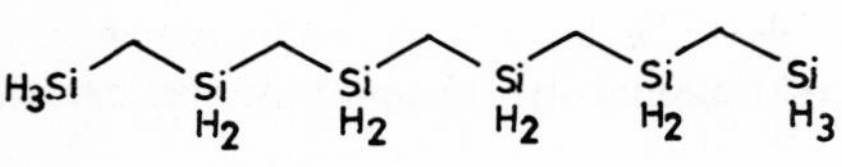

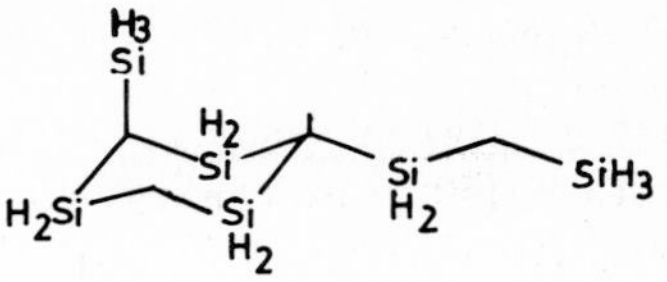

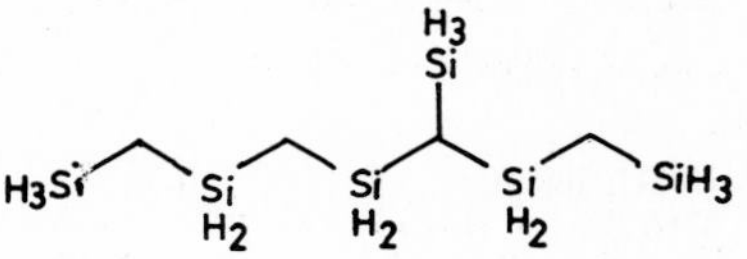

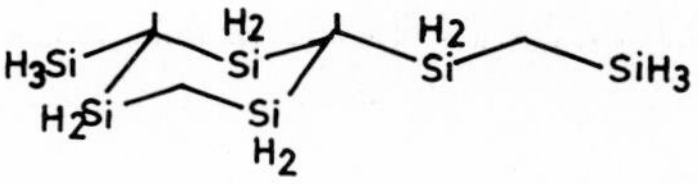

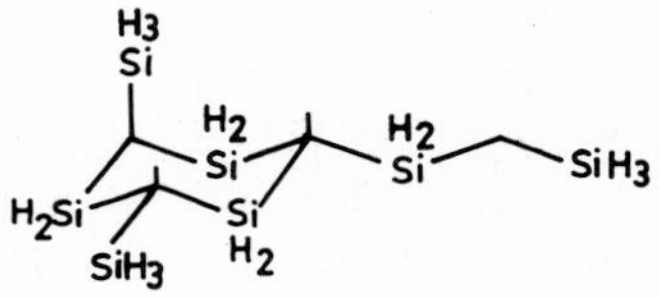

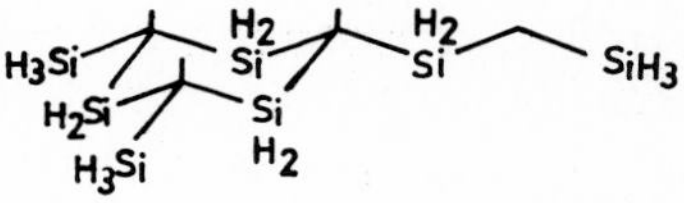

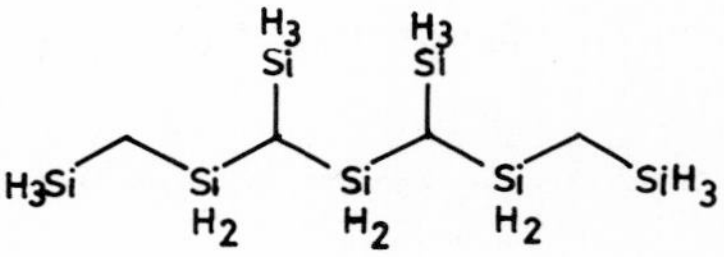

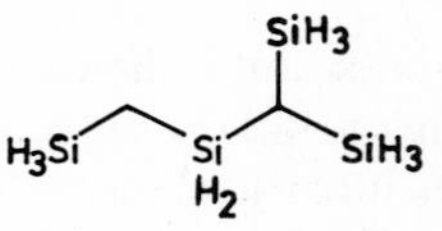

It is worthwile mentioning that carbosilane chains of varying lengths are linked to the skeletal C atoms of 1,3,5-trisilacyclohexane.

2.1.3 The Reaction with CCl_4

The reaction between silicon and CCl_4 has also been investigated by R. Müller and coworkers [50]. The reaction repeated at about 320 °C in a fluid bed yielded products which corresponded basically to what was previously found, namely, substituted derivatives of methane, ethylene and acetylene. These compounds are shown in Table 21. Some of them are accessible by no other way than by direct synthesis.

Table 21. Compounds obtained from the reaction of Si with CCl_4

Compounds	No.	Bp. (°C)	p (mm Hg)	%
CCl_4 C_2Cl_4		25	10^{-2}	20–30
C_2Cl_4 C_2Cl_6 $Cl_3Si-C \equiv C-SiCl_3$	172	40	10^{-2}	20
Cl_3Si $>C=C<$ $SiCl_3$ (Cl, Cl)	173	60	10^{-2}	30–40
$(Cl_3Si)_3CH$	174			
$(Cl_3Si)_3CCl$	175	90	10^{-2}	10–20
$(Cl_3Si)_4C$	176	Sublimation of Residue		
Cl_3Si, Cl_3Si $>C<$ $Si(Cl_2)$ $>C<$ $SiCl_3$, $SiCl_3$ (Cl_2)	177	100–120	10^{-3}	

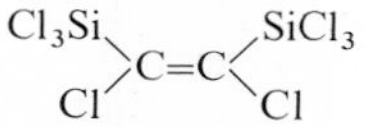

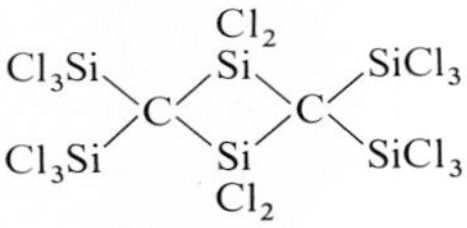

2.2 Advantages Associated with Forming Carbosilanes in a Fluid Bed

In summarizing these results it was established that the reactions between Si(Cu) and CH_2Cl_2 or $CHCl_3$ in a fluid bed will nearly exclusively produce carbosilanes. Reaction with CH_2Cl_2 produced linear carbosilanes in approximately 70–80 % yield, while from $CHCl_3$ carbon-branched carbosilane chains and carbon substituted trisilacyclohexane rings were obtained in similar yield. Bi- and polycyclic carbosilanes were not detected. It was also firmly established that as the reaction temperature increased, the amount of higher-molecular-weight compounds obtained in the product mixture decreased.

2.3 Comments on Mechanism of Formation

The reaction pathway followed by the reaction of CH_2Cl_2 with silicon can be understood by considering the mechanism developed by Klebanski and Fikhtengol'ts [51] for the reaction of Si(Cu) with MeCl. The first stage involves adsorption of CH_2Cl_2 molecules on the surface of the catalyst, whereby the CH_2 group and the Cl atom of CH_2Cl_2 are directed towards the Si atom and Cu of the catalyst respectively according to their charge distribution: $\overset{\delta-}{Cl}-\overset{\delta+}{CH_2}-\overset{\delta-}{Cl}$ and $\overset{\delta+}{Cu}-\overset{\delta-}{Si}$. The next stage involves the formation of an $Si-CH_2-Si$ group and of CuCl. This occurs by CH_2Cl_2 giving up its Cl atoms, the remaining CH_2 group being then coordinated by two Si atoms. Subsequently, transfer of the Cl atoms onto Si occurs, whereby at first a transition state is assumed to exist from which emerges simultaneous desorption of the molecule from the surface of the catalyst. The uptake of Cl must be balanced with that of the CH_2 group. When this is achieved, reaction will proceed via extension of the length of the molecular side chain, and cease only when Cl uptake outweighs its counterpart. In this case, the valencies of silicon will be satisfied by bonding three Cl atoms to it. Increasing the temperature accelerates the successive transfer of chlorine atoms onto silicon, and as a consequence of pyrolysis a decrease in uptake of CH_2-units is observed. Simultaneous surrender of two Cl atoms from CH_2Cl_2 is confirmed by the fact that the otherwise expected intermediate $ClCH_2-SiCl_3$ is not found in the reaction mixture. The $Si-Cl$ bond which forms once in this proposed mechanism does not take part in further reactions. This is also shown by the failure of $MeSiCl_3$ to react with Si(Cu) under the same conditions.

This suggested mechanism clarifies why unbranched carbosilane chains are formed in the reaction with CH_2Cl_2, while chain branching occurs in the reaction involving $HCCl_3$. It is also evident that linear compounds will not form during the reaction of Si with CCl_4.

These results generated some questions. It is conceivable that the reaction initiated on the catalytic surface will continue in the gas phase, producing eventually the observed compounds (carbosilane chains and 1,3,5-trisilacyclohexanes with carbosilane chains as substituents). Therefore, isolated carbosilane chains such as $(Cl_3Si-CH_2-SiCl_2)_2CH_2$ were heated to the usual reaction temperature without a catalyst, and in doing so generated no reaction even after an extended time. Similarly, by passing gaseous carbosilanes over the activated catalyst under the original conditions of reaction (350 °C, fluid bed) no reaction occurred. It can be concluded, therefore, that the overall reaction occurs on the surface of the catalyst and that the molecules desorbed from the surface of the catalyst under the given conditions do not react further in the gas phase.

Only at temperatures of about 500 °C and reaction times of at least 3 days can changes in the carbosilanes be observed. Linear carbosilanes containing 4 or more silicon atoms produce $(Cl_2Si-CH_2)_3$ through degradation of the $Si-C$ chain. However, this is not achieved starting with $(Cl_3Si-CH_2)_2SiCl_2$ [179].

2.4 The Reaction with $(Cl_3Si)_2CCl_2$

A novel group of carbosilanes is formed in the reaction of $(Cl_3Si)_2CCl_2$ with silicon (Cu catalyst) in a fluid bed at about 310 °C. Besides of $(Cl_3Si)_4C$ the following compounds were isolated [179].

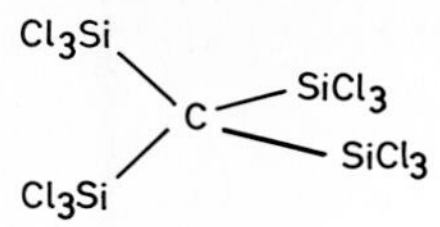

$$Si_4 C Cl_{12}$$

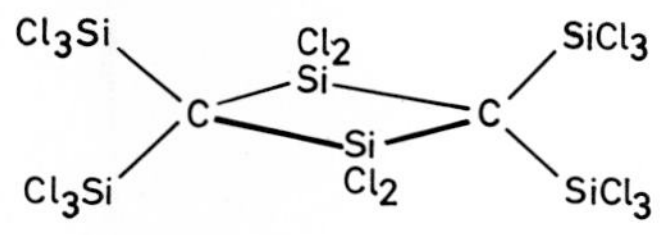

$$Si_6 C_2 Cl_{16}$$

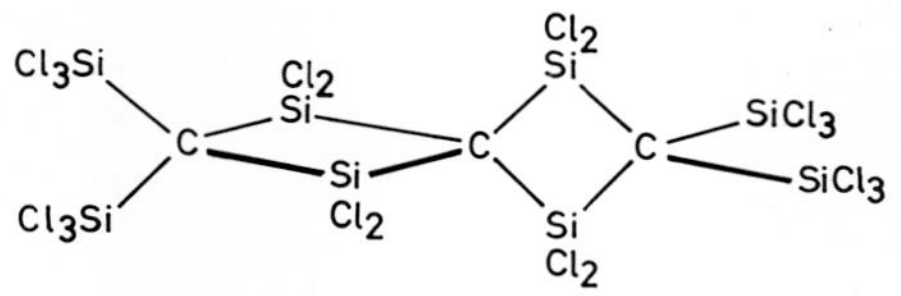

$$Si_8 C_3 Cl_{20}$$

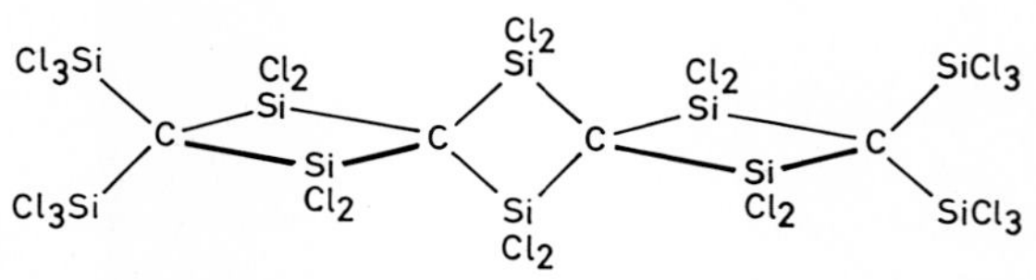

$$Si_{10} C_4 Cl_{24}$$

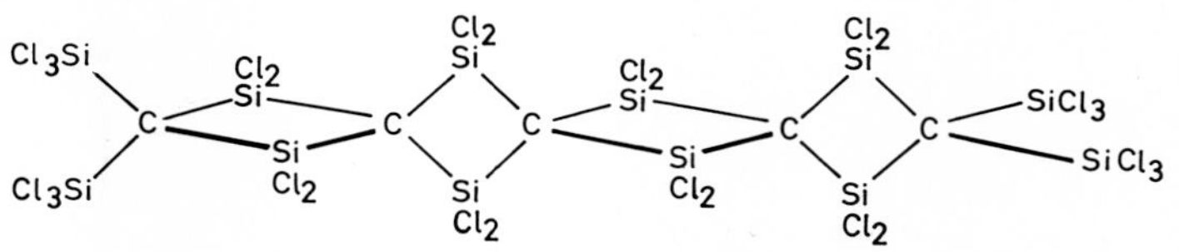

$$Si_{12} C_5 Cl_{28}$$

Spiro-connected four-membered ring structures with a common carbon atom were obtained. Small amounts of even higher molecular carbosilanes of the same structure type seem also to be present. The isolated compounds form white soluble crystals.

3. Formation of Polycyclic Molecular Skeletons through Rearrangement of Carbosilanes with AlBr$_3$ or AlCl$_3$

3.1 Introduction

The formation of carbosilanes through rearrangement reactions using AlCl$_3$ is already known. This process is best illustrated by considering the transition of Si—Si—C to

Si—C—Si by $AlCl_3$. This transition is accelerated by $AlCl_3$, since the Si—C—Si system is an energetically favored one:

$$Me_3Si-SiMe_2(CHCl_2) \xrightarrow[70-80\,°C]{AlCl_3} Me_2ClSi-CHCl-SiMe_3$$

$$\downarrow 140-150\,°C$$

$$Me_2ClSi-CHMe-SiClMe_2$$

In this case, the Si—C—Si moiety is even more favored by the formation of another Si—Cl group [52, 53].

Frye, Weyenberg and Klosowski [54] were the first to report on the formation of 1,3,5,7-tetramethyl-1,3,5,7-tetrasilaadamantane by influence of aluminium halides on $(Me_2Si-CH_2)_3$. Later, it was shown that linear carbosilanes like $(Me_3Si-CH_2)_2SiMe_2$ on reaction with $AlBr_3$ produced 1,3,5,7-tetrasilaadamantanes, proceeding through the cyclic intermediate $(Me_2Si-CH_2)_3$ [55]. Smaller amounts of monobrominated carbosilanes such as $Me_3Si-CH_2-SiMe_2Br$, $Me_3Si-CH_2-SiMe_2-CH_2-SiMe_2Br$ and Si-brominated 1,3,5-trisilacyclohexane were present. The next step was to investigate whether an additional ring-closing reaction could be induced involving carbosilane side chains present on the Si—C skeleton of, for example, 1,3,5,7-tetrasila-adamantane.

3.2 The Reactions of Structurally Different Carbosilanes

A carbosilane chain present on a skeletal carbon atom containing three Si atoms was not cleaved. Instead, ring closure was initiated through elimination of $SiMe_4$. The tertiary CH groups here are not attacked. The overall validity of this ring-building was established by considering the following examples:

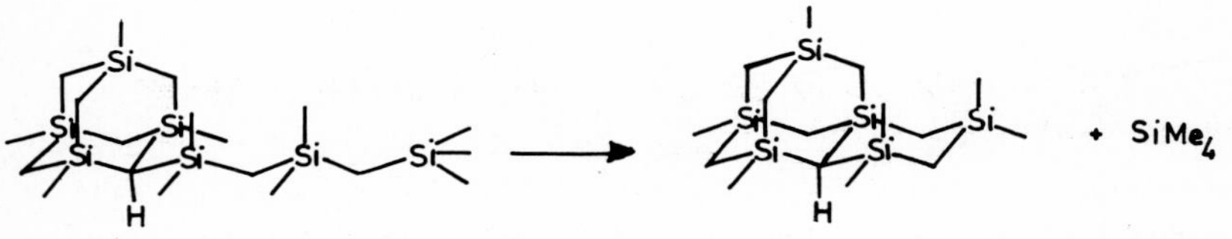

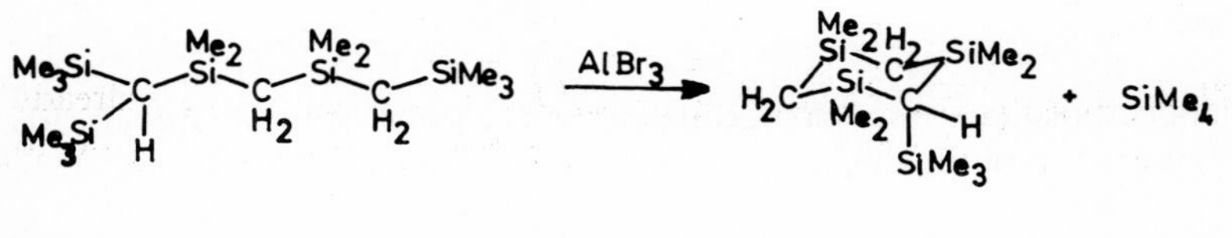

$+ \;\; SiMe_4$

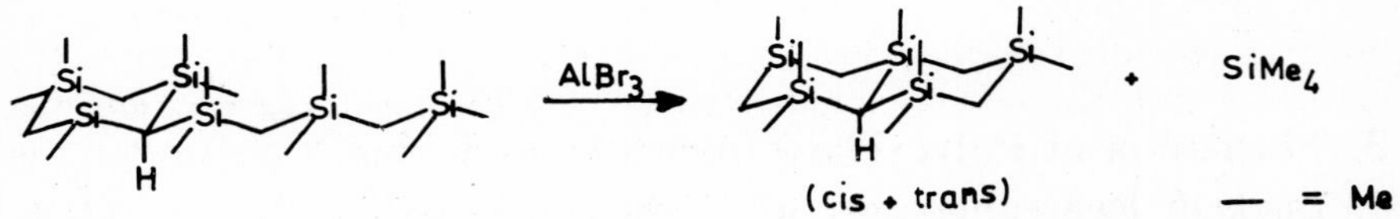

(cis + trans)

$—\; = \; Me$

The byproducts which appeared from this reaction consisted essentially of Si-brominated derivatives of the main cyclic compounds produced [55]. These reactions proceed readily at room temperature. Also, carbosilanes which have a quaternary carbon atom and an appropriate chain length react further to produce corresponding cyclic structures:

$$
\begin{array}{c}
\mathrm{Me_3Si} \\
\mathrm{Me_3Si} \\
\mathrm{Me_3Si}
\end{array}\!\!\!\!C\!-\!\underset{Me_2}{Si}\!-\!CH_2\!-\!\underset{Me_2}{Si}\!-\!CH_2\!-\!SiMe_3
\xrightarrow{\ \mathrm{AlBr_3}\ }
\ \mathrm{Me_3Si-}\underset{\underset{Me_3Si}{|}}{C}\!\underset{\underset{H_2}{}}{\overset{\overset{Me_2}{Si}}{\diagdown}}\!\!\begin{array}{c}H_2\\Si-C\\Me_2\end{array}\!\!\!\!SiMe_2\ +\ SiMe_4
$$

Out of these results arose a number of questions which gave rise to the following investigations.

3.2.1 The Effect of the Length of Skeletal C-Bonded Side Chains on Ring Formation

The reactions described in Fig. 9 (Eqs. 1 to 6) show the effect of the variation of chain length on ring building. It is seen that under the same mild conditions, $SiMe_4$ and $(Me_3Si)_2CH_2$ are eliminated on ring closing (Eqs. 1 to 3). Compound <u>188</u> will yield <u>209</u> only under very specific conditions (Eq. 16), otherwise <u>179</u> + <u>189</u> are formed (Eq. 4) as adamantane derivatives. Compound <u>190</u> can form a six-membered ring by elimination of $SiMe_4$ as well as CH_4, as is shown by Eq. 5. Compounds <u>191</u> and <u>181</u>, expected as products, were produced in the ratio 3:2. A number of more complicated reactions occur also. Starting from compound <u>192</u> no simple ring closing reaction is possible, nor does elimination of methane produce compound <u>191</u>. Only under harsher conditions will a complicated regrouping occur in compound <u>192</u>, which results in the formation of compound <u>40</u>. The compound produced in the largest yield on conversion of compound <u>192</u> is not compound <u>40</u>, but tris(trimethyl-silyl)methane <u>174</u> (Eq. 6).

3.2.2 Reactions of Carbosilanes Containing Side Chains Bonded to Si Atoms in the Molecular Skeleton

In contrast to the Si-substituted tetrasilaadamantanes such as compound <u>193</u>, the Si-substituted trisilacyclohexanes <u>194</u>, <u>195</u> and <u>196</u> react with $AlBr_3$ to produce poly-cyclic systems as shown in Fig. 10. However, if the first six-membered ring is formed through elimination of methane, a degradation reaction yielding compound <u>174</u> outweighs the former reaction. The fact that the tetrasilaadamantane derivative <u>193</u> does not produce any corresponding reaction could well mean that the methylene carbon atoms in the adamantane skeleton adjacent to the side chain are not attacked by $AlBr_3$ to from a new ring. To date, there has been no example presented in which a $(\!\!\geq\!\!Si)_3CH$ group forms by influence of $AlBr_3$ on $(\!\!\geq\!\!Si)_2CH_2$.

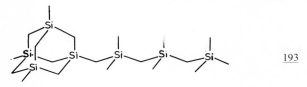

193

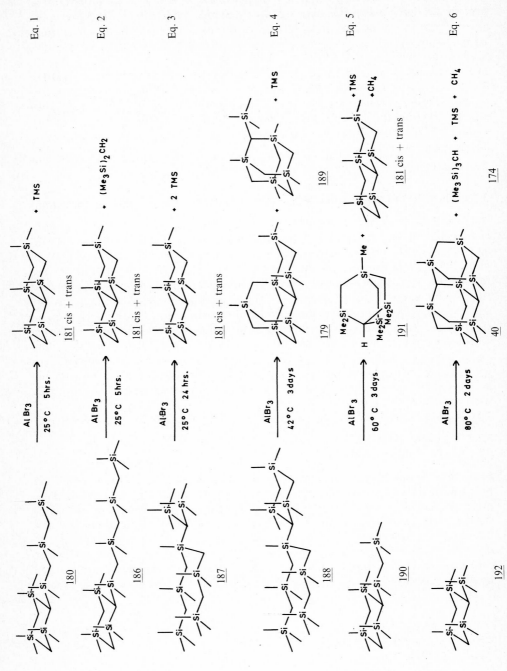

Fig. 9. Influence of the length of the Si—C—Si side chain attached to a skeletal C atom on reaction with AlBr$_3$

On the other hand, the trisilacyclohexane ring can reorient itself in such a way that a methylene group is available for attack as a prerequisite for subsequent ring closure.

3.2.3 Ring Closure by Elimination of Methane

The conversion of compound _197_ to produce compound _198_ according to Eq. 10 in Fig. 10 was already known [55, 56]. If one elevates the degree of silylation of compound _197_ and starts with compound _201_ and _202_, then no further reaction similar to Eq. 10 will occur [57]. In this case, only derivatives of the starting material were isolated in which the methyl groups were exchanged for Br atoms. The steric requirements of the molecules account for this behavior.

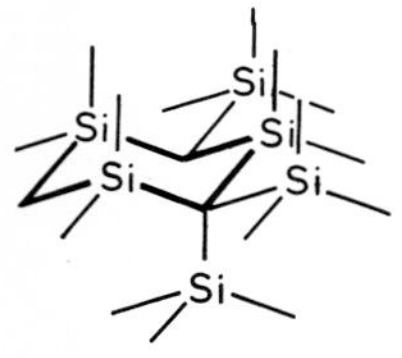

_201_ 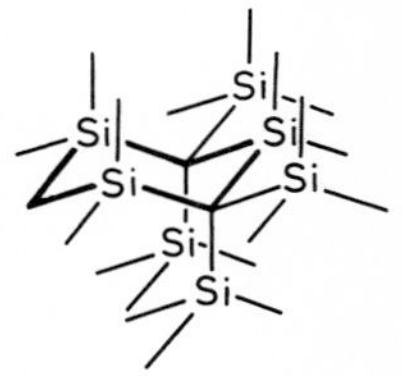   _202_

Under more favorable steric conditions, as in compound _199_ two molecules of CH_4 are eliminated according to Eq. 11, yielding the hexasilahexascaphane _200_ together with twice as much tetrasilascaphane _198_.

Methane can be eliminated only at elevated temperatures, due to the requirement of the cleavage of a very stable C—H bond, whereas in cases where $SiMe_4$ or $(Me_3Si)_2CH_2$ exist as byproducts of a reaction, the Si—C bond was cleaved.

3.2.4 Rearrangements Leading to the Formation of Larger Scaphanes or Carborundanes

The largest member of the scaphane class synthesized to date by this method is the hexascaphane _200_. It is synthesized from compound _199_ according to Eq. 11 [56], as well as by the reactions in Eq. 12 and Eq. 13. Equation 13 in Fig. 11 shows compound _212_ to be a direct precursor of compound _200_.

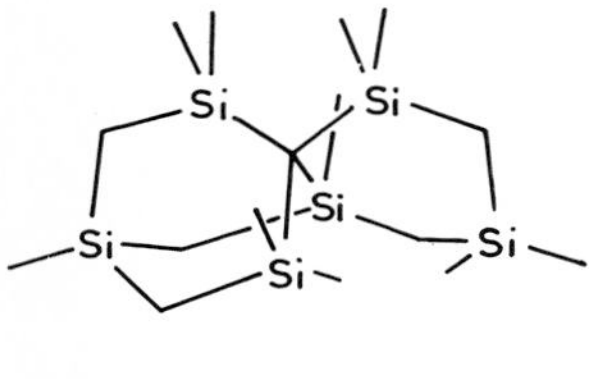

_212_

The expected construction of octasiladodecascaphane from compound _204_ through repeated elimination of methane failed to appear. Instead, the two remaining side chains were cleaved as soon as the molecular skeleton of _200_ was formed.

The largest carborundane to be synthesized to date by this method is _40_. It is formed according to Eq. 6 and Eq. 14.

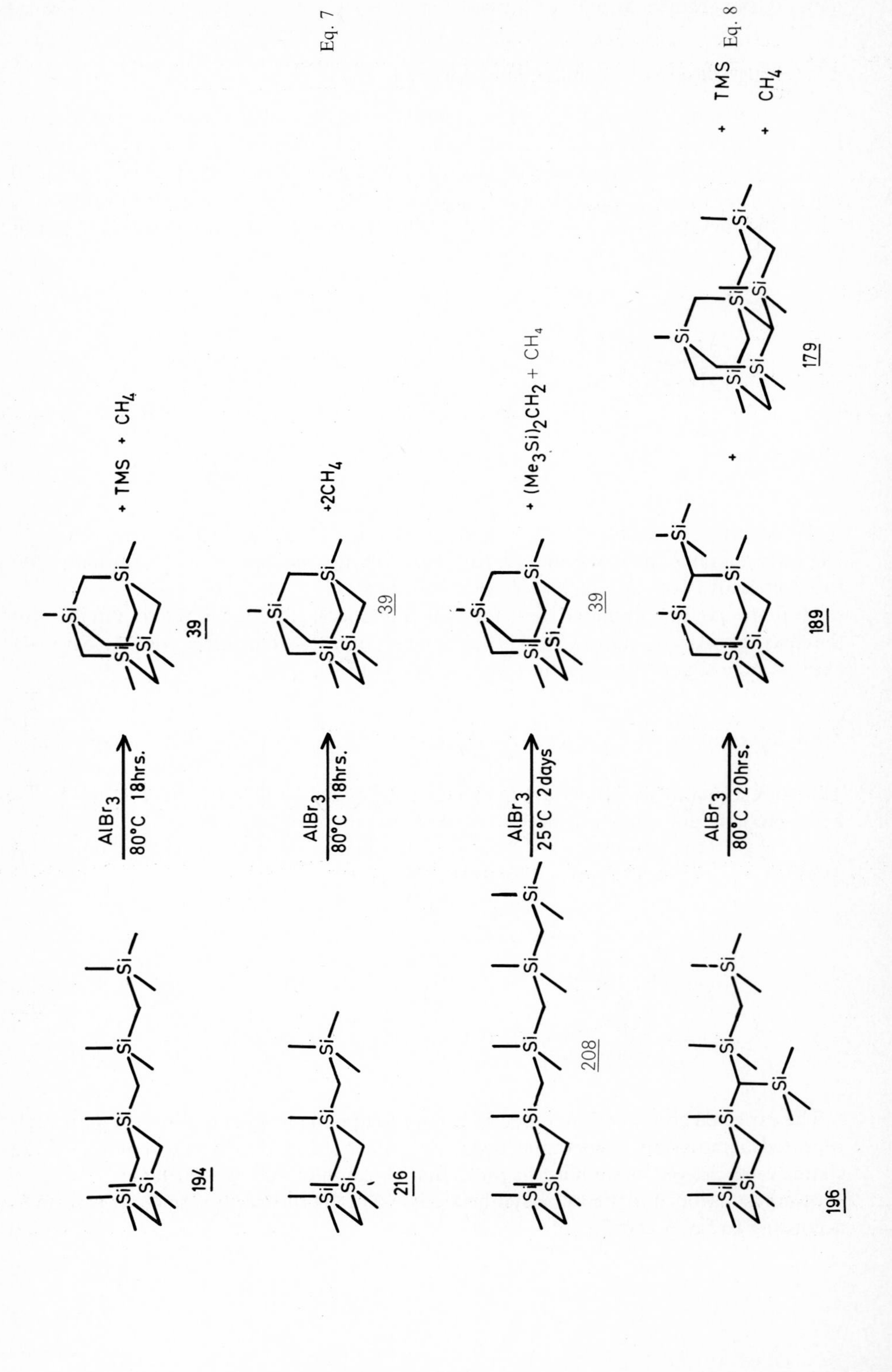
194
AlBr3
80°C 18hrs.
39
+ TMS + CH4
Eq. 7
216
AlBr3
80°C 18hrs.
39
+2CH4
208
AlBr3
25°C 2days
39
+ (Me3Si)2CH2 + CH4
196
AlBr3
80°C 20hrs.
189
+
179
+ TMS
+ CH4
Eq. 8

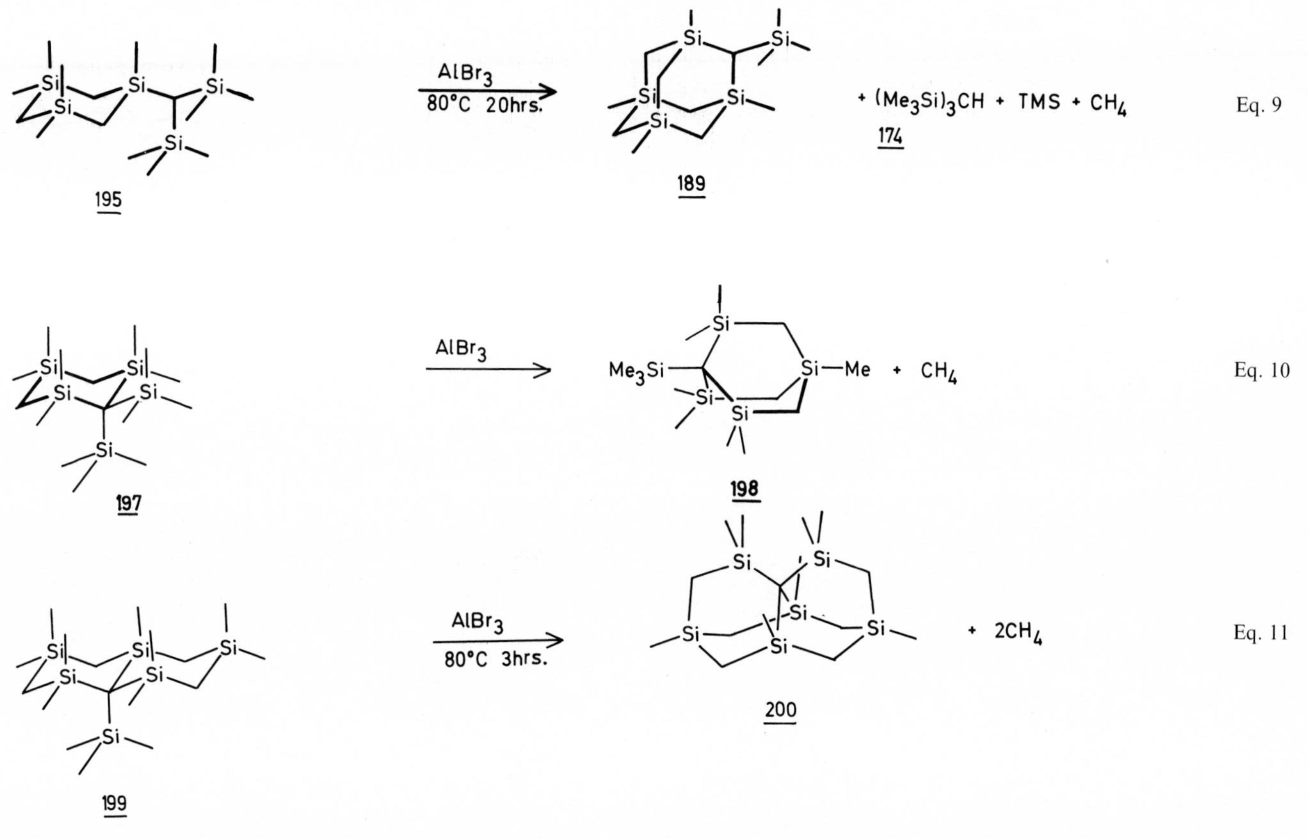

Fig. 10. Influence of the carbosilane side chain attached to a skeletal Si atom on reaction with AlBr$_3$

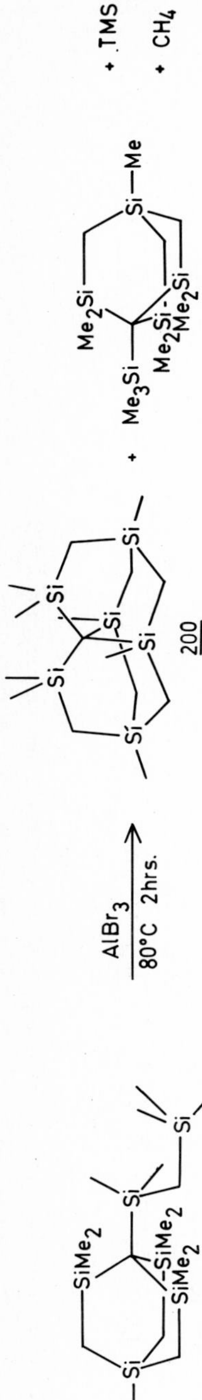

Eq. 12

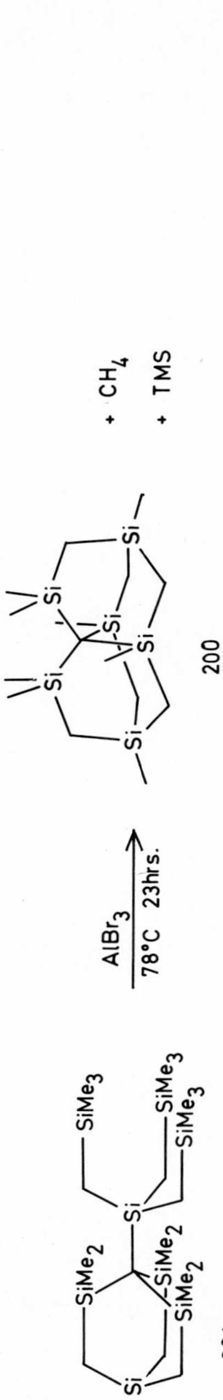

Eq. 13

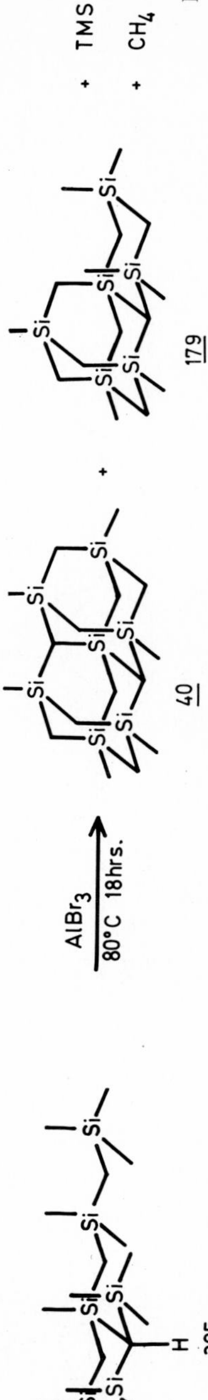

Eq. 14

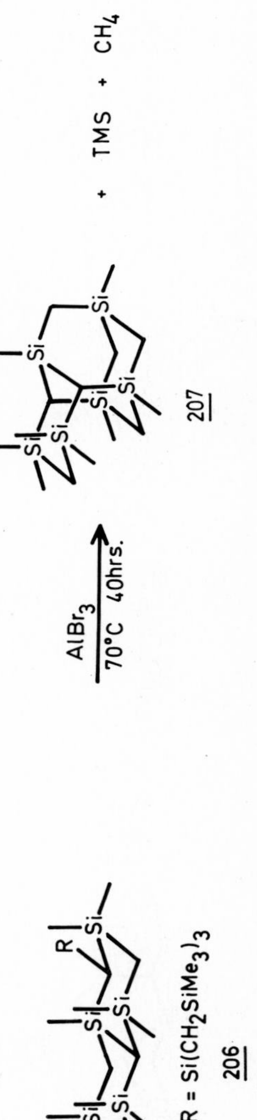

Eq. 15

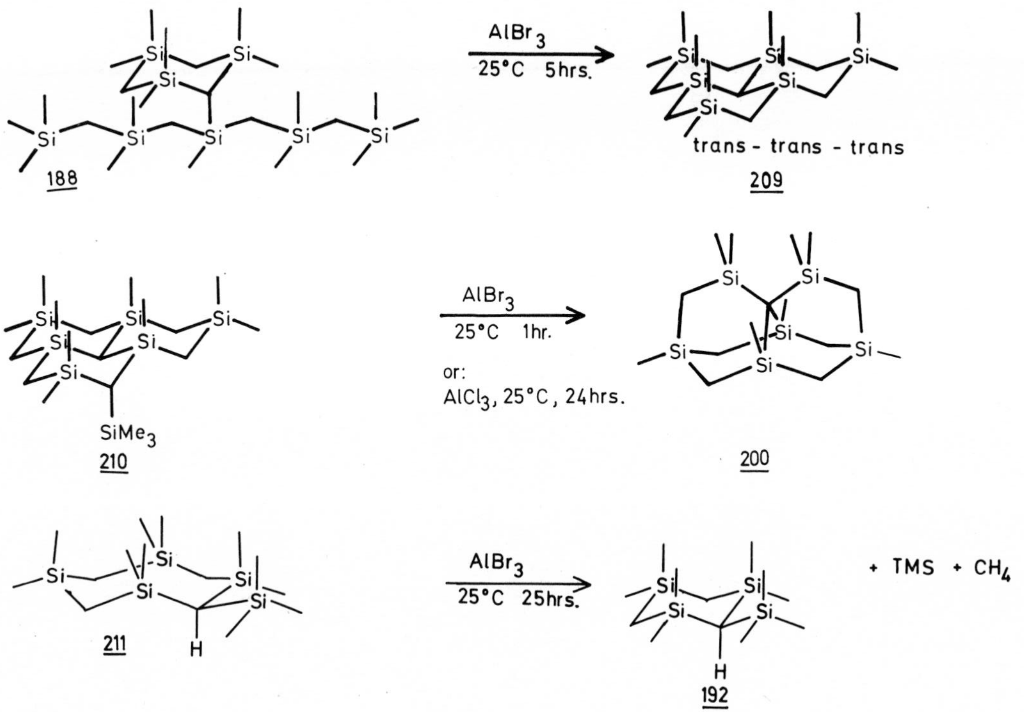

Eq. 16

Eq. 17

Fig. 11. Possibilities of ring closure via CH_4 elimination through the influence of $AlBr_3$

Synthesis of a further carborundane 207, known already from $SiMe_4$ pyrolysis, could be obtained through reaction of a mixture of C-substituted isomers of penta-siladecalin 206 with $AlBr_3$, as shown by Eq. 15 in Fig. 11.

3.2.5 Rearrangement of Larger Rings

The reactions described heretofore have all involved the formation of a six-membered ring. What about the behavior of larger substituted ring systems? By rearrangement of compound 211 again the preferred six-membered ring in compound 192 forms via ring contraction (Eq. 17, Fig. 11). In all the rearrangement reactions investi-gated, Si-brominated derivatives existed as byproducts. The synthesis of still larger representatives of the scaphanes and carborundanes is expected whenever suitable starting materials become available.

3.3 Investigation into Ring Closure Reactions During Rearrangement Reactions of Carbosilanes

Before the question of the pathways of such reactions is discussed, a few experimental findings must be mentioned:

1) The quaternary carbon atom group or even the tertiary substituted CH group will not be attacked on reaction of the corresponding carbosilane with $AlBr_3$. In simple cases, ring formation proceeds via elimination of $SiMe_4$ or CH_4.
2) Apart from cyclic Si-methylated main products, SiBr-containing compounds are formed to a limited extent.
3) The investigated carbosilanes will not react in the given way if $AlBr_3$ is replaced by organoaluminium compounds such as $RAlBr_2$ $(R = CH_2-SiMe_2-CH_2-SiMe_3)$.

Investigations into the mechanism of these reactions were originally restricted to simple cases of ring closure. The following reactions were chosen.

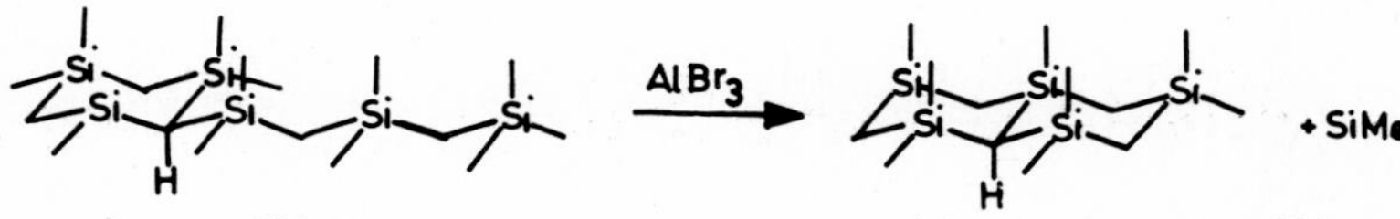

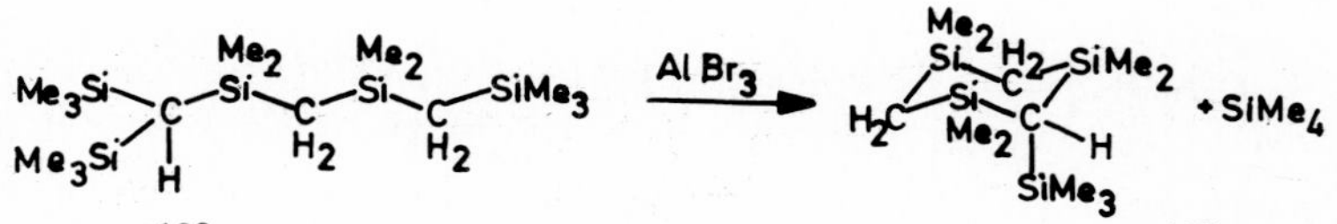

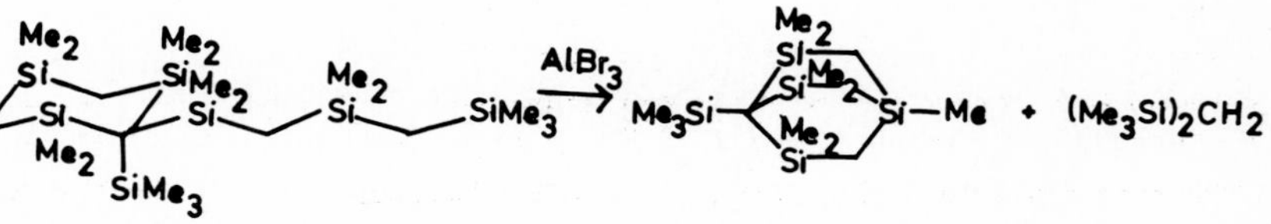

At the start, conditions were ascertained under which reactions proceeded according to the equations as given, and such that subsequent reactions could be excluded. The results of the formation of 1,3,5,7,9-pentasiladecalin from compound $\underline{180}$ are shown in Fig. 12 and Table 22. After 2 hours, 90% of this compound was present in the reaction mixture. Subsequently further complicated reactions took place. The exact conditions for the execution of these reactions have been established.

Table 22. Influence of reaction time on the reaction of compound $\underline{180}$ with $AlBr_3$. Conditions: 25 °C, no solvent used

Reaction Time	Compounds in reaction mixture in mol%					
(hrs)	6	181	192	35	189	179
2		89.0	2.3	5.7		
5		87.4	3.6	6.0		
10	approx	81.3	11.1	4.7		
15	30%	66.5	12.1	3.0	9.1	5.6
20		40.6	28.3	3.1	16.1	8.8
26		43.5	24.1	4.1	15.6	9.7
72		17.2	35.3	3.4	25.3	15.9

Yields are calculated referring to the isolated product mixture after hydrolytical work up and prepurification on a silica gel column.

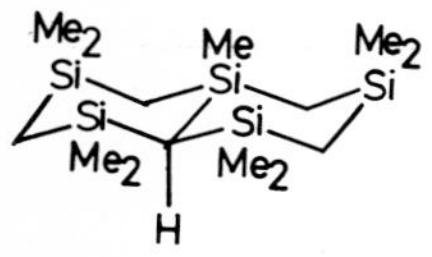

cis + trans
$\underline{181}$

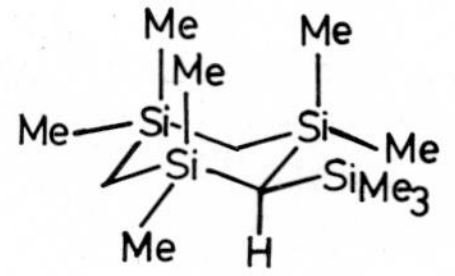

$\underline{192}$

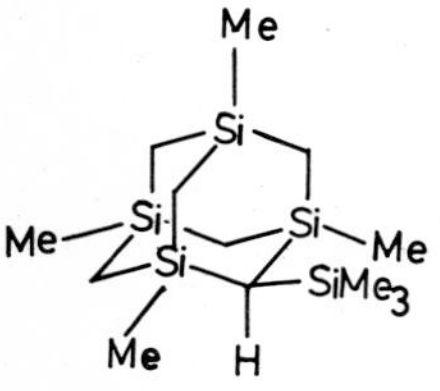

$\underline{189}$ 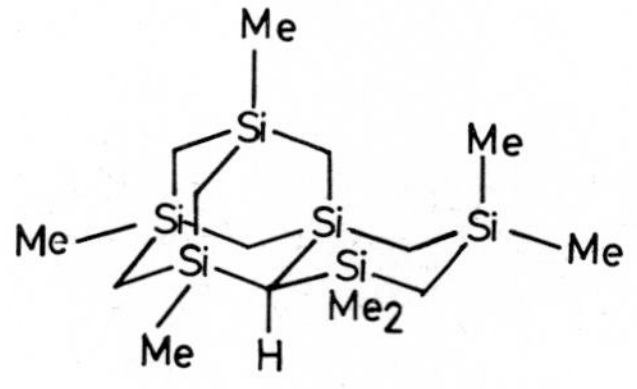 $\underline{179}$ 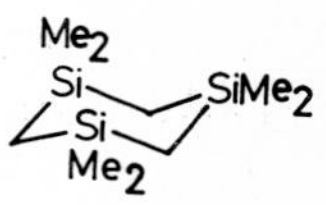 $\underline{35}$

The following considerations apply to the investigation of a suitable reaction mechanism: The formation of Si-brominated byproducts could occur to a limited extent either through combined action of Br_2 or HBr with $AlBr_3$, or through direct action of $AlBr_3$

$$Me_4Ad + Br_2 \xrightarrow{AlBr_3} Me_3BrAd + MeBr$$

$$(Me_3Si-CH_2)_2SiMe_2 + AlBr_3 \rightarrow Me_3Si-CH_2-SiMe_2Br + Me_3Si-CH_2-AlBr_2$$

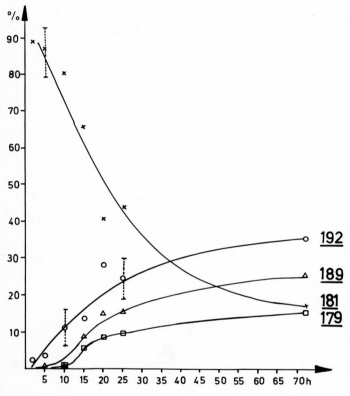

Fig. 12. Influence of reaction time on the reactions of compounds 180 and 181 with $AlBr_3$

Rearrangement could then conceivably occur by two possible pathways:

Mechanism 1: In the first reaction step, a Si—C bond is cleaved followed by ring formation supported by the initially-appearing Al compound.

Mechanism 2: $AlBr_3$ acts as a Lewis acid and initiates rearrangement through polarization to produce an energetically stable six-membered ring and $SiMe_4$

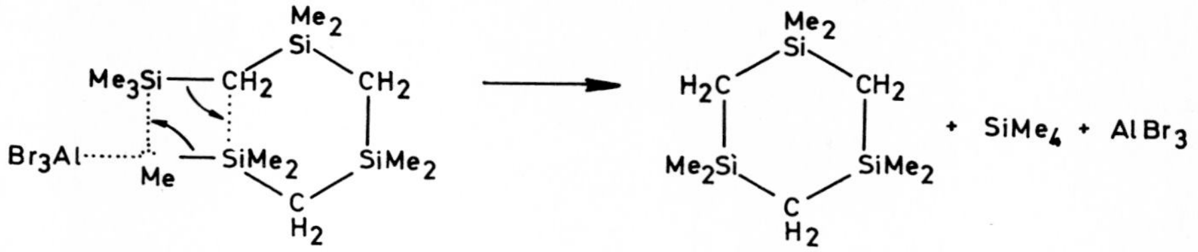

Both reaction pathways explain the ring building as well as the formation of $SiMe_4$ and the recovery of $AlBr_3$. The decision between these two mechanisms was made possible by rearrangement of the partly deuterated compound 180a.

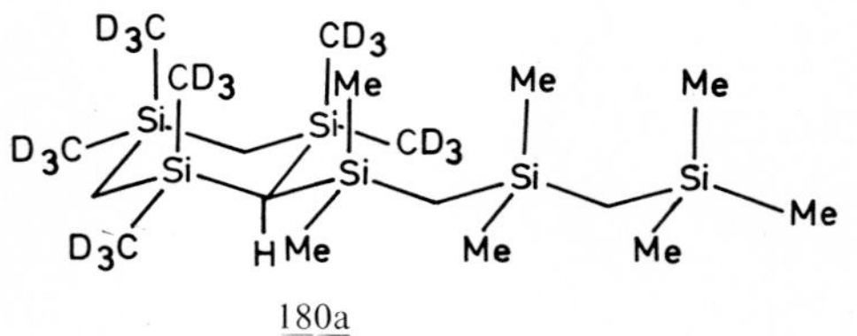

180a

According to mechanism 1 one would expect a C—D bond and a Si—C bond to disrupt, thus forming a tetramethylsilane with a $SiCH_2D$ group. The second mechanism, based on the influence of $AlBr_3$ as a Lewis acid, allows for the formation of a coordinated intermediate, thus producing through polarization a weakening of the Si—C bond. This may also be shown diagramatically:

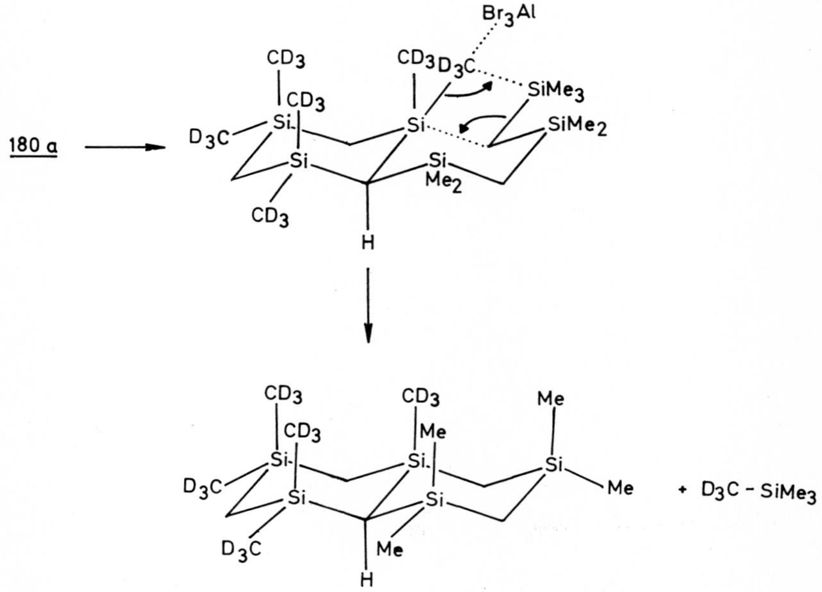

It is seen here that only Si—C bonds and no C—D bonds are broken. The reaction product is a 1,3,5,7,9-pentasiladecalin with five CD_3 groups and a D_3C—$SiMe_3$ molecule. Actually only reaction products with Si—CD_3 groups form, and through the influence of $AlBr_3$ a redistribution of CH_3 or CD_3 groups within the reaction products results [58]. This may be shown similarly:

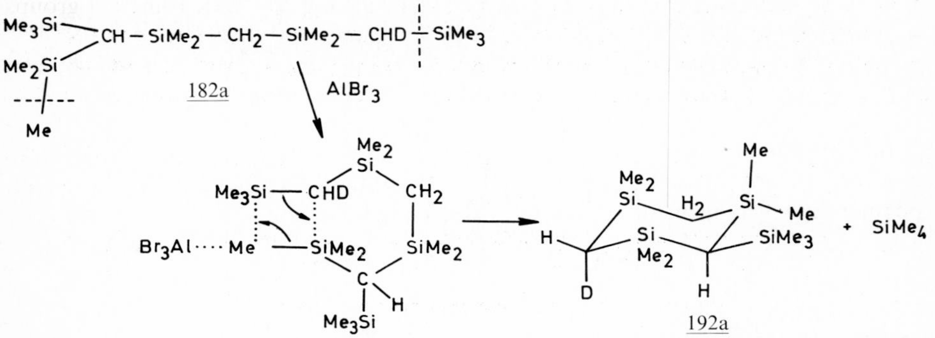

Me_3Si—CH—$SiMe_2$—CH_2—$SiMe_2$—CHD—$SiMe_3$
Me_2Si

--|--

Me   182a $AlBr_3$

Me_3Si—CHD Me_2 Si CH_2

Br_3Al····Me —$SiMe_2$ $SiMe_2$
C
Me₃Si H

Me_2 Si   H₂   Me   Si—Me
H— Si SiMe₃   + $SiMe_4$
Me₂
D   H
192a

In order that mechanism 2 be substantiated, a D-atom contained within a CHD group must be bound in a six-membered ring, thereby releasing deuterium-free $SiMe_4$. This was confirmed by experimental investigation [58]. The rearrangement is carried out by breakage of two Si—C bonds in compound 182a.

To illustrate a comparable mechanism in which $AlBr_3$ initiates the cyclization reaction of carbosilanes upon exclusion of Me_3Si—CH_2—$SiMe_3$, the following example is given:

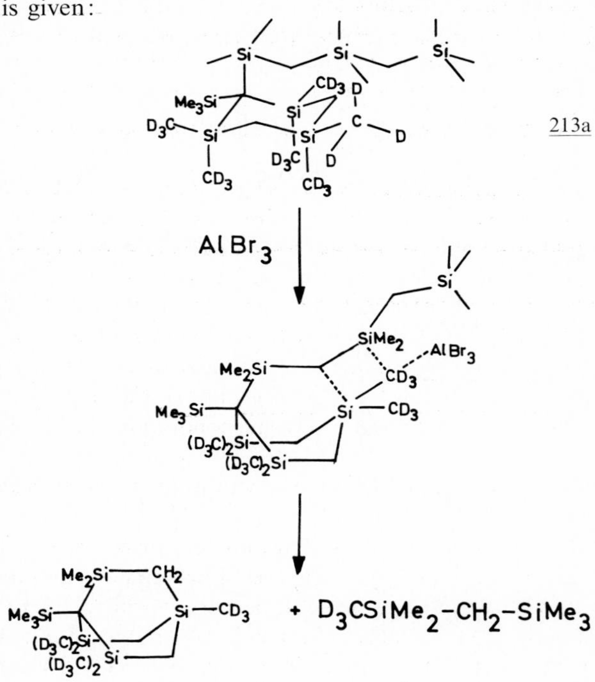

Si   Si   Si
Me_3Si CD_3 D
D_3C Si   Si   C—D   213a
D_3C D
CD_3 CD_3

$AlBr_3$

Si
$SiMe_2$ $AlBr_3$
Me_2Si CD_3
Me_3Si Si —CD_3
$(D_3C)_2Si$
$(D_3C)_2Si$

Me_2Si—CH_2
Me_3Si Si—CD_3 + D_3CSiMe_2—CH_2—$SiMe_3$
$(D_3C)_2Si$
$(D_3C)_2Si$

198a   6a

Mass spectrometric investigations of the products after reaction of the partly deuterated compound 213a provide evidence for the mechanism in which $AlBr_3$ as a Lewis acid coördinates with the carbosilane and through polarization causes a corresponding weakening of the Si—C bond. Cyclization then results exclusively through cleavage of Si—C bonds, and not of C—H bonds. Organoaluminium compounds are not observed (Tab. 23). Nevertheless, methyl and deuteromethyl groups are found to be statistically distributed over the reaction products.

The reaction of the partly deuterated compound 197a with $AlBr_3$ was investigated as an example of a cyclization reaction based on the elimination of methane.

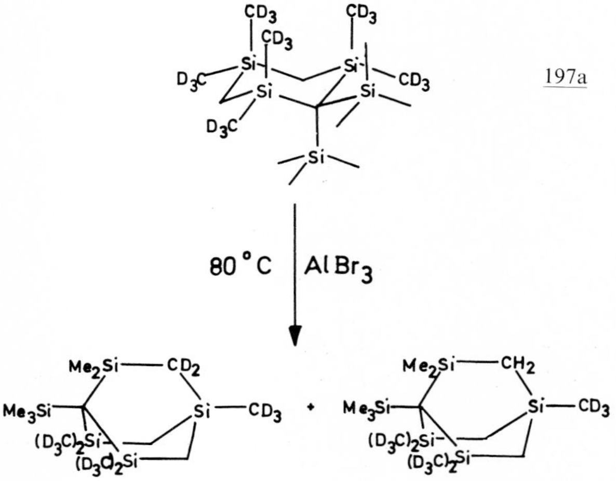

+ CH_3D + CD_3H + Me_4Si + $(Me_3Si)_4C$

The resulting compounds were analysed by mass spectrometry or IR spectroscopy (methane). Methane elimination always requires breakage of C—H bonds, meaning of course that deuteration experiments would not lead to a distinction between different mechanisms. Employing an analogous reaction pathway as established previously for the cyclization through elimination of $SiMe_4$ or $Me_3Si-CH_2-SiMe_3$, a close inspection of the partly deuterated products enables a statement to be made on the stereochemical course of the reaction. It seems clear that Si—C cleavage was initiated by $AlBr_3$, preferably acting on the methyl groups of the ring. Where C—H bond breakage was necessary, this occurred on corresponding methyl groups of the side chain.

A statistical distribution of CH_3 and CD_3 groups within the products is also found in this example.

^{27}Al- and ^{1}H-NMR spectroscopic investigations confirm these mechanism [58]. An alternative reaction pathway through the Si-brominated compound 214 and the Al-organic compound 215 can be excluded because the independent synthesis of these compounds as well as the subsequent reaction of the pure compounds 214 and 215 yielded other products. Table 23 demonstrates that the two compounds do not occur as intermediates in the reaction of compound 213 with $AlBr_3$.

Table 23. Products from the reactions of compound 213 as well as compounds 214 and 215 with AlBr$_3$

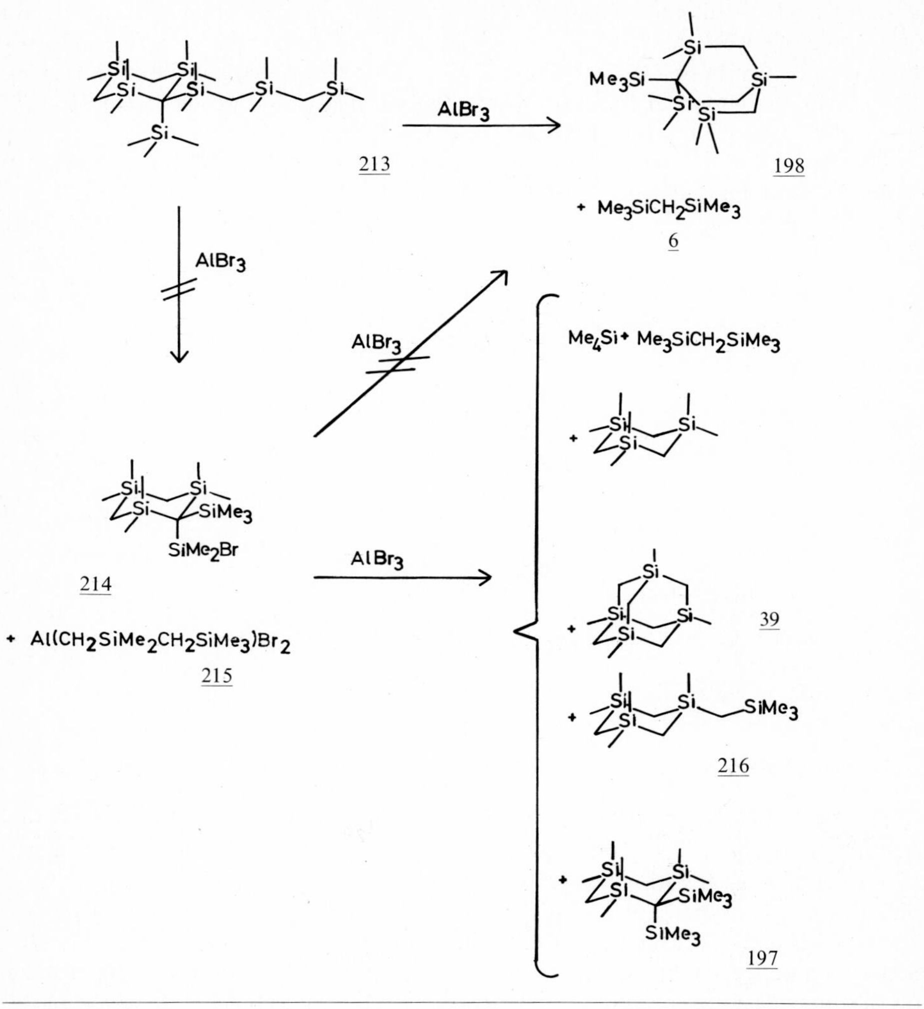

3.4 Investigations into the Stability of Carbosilane Skeletons Towards AlBr$_3$ and AlCl$_3$

As a result of the previous investigation it was recognized that reaction conditions will influence the products isolated from the reaction of carbosilanes with AlBr$_3$, because the primary reaction products will react further. This is distinctly clear from inspection of the reactions shown in Table 22 and Fig. 12. One poses the question then: Which carbosilane skeletons, if any, are stable against the effect of AlBr$_3$?

3.4.1 The Formation of CH_2-Linked 1,3,5,7-Tetrasilaadamantanes

The 1,3,5,7-tetrasilaadamantane structural unit itself has proved to be relatively stable against the effects of AlX_3 (X=Br, Cl). Reaction between Si-methylated 1,3,5,7-tetrasilaadamantane 39 and $AlBr_3$ occurs very slowly in benzene (30% converted in 8 days at 80 °C), but without solvent, the rate was much faster (30% converted after 4 hrs at 90 °C). Apart from unconverted compound 39, mono-brominated 1,3,5,7-

Table 24. Compounds isolated from the reaction of tetramethyltetrasilaadamantane 39 with $AlBr_3$

39

217 218

m.p. 186 - 189 °C

219

m.p. 192 °C

220

m.p. 218 - 221 °C

221

222

m.p. 220 - 229 °C

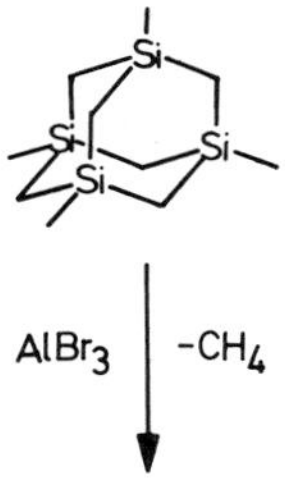

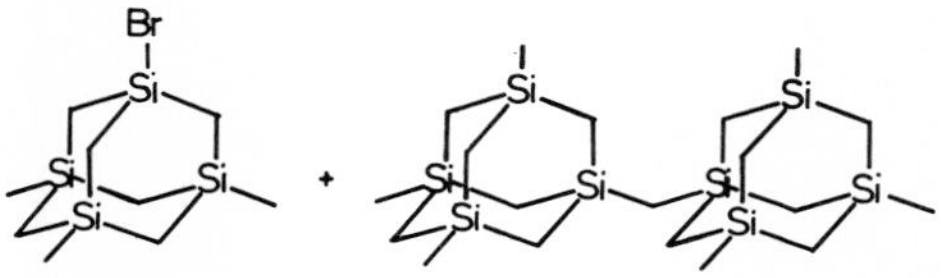

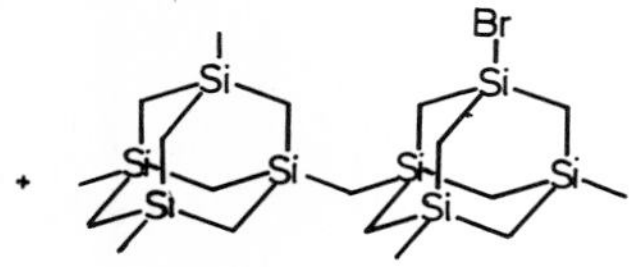

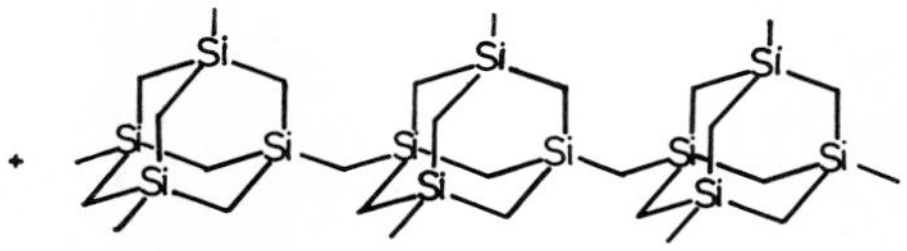

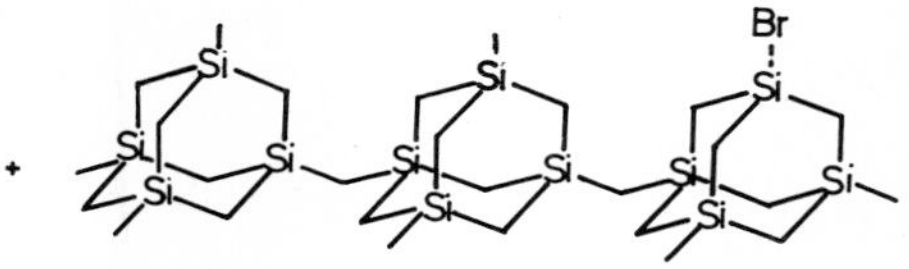

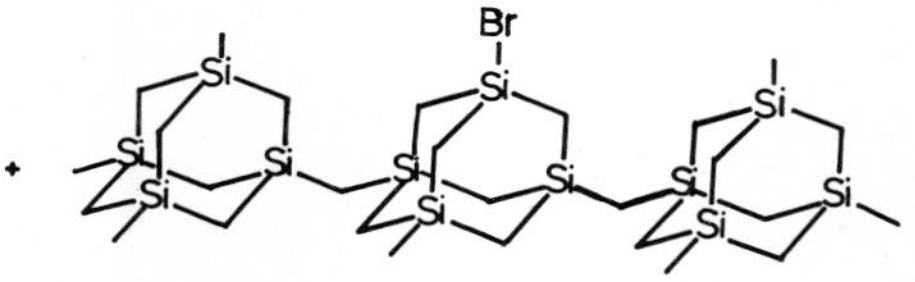

(Continued)

Table 24 (cont.)

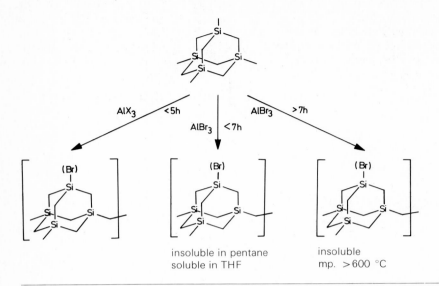

insoluble in pentane
soluble in THF

insoluble
mp. >600 °C

tetrasilaadamantane <u>217</u>, CH$_4$ and a small amount of SiMe$_4$, some methylene-bridged tetrasilaadamantane chains were isolated from the reaction mixture by means of HPLC (Table 24) [59].

Through ^{1}H-NMR spectroscopy and mass spectrometry, the methylene-linked tetrasilaadamantane chains in compounds <u>218</u> and <u>220</u>, as well as their brominated analogues in <u>219</u>, <u>221</u> and <u>222</u> were characterized.

An extension of time for the reaction between 1,3,5,7-tetrasilaadamantane <u>39</u> and AlBr$_3$ showed at first the formation of pentane-insoluble oligomers which were soluble in THF or benzene (reaction time: 4 to 7 hrs at 90 °C). Reaction times exceeding 7 hrs result in the synthesis of polymers insoluble in commonly used solvents, polymers which are constructed in the same way as those previously encountered. An IR spectroscopic investigation revealed the presence of absorption bands which are comparable to those of 1,3,5,7-tetrasilaadamantane or which are characteristic for oligo-tetrasilaadamantanes containing two to four tetrasilaadamantane structural units.

As a result of this investigation, the insoluble polymers were shown to contain three-dimensional methylene-bridged tetrasilaadamantane units. Their skeletal structure was almost entirely resistant to attack by AlBr$_3$, and on preservation of their skeletal structure a partial CH$_3$/Br exchange on silicon resulted, producing brominated derivatives [59].

The formation of the CH$_2$-bridged oligo-tetrasilaadamantanes can be clarified by consideration of mechanism 2 formulated for cyclization reactions. Here, methane is released by action of AlBr$_3$ through an intermolecular and not through an intra-molecular process. The presence of methane was proved experimentally [59].

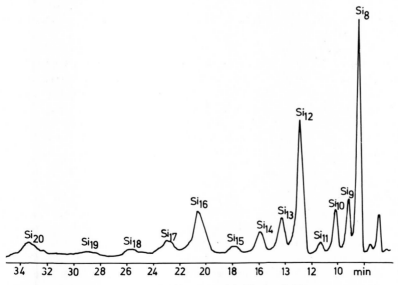

3.4.2 Reactions of 1,3,5,7-Tetrasilaadamantanes with AlCl$_3$

The use of the weaker Lewis acid AlCl$_3$ meant a reduction in reaction rate without a change in the mechanism of reaction. Although the reaction of Si-methylated 1,3,5,7-tetrasilaadamantane with AlCl$_3$ in the absence of solvent at 90 °C yielded

Fig. 13. HPLC Chromatogram. Packing material: Nucleosil-C18. Eluent: CH$_3$OH/C H$_6$ = 60/40 Vol.%

a THF-insoluble product, by reducing the reaction time to 4 hrs, a pentane-soluble product mixture was obtained [59]. After hydrolytic work-up and removal of solvent through distillation, unreacted tetrasilaadamantane and its monochloro derivative were separate^1 by sublimation at 90 °C under vacuum. The bulk of the dimer $Si_8C_{19}H_{44}$ (218) was then isolated by extraction with an 80:20 mixture of methanol/benzene. The residue was filtered and then prepurified by silica gel column chromatography. Figure 13 shows the HPLC chromatogram obtained for this mixture.

The methanol/benzene mixture to be used as mobile phase in the HPLC separation was tried at five ratios, 90:10; 80:20; 70:30; 60:40 and 50:50. The optimum conditions were achieved using a ratio of 60:40 vol%. With a higher methanol percentage in the mixture, the solubility of compounds of higher molecular weight was too small. If the polarity of the solvent mixture was decreased by increasing the proportion of benzene, then resolution of the chromatogram in the region of up to twelve silicon atoms became progressively worse.

In this way, 1,3,5,7-tetrasilaadamantane units connected by CH_2 groups could be isolated containing up to twenty silicon atoms. The chromatogram shows especially strong signals for compounds containing 8, 12, 16 and 20 silicon atoms, which corresponds to compounds containing 2 to 5 methylene-bridged tetrasilaadamantane units.

An ^1H-NMR investigation of the mass spectrometrically-identified compounds showed that an isomeric mixture of straight and branched-chain tetramers as well as n-, iso- and neo-pentane analogues of pentamers existed. The isomers of tetrameric tetrasilaadamantane could be separated and eventually isolated by HPLC.

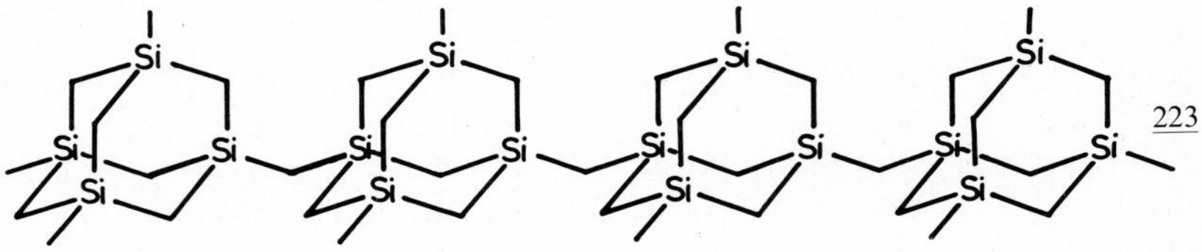

223

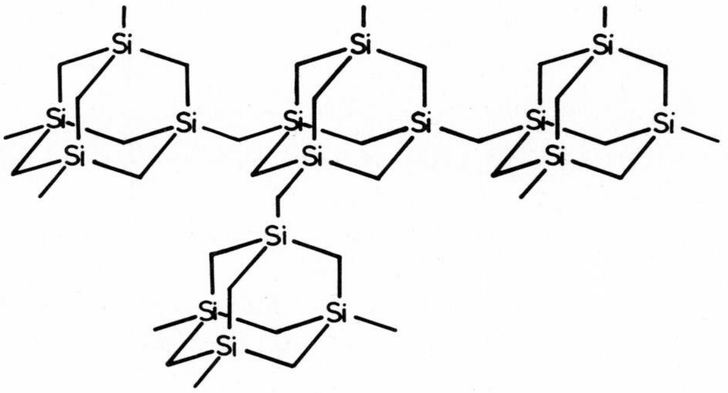

224

Between each pair of main peaks (Fig. 13) appeared three smaller signals, representing the compounds with 9, 10 and 11; 13, 14 and 15; and 17, 18 and 19 Si atoms respectively.

These correspond formally to an extension of the tetrasilaadamantane oligomers containing 1, 2 or 3 SiC_3H_8 structural units, as proved by a mass-spectrometric investigation of fractions Si_{10} and Si_{13}.

One- to three-fold addition of SiC_3H_8 units to the 1,3,5,7-tetrasilaadamantane chain can be explained by consideration of two structural alternatives:

a) Substitution of a proton on the 1,3,5,7-tetrasilaadamantane skeleton by a Me_3Si group.

b) Expansion of the methylene bridge between the 1,3,5,7-tetrasilaadamantane units by insertion of $-CH_2-SiMe_2-$ groups. This alternative, for example, can account for compounds containing nine silicon atoms. If in the underlying tetrasila-adamante chains fewer methylene bridges were present than SiC_3H_8 structural units incorporated, a corresponding extension of the tetrasilaadamantane chain by several $-CH_2-SiMe_2-$ groups would arise.

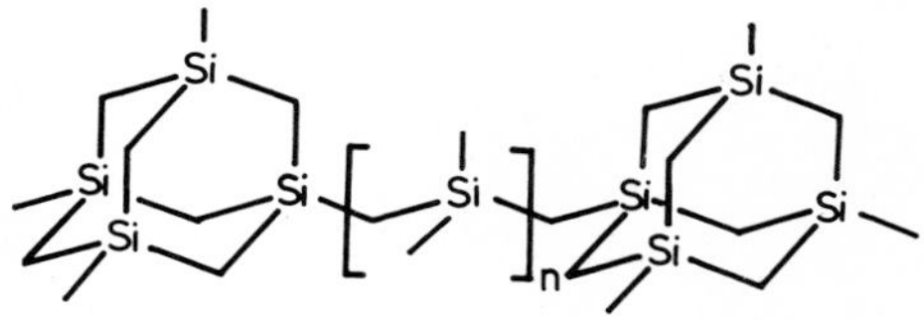

These compounds, amounting to 10–15% of the product mixture, require for their formation the dismantling of the 1,3,5,7-tetrasilaadamantane skeleton. Similar degradation reactions produce the $SiMe_4$ found in the reactions involving $AlBr_3$. However, because insurmountable difficulties arose in assigning peaks in the 1H-NMR spectra of fractions Si_{10} and Si_{13}, still no real distinction could be drawn between the two possible mechanisms.

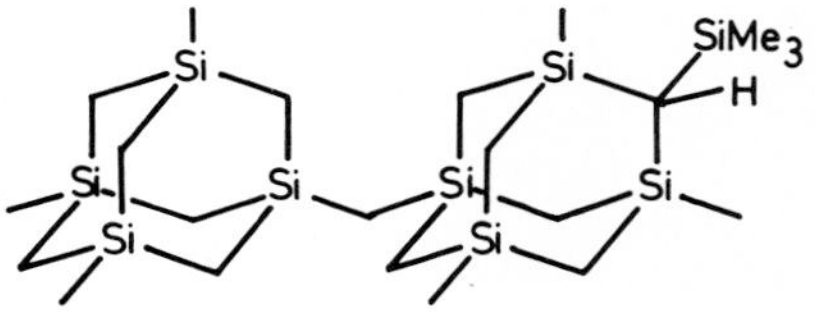

3.4.3 The Behavior of Heptasiladiadamantane Towards AlBr$_3$

It was shown by the reaction of heptasiladiadamantane $\underline{40}$ with AlBr$_3$ that this structural unit is also relatively stable. Even after a long reaction time (13 days at 90 °C), a reaction mixture was obtained which contained a large amount of starting material besides some deterioration products. These particular compounds were isolated with the help of HPLC, and are shown in Table 25. A pathway leading to the higher polycyclic system $\underline{61}$ was observed to only a limited degree [59].

3.4.4 The Behavior of Tetrasilatriscaphanes Towards AlBr$_3$

The boat conformation of the trisilacyclohexane building block in tetrasilabarrelane $\underline{191}$ is not maintained after reaction with AlBr$_3$. After a short reaction time (5 hrs at 80 °C) and also at lower reaction temperature (24 hrs at 20 °C), tetrasilabarrelane $\underline{191}$ could no longer be detected. The products, isolated by means of HPLC and shown in Table 26, arise from deterioration of the tetrasilabarrelane as well as from con-

Table 25. Compounds formed by reaction of hexamethylheptasiladiadamantane 40 with $AlBr_3$

40

192 189 55

40 (or isomers)

61

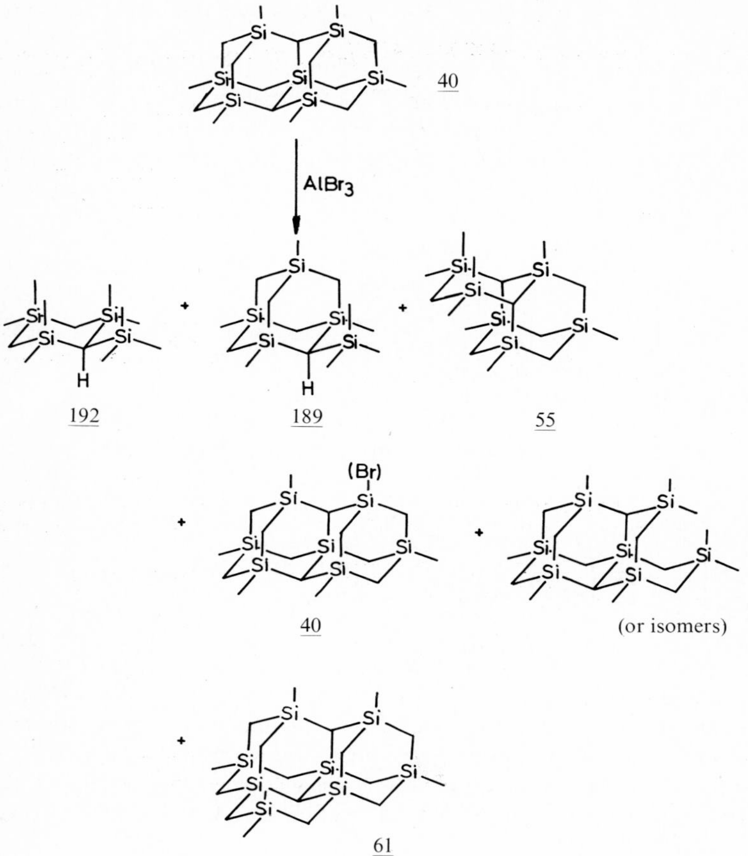

struction of larger carbosilane systems. All trisilacyclohexane structural units contained in these products exist in the chair conformation. The previously unknown bis(trimethylsilyl)tetrasilaadamantane 225 was characterized by mass spectrometry and ^{1}H-NMR spectroscopy [59].

225

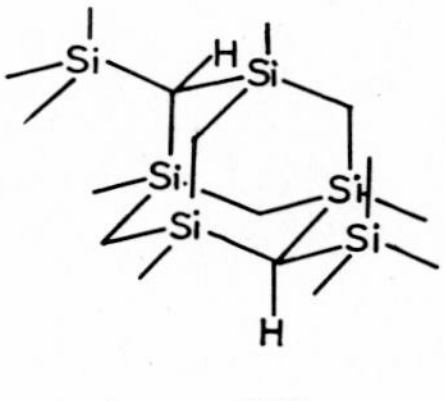

Table 26. Reaction products obtained by reaction of tetrasilabarrelane 191 with AlBr$_3$

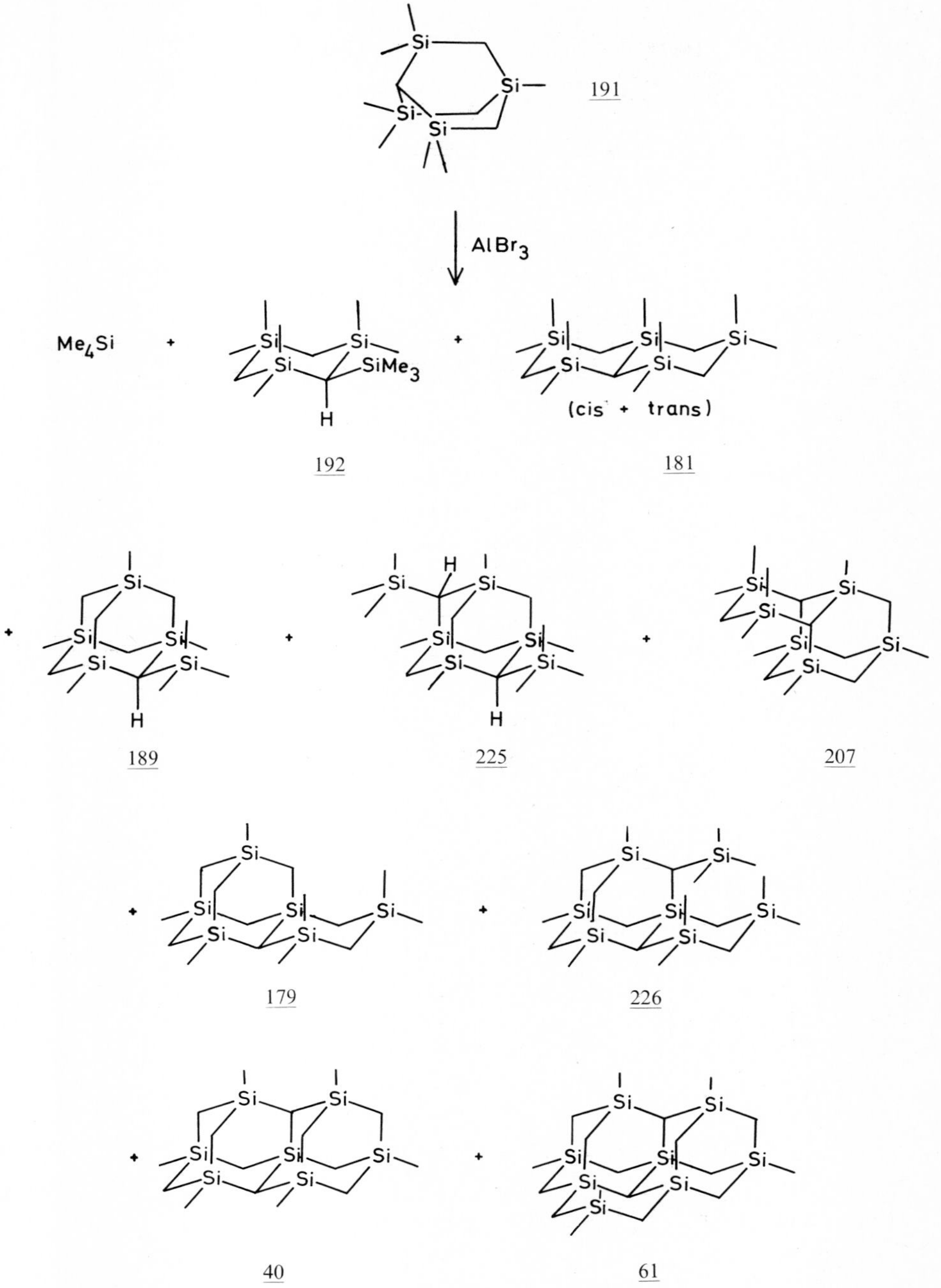

3.4.5 The Effect of the Lewis Acids BCl₃, PCl₃ and SbCl₃ on Unstrained Carbosilanes

No reaction was established by ^{1}H-NMR spectroscopy between the Lewis acids BCl_3, PCl_3 or $SbCl_3$ and 213 even after a reaction time of 14 days at 80 °C in benzene solution:

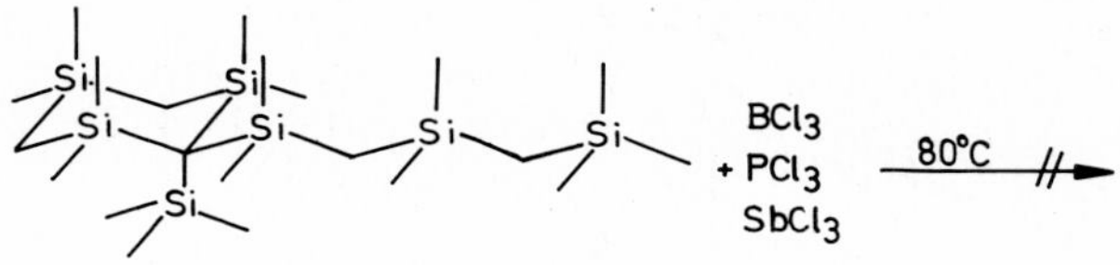

213

3.4.6 The Behavior of Si-Halogenated Carbosilanes Towards AlBr₃

Already by introducing one halogen atom onto silicon, carbosilanes become more stable against the effect of AlX_3 (where $X = Cl$, Br). Compounds in which all silicon atoms are chlorinated, such as 83 and 80, after a longer reaction time (6 days at 80 °C, in absence of solvent), show no change in the molecular skeleton. Employment of $AlBr_3$ only resulted in a mere exchange of chlorine for bromine:

$$(Cl_2Si-CH_2)_3 \qquad\qquad + \; AlX_3 \xrightarrow{\;80\,°C\;} \!\!/\!\!/\!\!\to$$

83

$$Cl_3Si-CH_2-SiCl_2-CH_2-SiCl_3 + AlX_3 \xrightarrow{\;80\,°C\;} \!\!/\!\!/\!\!\to$$

80

The replacement of one methyl group by a halogen impedes the cyclization of carbosilanes in comparison to their methylated analogues. Reaction of $Si_5C_{14}H_{37}X$ (X = Br, Cl) 214 with $AlBr_3$ causes a methyl/halogen exchange without transformation of the molecular skeleton:

$$\xrightarrow{\text{AlBr}_3} \quad CH_3/X\text{- exchange}$$
$$(X = Br, Cl)$$

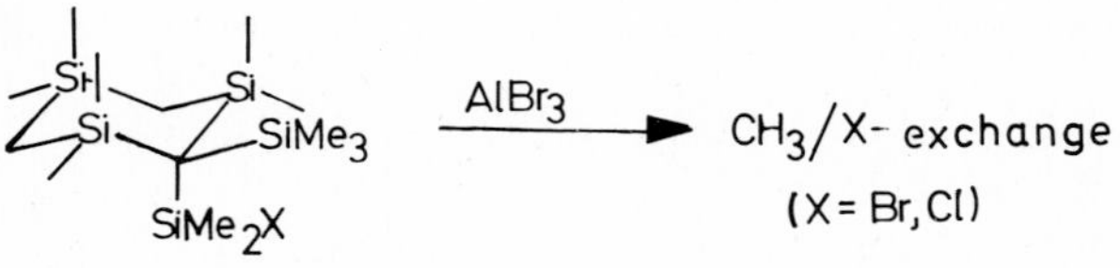

214

The reaction of $(Me_3Si)_2CH(SiMe_2CH_2)_2SiMe_2Br$ with $AlBr_3$ produced brominated trisilacyclohexane derivatives. The compounds underwent a primary Me/Br redistribution resulting in the methylated compound 182 and its cyclization product 192, followed by renewed Me/Br exchange:

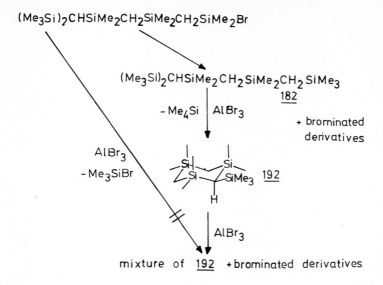

The analysis of the volatile compounds confirmed this reaction pathway. Tetramethylsilane was detected exclusively by ^{1}H-NMR spectroscopy; no proof for the existence of Me$_3$SiBr was forthcoming. These results substantiate the assumption that halogenated products resulting from cyclization of methylated carbosilanes are generated through Me/Br exchange only after ring closure.

3.4.7 Summary

From the results on the effect of AlBr$_3$ on carbosilanes presented here, it is recognized that these reactions show potential for development. After synthesis of suitable starting materials, a further insight could be gained into the several reaction pathways. This is often a toilsome undertaking, but it should not be overlooked that only in this way can an increased number of compound types be identified, substances that cannot be revealed under present circumstances by ordinary organometallic syntheses.

It is surprising that the basic structures of carbosilanes obtained from the pyrolysis of SiMe$_4$ and other methylsilanes and from the reaction of linear and cyclic carbosilanes with AlBr$_3$ are in principle the same. This finding has not yet been satisfactorily explained. Without doubt it can be said that the mechanisms pertaining to these reaction pathways are different. However, it must be considered that the high molecular-weight polycyclic carbosilanes and scaphanes from the pyrolysis of SiMe possibly result simply through thermal effects. It is not yet known from which concrete intermediates these compounds are built. The chemical pathways are better defined in reactions involving AlBr$_3$. The basic structures of the compounds formed in these two different pathways will naturally correspond, because these are the most energetically-stable representatives of these classes of compounds. Obviously, reactions made possible by AlBr$_3$ proceed between 20° and 100 °C, while the same products are produced by thermal treatment only at approximately 500 °C or above.

4. Organometallic Syntheses of Carbosilanes

In the investigations into the organometallic synthesis of carbosilanes, several different objectives were pursued:

a) It was intended to acquire generally-applicable methods for the step-by-step synthesis of carbosilanes, whereby the incorporation of appropriate protecting groups is a vital requirement. These syntheses make use of Si-functional groups. It was also intended to synthesize carbosilanes with structures not previously known from pyrolysis.

b) New possibilities of synthesis should be opened up in order to achieve carbosilanes with silicon and carbon occupying just the inverse positions in the molecular skeleton, and in order to employ a carbon moiety as the final linking unit in the synthesis of special carbosilane skeletons. This structural type in fact appears in methylsilane pyrolysis products only in exceptional cases.

c) Some of the syntheses carried out also function as a means to obtain starting materials for further investigations. This includes the clarification of the effect of Si-methylation or chlorination respectively on the reaction of 2,2-dichloro-1,3-disilapropanes with MeMgCl (Chapt. III.3.2) or the synthesis of suitable starting compounds for rearrangement and ring closure reactions of carbosilanes with $AlBr_3$ (Chapt. II.3).

4.1 SiPh Groups as Important Protecting Groups in the Organometallic Synthesis of Carbosilanes

The organometallic syntheses of cyclic and polycyclic carbosilanes have already been summarized [60]. These syntheses are made possible partly by initial incorporation of a shielding group which will be transferred into a functional group before the last reaction stage. Especially suitable in such circumstances are Si—Ph groups, which are easily cleaved:

$$\geqslant Si-C_6H_5 + HBr \rightarrow \geqslant SiBr + C_6H_6$$

Because the tendency for Si—Ph groups to be cleaved is less when more electronegative substituents are present on silicon, graduated reaction rates can be achieved.

4.1.1 Synthesis of Pentasiladecalin

The synthesis of cis and trans 1,3,5,7,9-pentasiladecalin through (PhSiMe$_2$—CH$_2$—SiMe$_2$—CH$_2$)$_2$SiPhMe serves as an example of this principle. The subsequent cleavage of Si—Ph groups causes a transfer of substituents, forming (BrMe$_2$Si—CH$_2$—SiMe$_2$—CH$_2$)$_2$SiBrMe followed by ring closure by means of CHBr$_3$ and lithium [61].

4.1.2 Synthesis of Hexasilaperhydrophenalene

Further examples are provided by the syntheses of cis-trans and trans-trans hexa-silaperhydrophenalene. After the synthesis of the chain $Ph_2MeSi—CH_2—(SiMe_2—$ $—CH_2—SiMePh—CH_2)_2—SiMe_2—CH_2Br$, cleavage of the Si—Ph bond by bromine produces $MeBr_2Si—CH_2—(SiMe_2—CH_2—SiMeBr—CH_2)_2—SiMe_2—CH_2Br$. Reaction of this substance with $LiAlH_4$ to produce $MeH_2Si—CH_2—(SiMe_2—CH_2—$ $SiMe—CH_2)_2—SiMe_2—CH_2Br$, followed by ring closure after metallation by lithium of the $—CH_2Br$ group, results in $HMeSi(CH_2—SiMe_2—CH_2—SiHMe—$ $CH_2)_2SiMe_2$. After subsequent bromination of the Si—H groups to produce $MeBrSi(CH_2—SiMe_2—CH_2—SiBrMe—CH_2)_2SiMe_2$, reaction with $HCBr_3$ and Li will cause a partitioning of the ring which enables the isolation of the isomers A and B [62]. For the purpose of clear assignment of the individual atoms in the molecular skeleton, this compound is also named 1,3,5,7,9,11,11-nonamethyl-1,3,5,7,9,11-hexasilatricyclo[7.3.1.0^{5,13}]tridecane $Si_6C_{16}H_{36}$ in Chapt. IV.1.5 dealing with the structural investigation of the trans-trans isomer:

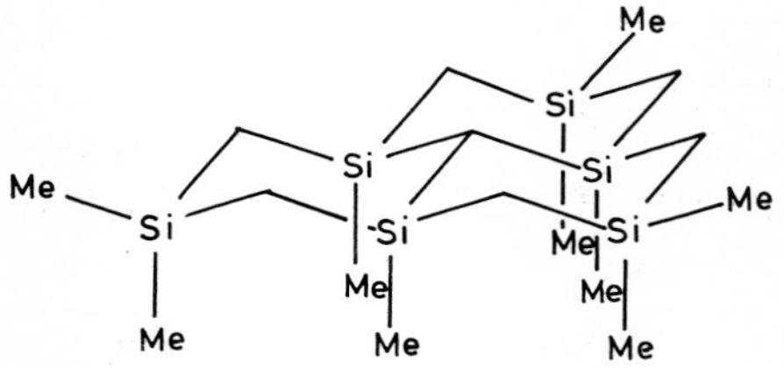

A
trans-trans

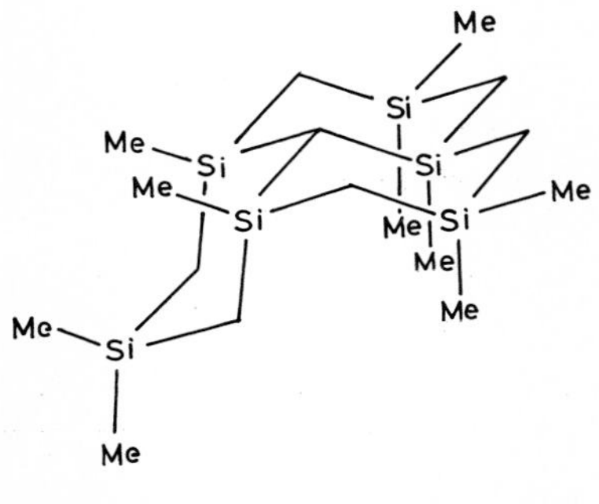

B
cis-trans

4.1.3 Synthesis of Heptasila[4.4.4]propellane

In an analogous way the synthesis of 1,3,5,7,9,12,14-heptasila[4.4.4]propellane was achieved [63]. After formation of $PhSi(CH_2—SiMe_2—CH_2—SiMe_2Ph)_3$, cleavage of the Si—Ph groups using bromine produced $BrSi(CH_2—SiMe_2—CH_2—SiMe_2Br)_3$ followed by ring closure with CCl_4 and Li to form

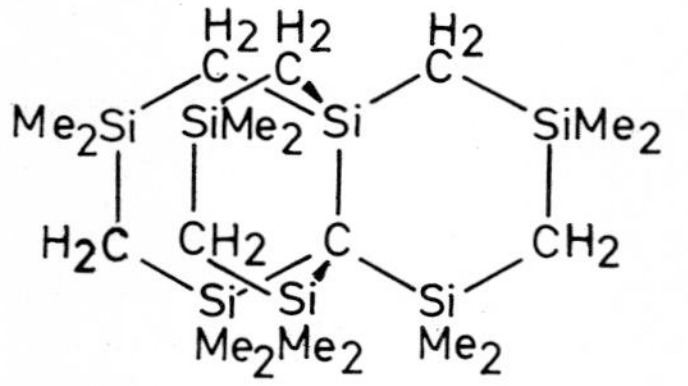

4.2 Investigations into the Synthesis of Tetrasilaadamantanes

We now can report the organometallic synthesis of 1,3,5,7-tetrasilaadamantane. This investigation was based on 1,3,5-trisilacyclohexane $(Cl_2Si—CH_2)_3$ obtained without difficulty by direct synthesis from Si. Starting with this basic compound, the synthesis appeared possible if sufficiently bulky groups could allign themselves in such away that the functional groups maintain the necessary axial positions on the ring. Synthesis of the suitable 1,3,5-trisilacyclohexanes therefore is an indispensable requirement.

4.2.1 Investigations of 1,3,5-Trisilacyclohexanes

4.2.1.1 Synthesis of $(t\text{-BuHSi}—CH_2)_3$

The syntheses of 1,3,5-triphenyl-1,3,5-trisilacyclohexane and 1,3,5-tri-t-butyl-1,3,5-trisilacyclohexane afforded the possibility of selective bromination of $(H_2Si—CH_2)_3$ to $(BrHSi—CH_2)_3$ and the reaction of the resulting cis-cis and cis-trans isomeric mixture with PhMgBr or t-BuLi, respectively [65]:

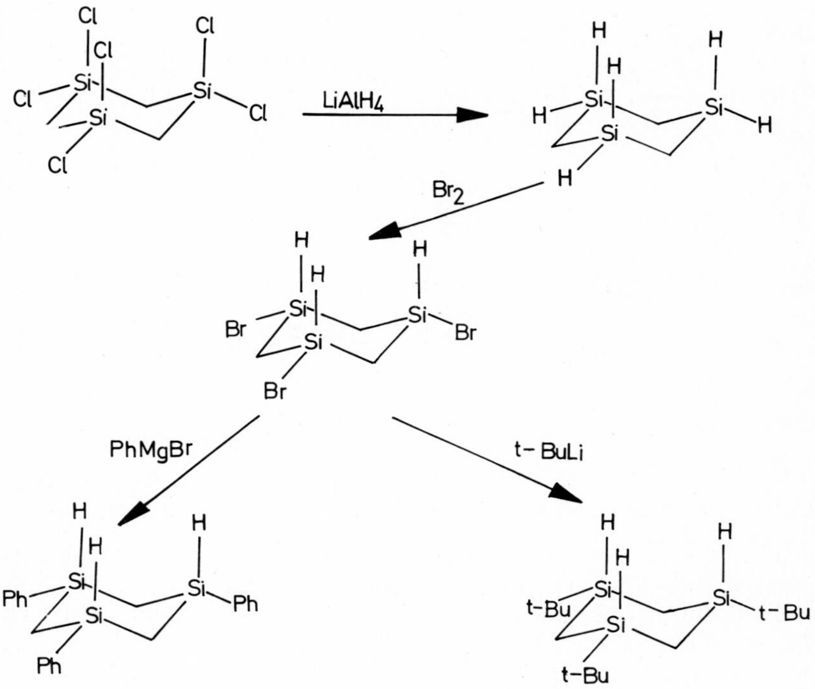

In the case of $(t\text{-BuHSi}—CH_2)$ the cis-cis isomer was separated by sublimation of the product mixture. The composition of the isomeric mixture of $(t\text{-BuHSi}—CH_2)_3$ depends on the reaction temperature. At $-20\ °C$, the amount of cis-cis and cis-trans isomers present were similar; however, at $0\ °C$ the contribution of the cis-cis isomer was only 23%.

4.2.1.2 Synthesis of $(PhHSi-CH_2)_3$

The reaction of $(BrHSi-CH_2)_3$ with PhMgBr generated $(PhHSi-CH_2)_3$ in a 38% yield, and through subsequent fractional crystallization of its corresponding isomers, the cis-cis (20%) and cis-trans (80%) isomers were separated.

4.2.1.3 Formation of $(RBrSi-CH_2)_3$ through Bromination

The bromination of the Si—H groups in $(RHSi-CH_2)_3$ (R = Ph or t-Bu) was possible at 20 °C without cleavage of existing Si—Ph groups, and so as a result $(PhBrSi-CH_2)_3$ as a viscous substance and $(t-BuBrSi-CH_2)_3$ as crystals could be obtained.

4.2.1.4 Synthesis of $(PhHSi-CH_2)_3$ via $(Ph_2Si-CH_2)_3$

The compound $(Ph_2Si-CH_2)_3$ is better prepared from the corresponding fluorosilane:

$$(F_2Si-CH_2)_3 + 6 PhLi \rightarrow (Ph_2Si-CH_2)_3$$

The use of $(F_2Si-CH_2)_3$ is very advantageous because of its relative stability against hydrolysis and because the Si—F group reacts readily with PhLi. The $(F_2Si-CH_2)_3$ is accessible through reaction of $(Cl_2Si-CH_2)_3$ with ZnF_2 [66]. In the solid state, $(Ph_2Si-CH_2)_3$ exists in a distorted boat form (Chapt. IV.1.1.2) [89]. This compound is also obtained under analoguous conditions by reaction of $(Cl_2Si-CH_2)_3$ with PhLi.

The synthesis of 1-fluoro-1,3,5,5-pentaphenyl-1,3,5-trisilacyclohexane is achieved through reaction of $(F_2Si-CH_2)_3$ with PhLi only by adherence to definite reaction conditions, because otherwise $(Ph_2Si-CH_2)_3$ forms:

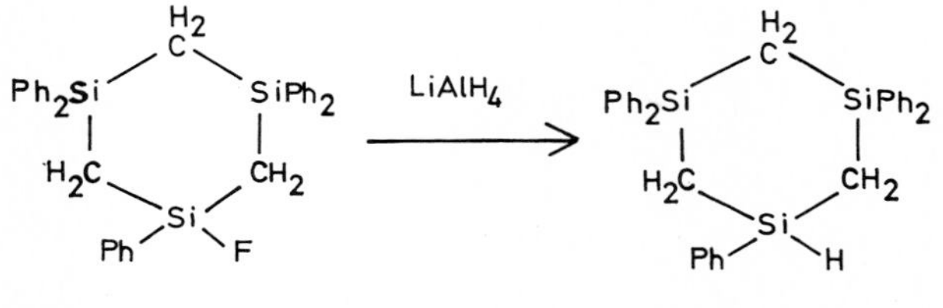

The reaction of different partly phenylated and fluorinated 1,3,5-trisilacyclohexanes with $LiAlH_4$ yields the corresponding SiHPh compounds which can then be separated by means of HPLC [65].

4.2.1.5 The Cleavage of Si—Ph Bonds in $(Ph_2Si-CH_2)_3$

The cleavage of the Si—Ph bond in such compounds was investigated as a means of introducing functional groups into the ring. This reaction, carried out in CCl_4 in the presence of Br_2, proceeds relatively sluggishly at first and is fully completed by refluxing the solvent. After subsequent hydrogenation with $LiAlH_4$ (the SiH-containing compounds are easier to isolate), $(PhHSi-CH_2)_3$ is obtained. The isomers of $(PhHSi-CH_2)_3$ from this reaction sequence appear in the same ratio (1:3) as by an organometallic synthesis. Also, by applying an excess of bromine, no further Si-Ph groups are cleaved from $(PhBrSi-CH_2)_3$.

This confirmed earlier observations that in the presence of Br_2 only one Si—Ph bond is cleaved, mainly because incorporation of more electronegative groups in the ring causes a remarkable stabilizing effect on the remaining Si—Ph bond [68]. On the one hand this enables a selective phenyl cleavage, but it also excludes simultaneous cleavage of all phenyl groups in the compound. Total cleavage becomes possible when $HCl/AlCl_3$ for instance is used, producing $(Cl_2Si—CH_2)_3$ [69].

4.2.2 Attempts to Synthesize Tetrasilaadamantanes

After 1,3,5-trisilacyclohexanes with functional groups in axial positions became available, it seemed logical to continue the organometallic synthesis of 1,3,5,7-tetrasilaadamantane through the following route:

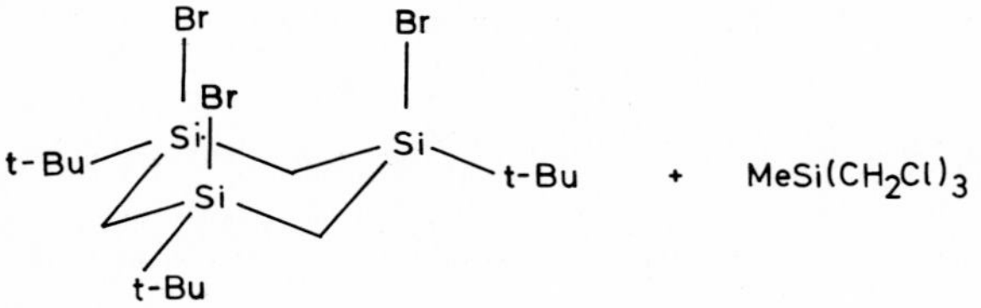

[chemical structure] + $MeSi(CH_2Cl)_3$

Indeed, coupling reactions with alkali metals yielded Si-methylated 1,3,5,7-tetrasilaadamantane, although the yield was very low. This is attributed to the fact that $MeSi(CH_2Cl)_3$ reacts with alkali metals to form C—C single bonds. The Si-compound remains essentially unchanged in the reaction mixture, and is converted by hydrolytic work-up to colorless crystals (mp. 250 °C) of the trisilanol:

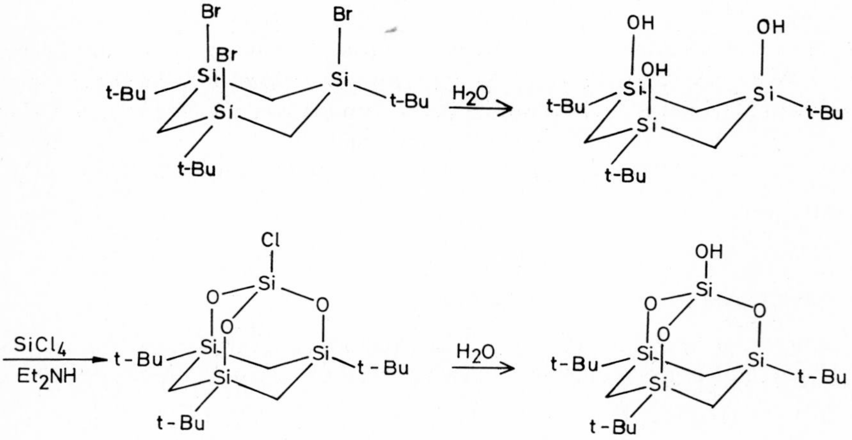

[chemical structures with H_2O, $SiCl_4 / Et_2NH$, H_2O]

This compound reacts with $SiCl_4$ to produce an adamantane derivative, forming thereby also a glassy residue as byproduct [70]. The hydrolysis produced then an adamantane skeleton with a silanol group.

4.3 Synthesis of Si-Substituted 1,3,5,7-Tetrasilaadamantane
Me$_3$(Me$_3$Si—CH$_2$—SiMe$_2$—CH$_2$—SiMe$_2$—CH$_2$) Ad

After a phenyl substituent is removed from Me$_3$(Ph—SiMe$_2$—CH$_2$)Ad* [68], it is possible then to link carbosilane chains of any desired length to the Si bridgehead. The synthesis could follow two possible routes, either through Me$_3$IAd [80] by reaction with LiCH$_2$—SiMe$_2$Ph in the presence of TMEDA in pentane (the solution had to be kept free of LiBr and ether mainly to prevent a halogen exchange between Me$_3$IAd and LiBr, because Me$_3$BrAd was found to behave in a divergent manner), or even better through Me$_3$HAd by means of the scheme

Me$_3$HAd + LiCH$_2$—SiMe$_2$Ph

$\downarrow$ THF, 50–60 °C; 2d

Me$_3$(Ph—SiMe$_2$—CH$_2$) Ad + LiH

$\downarrow$ + Br$_2$

Me$_3$(Br—SiMe$_2$—CH$_2$) Ad

Et$_2$O/C$_5$H$_{12}$ $\downarrow$ + LiCH$_2$SiMe$_2$Ph

Me$_3$(Ph—SiMe$_2$—CH$_2$—SiMe$_2$—CH$_2$) Ad

$\downarrow$ + Br$_2$

Me$_3$(Br—SiMe$_2$—CH$_2$—SiMe$_2$—CH$_2$) Ad

Et$_2$O/C$_5$H$_{12}$ $\downarrow$ + LiCH$_2$SiMe$_3$

Me$_3$(Me$_3$Si—CH$_2$—SiMe$_2$—CH$_2$—SiMe$_2$—CH$_2$) Ad   <u>193</u>

4.4 Synthesis of C-Bridged Cyclic Carbosilanes

The organometallic syntheses of polycyclic carbosilanes depicted so far have always made use of Si-functional groups. The synthesis of carbosilanes, either through pyrolysis of methyl and methylchlorosilanes or through rearrangement using AlBr$_3$ proceeds essentially to the same Si-bridged compounds. The 1,3,5,7-tetrasilaadamantane may be used as an example of this:

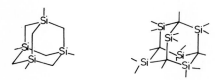

In the compounds which were isolated, silicon atoms nearly exclusively occupy the bridgehead position. Compounds containing carbon and silicon bound in a reverse manner, i.e. with carbon in the bridgehead position to give a Si$_6$C$_4$ adamantane

---

* The abbreviation Ad designates the 1,3,5,7-tetrasilaadamantane molecular skeleton. Substituents written in front of Ad are linked to the Si atoms, those written behind Ad are linked to the C atoms of the adamantane skeleton.

skeleton, were not formed. In order to obtain such compounds, the organometallic synthetic process must operate on the C-functional groups of the carbosilanes. Background investigations into such systems will now be described.

4.4.1 Syntheses via Metallation of CBr$_2$ Groups

4.4.1.1 C-Metallation and Reactions of 2,2-Dibromo-1,1,3,3,5,5-hexamethyl-1,3,5-trisilacyclohexane

Photobromination of $(Me_2Si-CH_2)_3$ enabled the formation of compound 227 [77] This reacted with n-BuLi at $-100\,°C$ in a solvent mixture of THF/ether/pentane (4:1:1) to give compound 228.

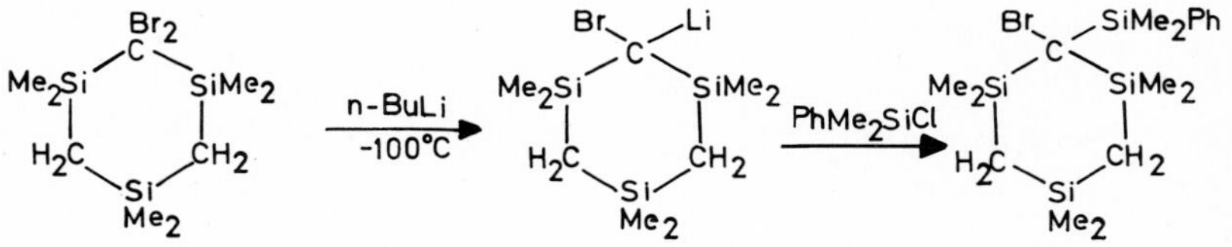

227    $\xrightarrow[-100°C]{\text{n-BuLi}}$ 228 $\xrightarrow{\text{PhMe}_2\text{SiCl}}$ 229

The formation of the lithiated compound 228 is recognized by its yellow-orange color in solution. It reacts subsequently with PhMe$_2$SiCl to form compound 229 [78]. Compound 229 is separated after hydrolytic work up of the reaction mixture through recrystallization from pentane at $-20\,°C$, followed by sublimation at $80\,°C$. It reacts with bromine at $20\,°C$ quantitatively to produce the transparent, highly viscous liquid compound 230:

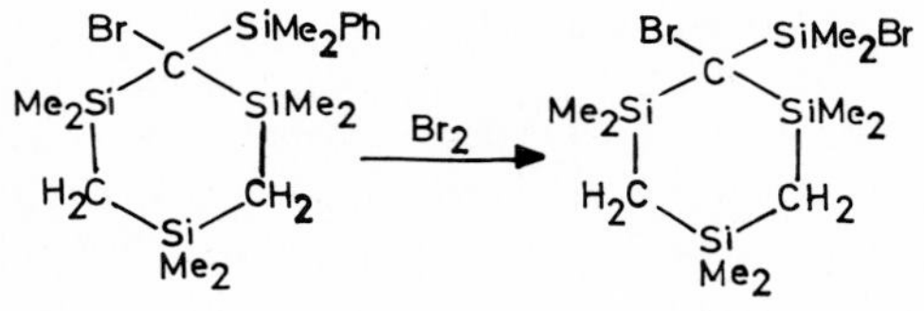

229    $\xrightarrow{Br_2}$ 230

On warming the reaction mixture from $-100\,°C$ to $20\,°C$, a butylation of compound 228 occurs due to existing butylbromide in solution. Compound 229 reacts with n-BuLi at first via an orange-colored intermediate 231 [78]. The trapping reaction between compound 231 and PhMe$_2$SiCl did not result in a second substitution. Rather a butylated compound 232 (60%) appeared as well as 233, containing a tertiary-substituted CH group. The formation of compound 232 is naturally explained by reaction of compound 231 and existing butylbromide in solution. Similarly, compound 233 arises from compound 231 through ether cleavage. After hydrolytic work-up, compounds 232 and 233 were purified over a silica gel column. Substitution of compound 231 with PhMe$_2$SiCl failed because of unfavorable steric conditions. Compound 229 can be metallated by reaction with MeLi. The intermediate compound 231 reacts with the MeBr just formed to generate compound 234.

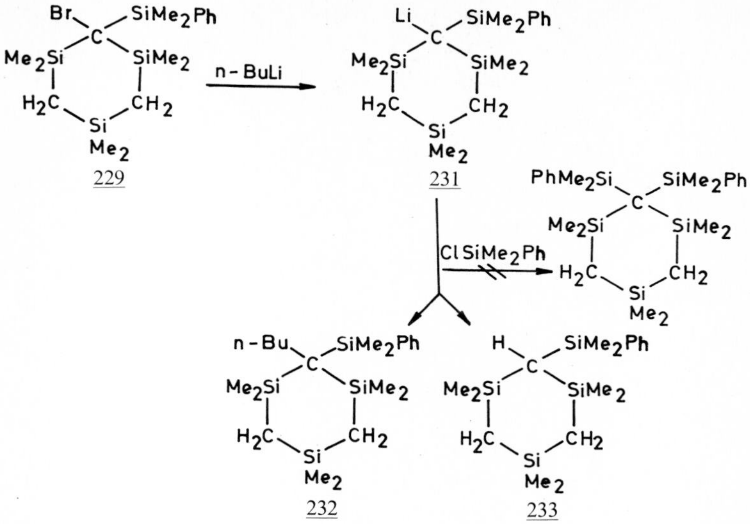

229 231 232 233

Trapping of compound 231 with $PhMe_2SiCl$ failed, and the trapping agent was quantitatively recovered:

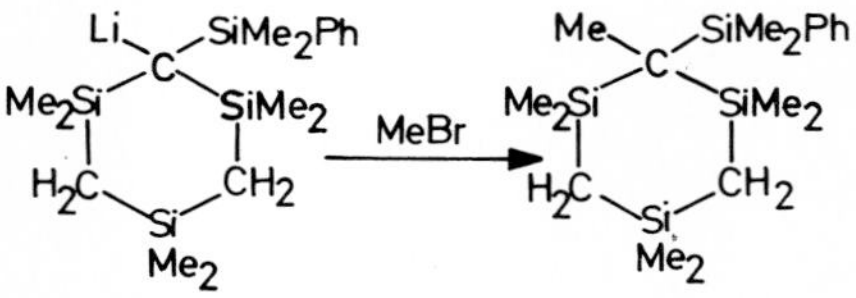

231 234

After recrystallization from pentane, compound 234 was obtained in 80% yield. Metallation of compound 229 to yield compound 231 also can be achieved with metallic Li. Subsequent treatment with MeI enables the formation of the same compound 234 in a 90% yield. Bromine attacks the SiPh group in this compound yielding 235 [78].

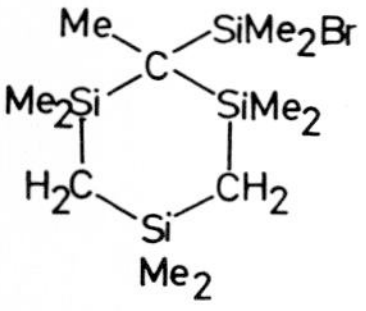

235

4.4.1.2 Synthesis of C-Bridged Spiro Carbosilanes

The comparably convenient access to compound 228 opens a way to the synthesis of spiro carbosilanes [78]:

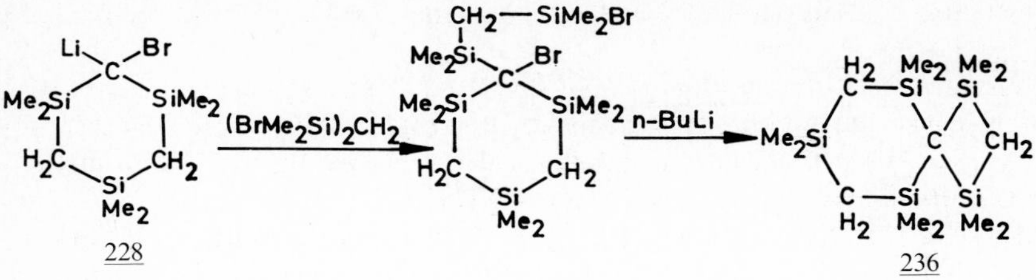

$$
\underset{228}{\overset{\displaystyle \text{Li}\diagdown\underset{\text{C}}{}\diagup\text{Br}}{\underset{\overset{\displaystyle \text{H}_2\text{C}\diagdown\underset{\text{Si}}{}\diagup\text{CH}_2}{\underset{\text{Me}_2}{}}}{\overset{\displaystyle \text{Me}_2\text{Si}\diagdown\diagup\text{SiMe}_2}{}}}
\quad\xrightarrow{(\text{BrMe}_2\text{Si})_2\text{CH}_2}\quad
\text{Me}_2\text{Si}\cdots
\quad\xrightarrow{n\text{-BuLi}}\quad
236
$$

Similarly, reaction of compound 228 with $(\text{BrMe}_2\text{Si}-\text{CH}_2)_2\text{SiMe}_2$ yielded the spiro compound 237 [78]:

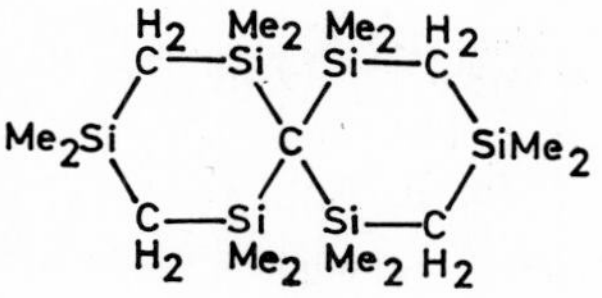

$$237$$

C-metallation of compound 230 is of interest in subsequent syntheses. The lithiated intermediate 238 reacts by abstraction of lithium bromide, forming a reactive intermediate 239 which then dimerizes to produce compound 240 [78]:

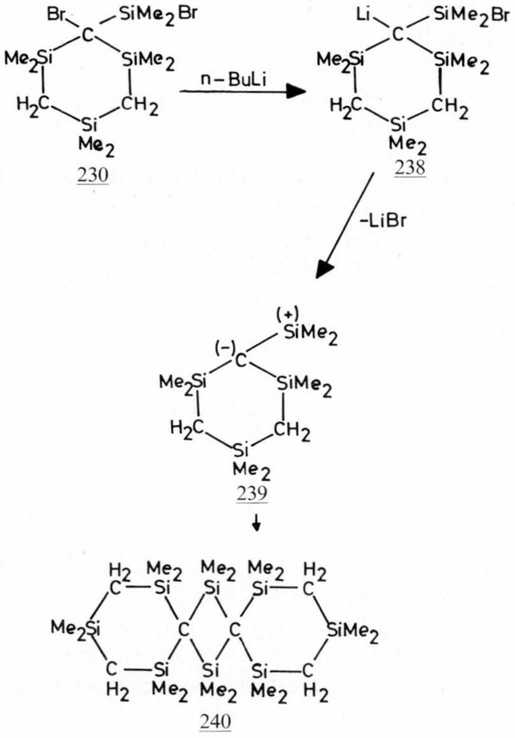

$$230 \xrightarrow{n-\text{BuLi}} 238$$

$$\xrightarrow{-\text{LiBr}}$$

$$239$$

$$\downarrow$$

$$240$$

The formation of compound <u>240</u> occurs as a result of a very selective reaction path, resulting in an 85% yield.

The four-membered ring in compound <u>240</u> displays an unusual stability. It is cleaved neither by HBr nor Br_2, whereas the same ring in $(Me_2Si—CH_2)_2$ readily undergoes such cleavage at $-78\ °C$. This unsual stability is also displayed by 1,1,3,3-tetramethyl-2,2,4,4-tetrakis(trimethylsilyl)-1,3-disilacyclobutane [79].

A similar reaction sequence was used by N. Wiberg and coworkers. Supported by stabilization through bulky groups they succeeded in preparing $Me_2Si=C(SiMe_3)—$ $—SiMe(t\text{-}Bu)_2$ by lithiating a carbon neighboring a Si-halogen group, and by subsequently eliminating lithium halide [75].

4.4.2 Syntheses via Metallation of CH$_2$ or CH Groups

The underlying principle in this method is the metallation of a skeletal C atom in carbosilanes by organolithium compounds. As examples, $(Me_2Si—CH_2)_3$ and the Si-methylated 1,3,5,7-tetrasilaadamantane were investigated (Chapt. III.4).

In this way, $(Me_2Si—CH_2)_3$ is caused to react with BuLi to produce the monolithiated compound:

$$
\begin{array}{ccc}
\underset{\substack{\\ \text{Si}}}{\text{Me}_2} & & \underset{\substack{\\ \text{Si}}}{\text{Me}_2}\\
\text{H}_2\text{C} \quad \text{CH}_2 & & \text{H}_2\text{C} \quad \text{C}{<}^{\text{H}}_{\text{Li}}\\
\text{Me}_2\text{Si} \quad \text{SiMe}_2 & \xrightarrow{\ \ \text{BuLi}\ \ } & \text{Me}_2\text{Si} \quad \text{SiMe}_2\\
\underset{\text{H}_2}{\text{C}} & & \underset{\text{H}_2}{\text{C}}
\end{array}
$$

4.4.2.1 Synthesis of $Me_4Ad(—SiMe_2—CH_2—SiMe_2Ph)$ and of $Me_4Ad(—SiMe_2—CH_2—SiMe_2—CH_2—SiMe_3)$

Selective metallation of skeletal C atoms is decided either by an increase in carbanion activity or by cation solvation, both achieved through the use of a good coordinating solvent. This investigation showed that the reagents n-BuLi/KOCMe$_3$ and $Me_3SiCH_2Li/KOCMe_3$ in THF were better metallation media than the combination of n-BuLi/TMEDA in hexane [55]. Dropwise addition of the organolithium reagent in hexane to a solution of Me_4Ad + $KOCMe_3$ in THF results immediately in the lithiation of a skeletal methylene group:

$$
\text{Me}_4\text{Ad}
\begin{cases}
\xrightarrow{\ \text{n-BuLi/KOCMe}_3\ } \text{Me}_4\text{AdLi} + \text{n-C}_4\text{H}_{10}\\[2ex]
\xrightarrow{\ \text{Me}_3\text{SiCH}_2\text{Li/KOCMe}_3\ } \text{Me}_4\text{AdLi} + \text{SiMe}_4
\end{cases}
$$

The reaction is so fast that it is practically complete at the end of the addition of the lithiating reagent. Trapping the Li-intermediate by means of suitable Si-functional compounds enabled the isolation of skeletal C-silylated compounds in approximately 70–80% yield. The use of KOCMe$_3$ in THF has the advantage of short reaction times and relatively simple work-up. However, the application remains limited to Me_4Ad and Si-methylated carbosilanes because of competing side reactions. For example, Me_3BrAd reacts to form $Me_3(Me_3C—O)Ad$.

The principal metallation reactions proceed without difficulty to the formation of the successive compounds

Me_4AdLi

$\downarrow$ ClSiMe$_2$Ph

$Me_4Ad-SiMe_2Ph$

$\downarrow$ Br$_2$

$Me_4Ad-SiMe_2Br$

$\downarrow$ LiCH$_2$SiMe$_2$Ph

$Me_4Ad-SiMe_2-CH_2-SiMe_2Ph$

(yield 75%)

$Me_4Ad(-SiMe_2-CH_2-SiMe_2Ph)$

$\downarrow$ Br$_2$

$Me_4Ad(-SiMe_2-CH_2-SiMe_2Br)$

$\downarrow$ LiCH$_2$-SiMe$_3$

$Me_4Ad(-SiMe_2-CH_2-SiMe_2-CH_2-SiMe_3)$ <u>178</u>

Compound <u>178</u>, an oily colourless liquid can also be obtained in a 73% yield in an alternative way:

$Me_4AdLi + BrSiMe_2-CH_2-SiMe_2-CH_2-SiMe_3$

$\downarrow$

$Me_4Ad(-SiMe_2-CH_2-SiMe_2-CH_2-SiMe_3)$ <u>178</u>

4.4.2.2 Synthesis of $(Me_3Si)_2CH-SiMe_2-CH_2-SiMe_2-CH_2-SiMe_3$

At first it appeared useful to build the appropriate branched unit containing an Si-functional group and then to extend the chain length according to [55]:

$$Me_3Si-CH_2-SiMe_3 \begin{array}{c} \text{n-BuLi/TMEDA} \\ \hline \text{t-BuLi/HMPTA} \\ \hline \text{THF} \end{array} \rightarrow (Me_3Si)_2CHLi \quad \begin{array}{c} + \text{ n-C}_4\text{H}_{10} \\ + \text{ t-C}_4\text{H}_{10} \end{array}$$

$Me_3Si-CH(Li)-SiMe_3$

$\downarrow$ Me$_2$PhSiCl

$(Me_3Si)_2CH-SiMe_2Ph$

$\downarrow$ Br$_2$

$(Me_3Si)_2CH-SiMe_2Br$

$\downarrow$ LiCH$_2$SiMe$_2$Ph

$(Me_2Si)_2CH-SiMe_2-CH_2-SiMe_2Ph$

$\downarrow$ Br$_2$

$(Me_3Si)_2CH-SiMe_2-CH_2-SiMe_2Br$

$\downarrow$ LiCH$_2$-SiMe$_3$

$(Me_3Si)_2CH-SiMe_2-CH_2-SiMe_2-CH_2-SiMe_3$ <u>182</u>

4.4.2.3 Synthesis of $(Me_3Si)_2CH-SiMe(CH_2-SiMe_2-CH_2-SiMe_3)_2$

The synthesis of this compound followed a reaction path similar to that in II.4.4.2.4:

$$(Me_2Si-CH_2)_2$$

$$\downarrow Br_2$$

$$BrMe_2Si-CH_2-SiMe_2-CH_2Br$$

$$\downarrow MeMgCl$$

$$Me_3Si-CH_2-SiMe_2-CH_2Br$$

$$\downarrow Li/Et_2O$$

$$Me_3Si-CH_2-SiMe_2-CH_2Li$$

$$\downarrow MeSiCl_3$$

$$ClMeSi(CH_2-SiMe_2-CH_2-SiMe_3)_2$$

$$\downarrow (Me_3Si)_2CHLi$$

$$(Me_3Si)_2HC-SiMe(CH_2-SiMe_2-CH_2-SiMe_3)_2 \qquad \underline{187}$$

4.4.2.4 Synthesis of $(Me_3Si)_3C-SiMe_2-CH_2-SiMe_2-CH_2-SiMe_3$

Originally, the simplest way to synthesize this compound appeared to be through metallation of the tertiary C atom of $(Me_3Si)_2CH-SiMe_2-CH_2-SiMe_2-CH_2-SiMe_3$ with MeLi, followed by reaction with Me_3SiCl. However, the reaction showed that apart from the desired monolithiation, dilithiation occurred. In addition to this, unchanged starting material remained. Reaction with Me_3SiCl therefore produced a mixture of:

$$(Me_3Si)_2CH-SiMe_2-CH_2-SiMe_2-CH_2-SiMe_3$$
$$+ (Me_3Si)_3C-SiMe_2-CH_2-SiMe_2-CH_2-SiMe_3$$
$$+ (Me_3Si)_3C-SiMe_2-CH_2-SiMe_2-CH(SiMe_3)_2$$

Difficulties were also encountered in the reaction below beginning with $(Me_3Si)_2CH_2$ as starting material:

$$(Me_3Si)_2CH_2$$

$$t\text{-BuLi/HMPTA} \downarrow THF, \ -78\ °C$$

$$(Me_3Si)_2CHLi + C_4H_{10}$$

$$\downarrow Me_3SiCl$$

$$(Me_3Si)_3CH$$

$$THF \downarrow MeLi$$

$$(Me_3Si)_3CLi$$

$$\downarrow Me_2SiCl_2 \ \text{(in excess)}$$

$$(Me_3Si)_3C-SiMe_2Cl$$

$$\downarrow LiCH_2SiMe_2Ph$$

$$(Me_3Si)_3C-SiMe_2-CH_2-SiMe_2Ph$$

$$\downarrow Br_2$$

$$(Me_3Si)_3C-SiMe_2-CH_2-SiMe_2Br$$

$$\downarrow LiCH_2-SiMe_3$$

$$(Me_3Si)_3C-SiMe_2-CH_2-SiMe_2-CH_2-SiMe_3 \qquad \underline{184}$$

These reactions proceeded smoothly in the initial stages of the reaction pathway until they reached $(Me_3Si)_3C—SiMe_2Cl$. This compound crystallizes upon concentration of the solution, and is isolated through sublimation at 80 °C/10^{-3} mm Hg. Further reaction with $LiCH_2—SiMe_2Ph$ occurs to a limited extent, obviously because the reactivity of the Si—Cl group is considerably restricted. As a result, either no subsequent reaction occurred, or under forced reaction conditions ether cleavage resulted. Carrying out the reaction under reflux in THF/TMEDA yielded in 3 days a mixture of $(Me_3Si)_3C—SiMe_2—CH_2—SiMe_2Ph$, Me_3SiPh, and ether-cleaved by-products. Further reaction time eventually yielded the desired compound, although it remained inpure.

4.4.3 Synthesis of C-Substituted Carbosilanes through Metallation of CH$_2$ or CH Groups

4.4.3.1 2-Trimethylsilyl-2-dimethyl(phenyl)silyl-1,1,3,3,5,5-hexamethyl-1,3,5-trisilacyclohexane and 2,2-Bis(trimethylsilyl)-1,1,3,3,5,5-hexamethyl-1,3,5-trisilacyclohexane

The synthesis of simple derivatives pertaining to this class of compounds is achieved in approximately 90% yield through metallation of a CH$_2$ group in hexamethyl-1,3,5-trisilacyclohexane with n-BuLi:

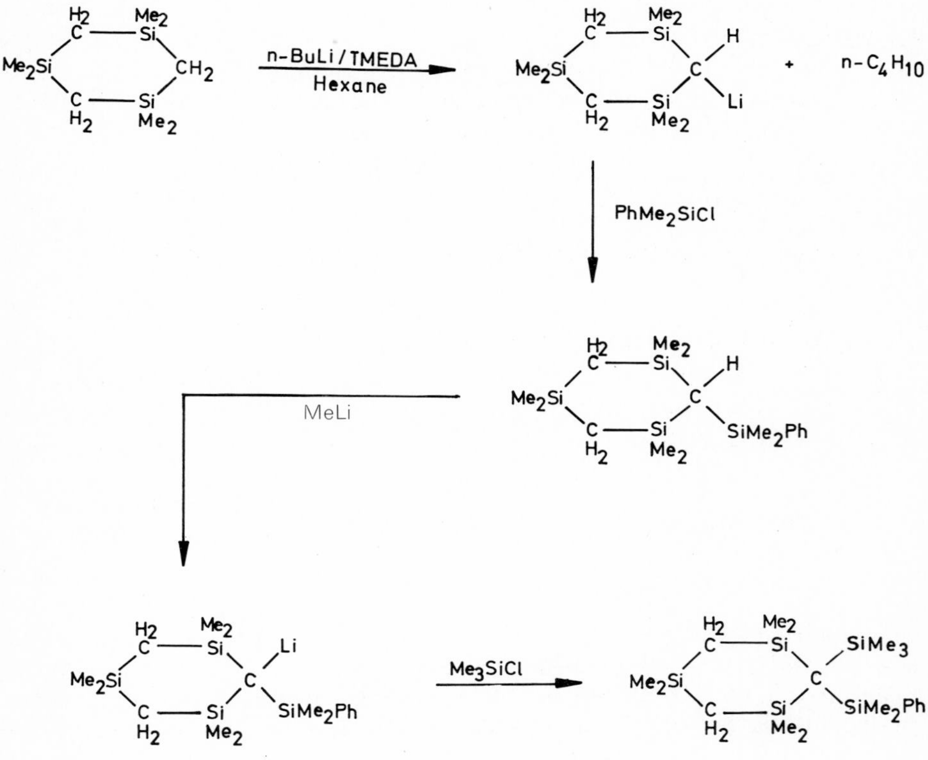

In an analogous way, compound 197 was synthesized:

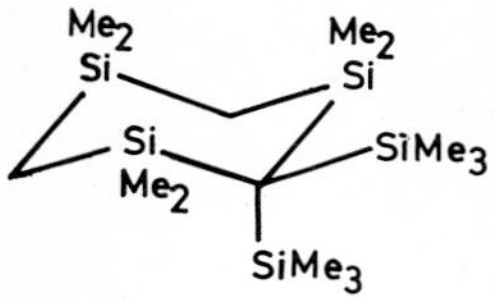

Attempts to extend the carbosilane chain in

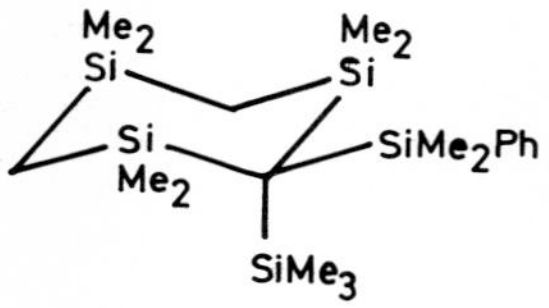

remained unsuccessful, mainly because conversion of the Si—Ph group to the corresponding Si—Br group by addition of HBr or Br_2 did not occur. In addition to this, attempts to increase the cleaving activity by addition of $AlBr_3$ to the reaction mixture resulted instead in unsurveyable side reactions apart from the desired SiPh cleavage [55].

The synthesis of 1,3,5-trisilacyclohexanes bearing longer carbosilane side chains on a ring carbon atom proceeds according to

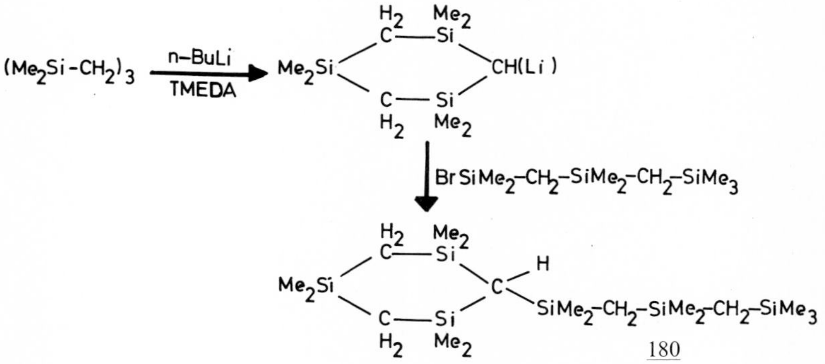

to form compound 180, an oily, transparent liquid of bp. 116–117 °C/10^{-3} mm Hg, in a 68% yield.

Similarly, C-metallation and further substitution of $(Me_2Si—CH_2)_3$ results in compounds 190, 186 and 188 (see p. 92).

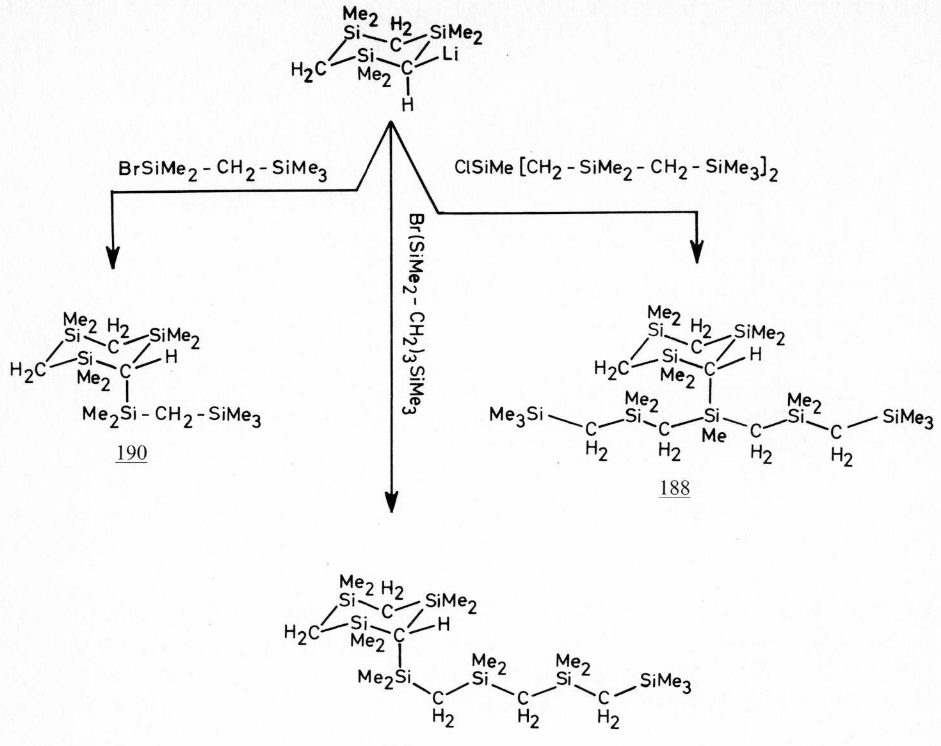

4.4.3.2 Synthesis of C-Silylated 6-Trimethylsilyl-1,3,5,7,9-pentasiladecalin

Synthesis could not be achieved by rearrangement of

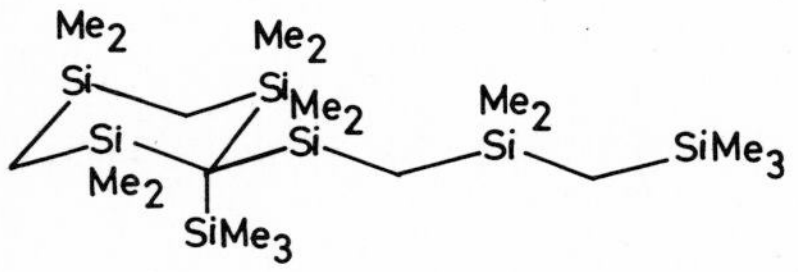

by AlBr$_3$, because the barrelane skeleton formed instead [55, 57]. The desired compound was made by lithiating 1,3,5,7,9-pentasiladecalin 181 with MeLi followed by reaction with Me$_3$SiCl, as shown in Fig. 14 [56].

Compound 181 existed as a mixture of 65 % trans and 35 % cis isomer. The investigation showed that the same lithiated compound resulted as an intermediate, regardless of which isolated isomer was used as starting material. This contrasts with the result that only the trans isomer of compound 199 is formed by reaction of compound 241 with Me$_3$SiCl, while hydrolysis of compound 241 produces the normal isomeric mixture [56].

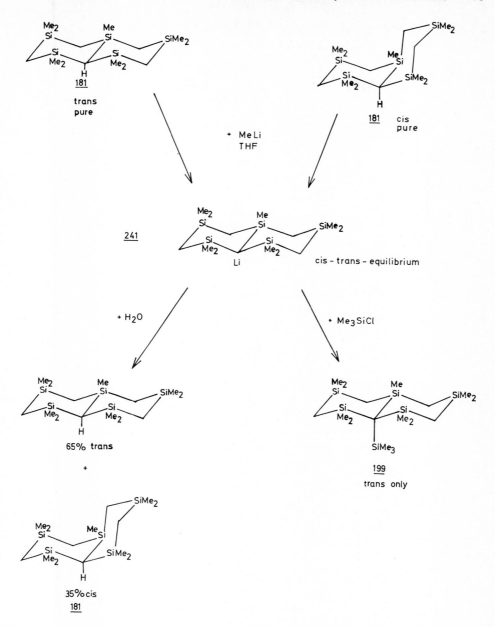

Fig. 14. Synthetic route leading to the formation of 6-trimethylsilyl-1,3,5,7,9-pentasiladecalin 199

4.4.3.3 Synthesis of C-Bridgehead Silylated Tetrasilatriscaphanes

These compounds were synthesized by initial C-metallation followed by reaction with $ClSi(CH_2-SiMe_3)_3$, as shown in the example below.

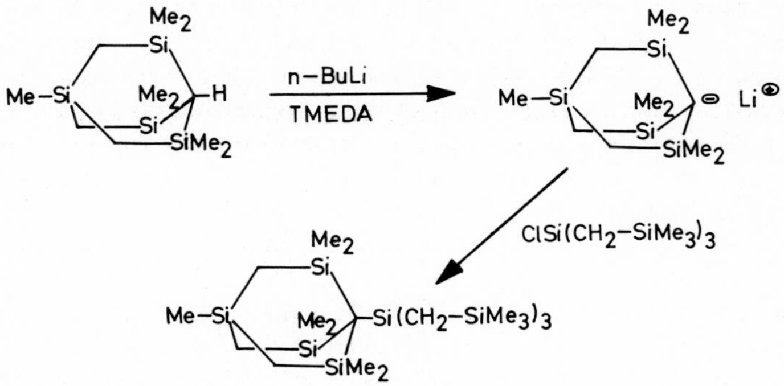

204

4.5 Synthesis of 1,3-Disilapropanes

4.5.1 Synthesis of 2,2-Dichloro-1,3-disilapropanes

The investigation into the reactions of various Si- and C-chlorinated 1,3-disilapropanes with MeMgCl and MeLi dealt with in Chapt. III.3.2.2 required the synthesis of such starting materials.

The organometallic synthesis of compounds $Me_3Si-CCl_2-SiMe_nCl_{3-n}$ ($n = 1$–3) started with the reaction of the carbenoid $Me_3Si-CCl_2Li$ with methylchlorosilanes. This reaction pathway was designed according to the synthesis of $(Me_3Si)_2CCl_2$ from $Me_3Si-CCl_2H$, n-BuLi and Me_3SiCl at -100 °C [119]:

$$Me_3Si-CCl_2H \xrightarrow[-\text{ n-BuH}]{+\text{ n-BuLi}} Me_3Si-CCl_2Li$$

$$\xrightarrow[-\text{ LiCl}]{+\text{ ClSiMe}_n\text{Cl}_{3-n}} Me_3Si-CCl_2-SiMe_nCl_{3-n} \qquad (n = 1-3)$$

$ClMe_2Si-CCl_2H$ was accessible through gas-phase photochlorination of Me_3SiCl [120]. Starting from $ClMe_2Si-CCl_2H$, two important intermediate stages were established by methylation or phenylation in the formation of 2,2-dichloro-1,3-disilapropanes:

$$ClMe_2Si-CCl_2H \left\lbrace \begin{array}{l} \xrightarrow{\text{MeMgCl}} Me_3Si-CCl_2H \\ \xrightarrow{\text{PhMgBr}} PhMe_2Si-CCl_2H \end{array} \right.$$

Then $Me_3Si-CCl_2-SiMe_2Cl$ was obtained according to

$$Me_3Si-CCl_2Li + Me_2SiCl_2 \rightarrow Me_3Si-CCl_2-SiMe_2Cl + LiCl$$

242

There were indications that transfer reactions occurred [119]:

$$Me_3Si-CCl_2Li + Me_3Si-CCl_2H \rightarrow (Me_3Si)_2CCl_2 + LiCCl_2H$$

The carbenoid acted as a nucleophile and attacked the Si—C bond of Me_3Si- $-CCl_2H$. If $Me_3Si-CCl_2H$ is added dropwise to n-BuLi, then this commutation reaction is restrained. This procedure guaranteed that as soon as $Me_3Si-CCl_2H$ comes into contact with n-BuLi, which is present in excess, it is lithiated, thus not allowing any chance for side reactions to occur. Regarding the reaction to produce compound $\underline{242}$, a twofold substitution of Me_2SiCl_2 through $Me_3Si-CCl_2Li$ to produce $(Me_3Si-CCl_2)_2SiMe_2$ had to be prevented. This was achieved by cooling the reaction mixture to -125 °C after obtaining the synthesis of $Me_3Si-CCl_2Li$. Then, a -78 °C solution of Me_2SiCl_2 in ether was added very quickly in a five-fold excess to the solution. After subsequent vigorous stirring, the mixture was maintained at -110 °C for 2–3 hours and then slowly allowed to warm to 20 °C. The resulting reaction mixture (80% conversion of starting materials) contained $Me_3Si-CCl_2-SiMe_2Cl$ in 85% yield and $(Me_3Si)_2CCl_2$ in 15% yield [114]. The $(Me_3Si)_2CCl_2$ was concentrated through fractional distillation and separated out in the pre-runs, so that the last fraction consisted of pure $Me_3Si-CCl_2-SiMe_2Cl$.

The synthesis of $Me_3Si-CCl_2-SiMeCl_2$, compound $\underline{243}$, was obtained by means of the pathway shown below:

$$Me_3Si-CCl_2Li + MeSiCl_3 \xrightarrow{-100\,°C} Me_3Si-CCl_2-SiMeCl_2 + LiCl$$

This was carried out under the same reaction conditions employed for the synthesis of $Me_3Si-CCl_2-SiMe_2Cl$ [114]. Side reactions again were successfully suppressed by cooling the reaction mixture and by addition of excess $MeSiCl_3$. The reaction products were obtained in a 75% yield, consisted after work-up of $Me_3Si-CCl_2-$ $-SiMeCl_2$ (88%) and $(Me_3Si)_2CCl_2$ (12%).

No $Me_3Si-CCl_2-SiCl_3$ could be formed in this way from $Me_3Si-CCl_2Li$ and $SiCl_4$. Rather, this reaction proceeded mostly to the formation of $(Me_3Si)_2CCl_2$, in approximately 50% yield, and to $Me_3Si-CCl_3$ together with $Cl_3Si-CCl_2H$ to a lesser degree.

Starting from $ClMe_2Si-CCl_2H$ and n-BuLi it was not possible to produce the corresponding carbenoid, which as originally thought would provide a means of synthesizing 1,3-disilapropanes with two functional silyl groups. The use of the known carbenoid $PhMe_2Si-CCl_2Li$ [102] opens up another pathway to this aim. By reaction with the respective methylchlorosilanes, Si—Ph bond cleavage using Br_2 and subsequent Br/Cl exchange, several compounds of the type $ClMe_2Si-CCl_2-SiMe_nCl_{3-n}$ were synthesized [114]. The method was tried out with $PhMe_2Si-CCl_2H$ and Me_3SiCl, a combination which was expected to give rise to the least difficulties

$$PhMe_2Si-CCl_2Li + Me_3SiCl \rightarrow PhMe_2Si-CCl_2-SiMe_3 + LiCl$$
$$\underline{243}$$

In order to avoid the formation of $(PhMe_2Si)_2CCl_2$ through trans-metallation, $PhMe_2Si-CCl_2H$ was slowly added dropwise to an equimolar solution of n-BuLi at -100 °C. Subsequent reaction with Me_3SiCl yielded compound $\underline{243}$ in an 85% yield. All attempts to cleave the Si—Ph bond in compound $\underline{243}$ by HCl, Cl_2 or HCl/ $AlCl_3$ to form a Si—Cl bond were unsuccessful. On the other hand, performing the reaction in the presence of excess bromine formed the known compound Me_3Si- $-CCl_2-SiMe_2Br$ [94] in very good yield. However, distillation of this compound resulted in a rearrangement, producing $Me_3Si-CBrCl-SiMe_2Cl$. By allowing

$Me_3Si-CCl_2-SiMe_2Br$ to react in PCl_3 solution at $-78\ ^\circ C$ with HCl, the desired chloro substitution was achieved.

$$Me_3Si-CCl_2-SiMe_2Br + HCl \xrightarrow[PCl_3]{-78\ ^\circ C} Me_3Si-CCl_2-SiMe_2Cl + HBr$$

Compound $ClMe_2Si-CCl_2-SiMe_2Cl$ 244 was synthesized in the same way [114].

$$PhMe_2Si-CCl_2Li + Me_2SiCl_2$$
$$\downarrow {-100\ ^\circ C}$$
$$PhMe_2Si-CCl_2-SiMe_2Cl + LiCl$$
$$\downarrow {Br_2}$$
$$BrMe_2Si-CCl_2-SiMe_2Cl + PhBr$$
$$\downarrow {HCl/PCl_3}$$
$$ClMe_2Si-CCl_2-SiMe_2Cl + HBr$$
$$\underline{244}$$

In the first reaction step above, namely, the reaction of $PhMe_2Si-CCl_2Li$ with Me_2SiCl_2, further substitution is prevented by using Me_2SiCl_2 in excess. Products resulting from side reactions, such as the formation of $(PhMe_2Si-CCl_2)_2SiMe_2$, were not observed. The Si—Ph cleavage resulted from the second reaction step, in the absence of solvent and by dropwise addition of excess Br_2. In order to avoid Si—Br/C—Cl exchange, the reaction temperature was maintained below 50 °C. After bromine and bromobenzene were distilled off under vacuum, purified $BrMe_2Si-$ $-CCl_2-SiMe_2Cl$ was obtained by vacuum condensation at 20 °C. Then Cl-substitution, the third reaction step of this pathway, was obtained by introducing HCl into a solution of $BrMe_2Si-CCl_2-SiMe_2Cl$ in PCl_3. The substitution was complete in 30 hrs.

The compound $ClMe_2Si-CCl_2-SiMeCl_2$ 245 was synthesized analogously by reaction of $PhMe_2SiCCl_2Li$ with $MeSiCl_3$ through the intermediate $PhMe_2Si-CCl_2-$ $-SiMeCl_2$ [114]. Here Si—Ph cleavage yielded $BrMe_2Si-CCl_2-SiMeCl_2$, which on reaction with HCl in PCl_3 produced the desired Br/Cl exchange.

Lastly, $Cl_2MeSi-CCl_2-SiCl_3$ is accessible by another method through methylation of $(Cl_3Si)_2CCl_2$ at 0 °C with MeMgCl in a mole ratio of 1:1 [112].

4.5.2 Synthesis of 2-Methyl-2-chloro-1,3-disilapropanes

A possibility for the synthesis of these compounds lies with halogen/metal exchange in the reaction of $(Me_3Si)_2CCl_2$ [119] with n-BuLi at -110 °C followed by reaction with MeI to produce $(Me_3Si)_2CMeCl$. This reaction path is appliable to compounds in which methyl or phenyl groups protect the silicon atom against nucleophilic attack by n-BuLi. For example, the formation of $(Me_3Si)_2CCl(CD_3)$ occurs by reaction of $(Me_3Si)_2CClLi$ with CD_3Br [113].

Another possibility for the synthesis of these compounds could lie with the metallation of $PhMe_2Si-CCl_2Me$ with n-BuLi to produce $PhMe_2Si-CClMeLi$, followed by reaction with methylchlorosilanes. Investigation showed that this pathway was not adhered to; rather, butylation products of methylchlorosilanes and polymers were formed. This is probably due to the instability of the carbenoid $PhMe_2Si-$

—CClMeLi in comparison to $PhMe_2Si—CCl_2Li$, caused mainly by the destabilizing influence of the adjacent methyl group.

Access to $Me_3Si—CMeCl—SiMe_2Cl$ $\underline{246}$ was possible through $Me_3Si—CCl_2—$ $—SiMe_2Ph$:

$$Me_3Si—CCl_2—SiMe_2Ph$$

$-100\,°C \Big\downarrow \text{n-BuLi}$

$$Me_3Si—CClLi—SiMe_2Ph + n-BuCl$$

$\Big\downarrow \text{MeI}$

$$Me_3Si—CClMe—SiMe_2Ph + LiI$$

$\Big\downarrow \text{HCl}$

$$Me_3Si—CClMe—SiMe_2Cl + PhH$$

Lithiation of the CCl_2 group in $Me_3Si—CCl_2—SiMe_2Ph$ was achieved by using a 10% excess of n-BuLi and proceeded quantitatively to formation of $Me_3Si—$ $—CClLi—SiMe_2Ph$. The high stability of the carbanion in this case is due to the mesomeric effect of the Cl atom and two further silyl groups present on carbon. The formation of the carbanion was made easier due to the delocalization of negative charge in the system of the Ph group. Reaction with MeI produces $Me_3Si—CClMe—$ $—SiMe_2Ph$ in 75% yield. The cleavage of Si—Ph groups occurred at $-90\,°C$ with liquid HCl. The reaction proceeded quantitatively and was complete after 24 hours [114]. Hence the stepwise formation of $ClMe_2Si—CMeCl—SiMe_2Cl$ $\underline{247}$ follows this pathway:

$$PhMe_2Si—CCl_2Li + PhMe_2SiCl$$

$\Big\downarrow$

$$(PhMe_2Si)_2CCl_2 + LiCl$$

$\Big\downarrow + \text{n-BuLi}$

$$(PhMe_2Si)_2CClLi + n-BuCl$$

$\Big\downarrow \text{MeI}$

$$(PhMe_2Si)_2CMeCl + LiI$$

$\Big\downarrow 2\,\text{Br}_2$

$$(BrMe_2Si)_2CMeCl + 2\,PhBr$$

$\Big\downarrow 2\,\text{HCl/PCl}_3$

$$(ClMeSi)_2CMeCl + 2\,HBr$$

Synthesis of $(PhMe_2Si)_2CMeCl$ is achieved in 86% yield by the reaction of $(PhMe_2Si)_2CClLi$ with MeI. The Si—Ph cleavage is possible in liquid HCl at $-90\,°C$, but this method cannot be used for the synthesis of larger amounts of product because the solubility of $(PhMe_2Si)CMeCl$ in HCl at $-90\,°C$ is too small. The cleavage of phenyl groups is more conveniently achieved with bromine in the absence of solvent. The reaction is complete in two days at $20\,°C$, using a double excess of bromine. Purification of $(BrMe_2Si)_2CMeCl$ was achieved through vacuum condensation at $20\,°C$ so as to avoid SiBr/CCl exchange. Conversion of Si—Br to Si—Cl groups occurred in a PCl_3 solution at $-78\,°C$ by HCl addition [114].

The synthesis of compounds such as $Cl_2MeSi-CMeCl-SiCl_3$, 248, where a larger chlorine content exists on silicon, cannot be achieved according to the above procedure. From investigations into the reaction of $(Cl_3Si)_2CCl_2$ with MeLi it is known that the reaction proceeds initially through methylation of the C atom, producing eventually $(Cl_3Si)_2CMeCl$ [112]. Regarding the reaction between $Cl_2MeSi-CCl_2-SiCl_3$ and MeLi in Et_2O, it must be ascertained that a deficiency of MeLi in the reaction will prevent a second metallation of the bridging C atom. The reaction carried out in a mole ratio of 1:0.8 at 20 °C produces a reaction mixture containing (besides unreacted starting compound and a smaller amount of $(Cl_2MeSi)_2CCl_2$ and $ClMe_2Si-CCl_2-SiCl_3$) only $Cl_2MeSi-CMeCl-SiCl_3$, which could be separated by means of gas chromatography [114].

III. Reactions of Carbosilanes

1. Introduction

There are some principal objectives in the study of the reactivity of carbosilanes: First, to what extent are the generally known reactions of Si-functional groups (Chapt. I.2) transferable to the silicon atoms of a carbosilane skeleton? Second, under what circumstances do changes such as Si—C cleavage and rearrangement occur in the molecular skeleton? These objectives overlap partly with those presented in Chapt. II.4 but require separate study. Of particular interest was an investigation into the effects of skeletal substituents on the behavior of the silicon and carbon atoms. Further, what about the stability of the molecular skeleton in compounds containing CH_2 or CCl_2 groups, respectively? What is the effect of SiCl, SiF and SiH groups? What are the reactions of the CCl_2 and CH_2 groups? Here the compounds $(Cl_3Si)_2CH_2$, $(Cl_2Si—CH_2)_3$, tetrasilaadamantane and their derivatives served as models for such investigations.

2. The Introduction of New Functional Groups on the Carbosilane Molecular Skeleton

The objective of the investigations now to be dealt with was to introduce functional groups on skeletal carbon atoms of the model carbosilanes, then to establish their influence on the Si—C bond, and hence to create a wider basis for the synthesis of carbosilanes.

2.1 C-Halogenation of Carbosilanes

2.1.1 C-Chlorinated Carbosilanes

Conversion of CH_2 groups to CCl_2 or CHCl groups in the carbosilane skeleton is expected to occur through photohalogenation, provided that cleavage of the Si—C bond does not occur. Because photochlorination affects usually the terminal methyl groups in Si-methylated carbosilanes, it was decided at first to investigate Si-chlorinated representatives of this class.

The first reaction carried out in this way was with $(Cl_3Si)_2CH_2$, and no $Si-C$ cleavage was observed [99]:

$$Cl_3Si-CH_2-SiCl_3 \xrightarrow[Cl_2]{hv} Cl_3Si-CCl_2-SiCl_3.$$

Similarly, chlorination of $(Cl_2Si-CH_2)_3$ occurs without cleavage [91]:

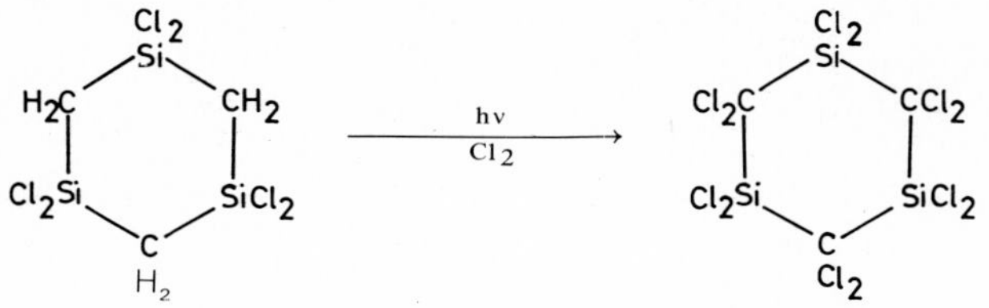

Carbon-chlorinated compounds are hydrolytically easier to cleave than their corresponding CH_2-containing analogues, as is shown in this example:

$$Cl_3Si-CCl_2-SiCl_3 \xrightarrow[OH^-]{H_2O} 2\equiv SiOH + H_2CCl_2 + 6\ HCl$$

$$(Cl_2Si-CCl_2)_3 + 6\ NaOH + 6\ H_2O \rightarrow 3\ Si(OH)_4 + 6\ NaCl + 3\ CH_2Cl_2$$

Through C-chlorination a considerable number of new compounds became accessible. From the molecular skeleton of 1,3-disilapropane the following compounds arose, differing only in the arrangement of Cl and H atoms in the skeleton. Ignoring for the present those compounds containing partially chlorinated groups, we found

1. $Cl_3Si-CH_2-SiCl_3$ 3. $H_3Si-CH_2-SiH_3$

2. $Cl_3Si-CCl_2-SiCl_3$ 4. $H_3Si-CCl_2-SiH_3$

The number of compounds obtained increases if one includes those in which the C and Si atoms are partially chlorinated:

5. $Cl_3Si-CHCl-SiCl_3$ 6. $H_3Si-CHCl-SiH_3$

7. $H_{3-n}Cl_nSi-CH_2-SiH_{3-n}Cl_n,$ $(n = 1-3)$

In carbosilanes with three or more silicon atoms just by different arrangement of the C-chlorinated groups a number of additional derivatives emerges, as shown in examples 8–11. Moreover, considering the various possible substitution patterns of hydrogen and chlorine on silicon, this number will considerably increase.

8. $\equiv Si-CHCl-Si-CH_2-Si\equiv$ 10. $\equiv Si-CHCl-Si-CCl_2-Si\equiv$

9. $\equiv Si-CCl_2-Si-CH_2-Si\equiv$ 11. $\equiv Si-CHCl-Si-CHCl-Si\equiv$

The formation of CHCl-containing compounds becomes progressively more difficult in the series $(Cl_3Si)_2CH_2$, $(Cl_3Si-CH_2)_2SiCl_2$ and $(Cl_2Si-CH_2)_3$. Only $(Cl_3Si)_2CHCl$ could be synthesized on a preparative scale through photochlorination in the gas phase exploiting the considerable difference in boiling points, that exist between the CH_2- and CHCl-containing compounds.

The photochlorination of $(Cl_3Si-CH_2)_2SiCl_2$ in the liquid phase proceeds faster in CCl_4 than in the absence of solvent. Indications of the course of the chlorination process, as well as clues to how these partially chlorinated products are formed, came from an analysis of the ^{1}H-NMR spectra taken during the chlorination. The compounds $Cl_3Si-CHCl-SiCl_2-CH_2-SiCl_3$, $Cl_3Si-CCl_2-SiCl_2-CH_2-SiCl_3$ and $Cl_3Si-CCl_2-SiCl_2-CHCl-SiCl_3$ appear step by step as intermediates, of which $Cl_3Si-CHCl-SiCl_2-CH_2-SiCl_3$ forms only at the beginning of the reaction, followed by $Cl_3Si-CCl_2-SiCl_3-CHCl-SiCl_3$, but not before the starting material has been fully consumed. Only $Cl_3Si-CCl_2-SiCl_2-CH_2-SiCl_3$ exists in the reaction mixture for an extended period. The fact that $(Cl_3Si-CHCl)_2SiCl_2$ fails to appear in the reaction mixture, while $Cl_3Si-CCl_2-SiCl_2-CH_2-SiCl_3$ appears in the reaction mixture for a considerable time, demonstrates that an already-formed CHCl group will be chlorinated faster than a still-intact CH_2 group. The relative ratios of the respective compounds present in this reaction mixture were determined by integrating the ^{1}H-NMR spectra; of course $(Cl_3Si-CCl_2)_2SiCl_2$ did not register. A preparative amount of $Cl_3Si-CCl_2-SiCl_2-CH_2-SiCl_3$ was accessible by appropriate termination of the chlorination reaction. Of course, $(Cl_3Si-CCl_2)_2SiCl_2$ as the final product of the chlorination is the most accessible compound in this reaction [92].

The C-chlorinated linear carbosilanes are liquids, while $(Cl_3Si-CCl_2)_2SiCl_2$ is a viscous compound which degrades on attempted distillation.

The photochlorination of $(Cl_2Si-CH_2)_3$ in CCl_4 yields as the end product fully chlorinated $(Cl_2Si-CCl_2)_3$. The reaction proceeds in a manner similar to that of its linear analogues, so the termination of the chlorination raction at a particular point enabled the isolation of crystalline compound 249 in 60% yield. Continuation of the chlorination enabled the incorporation of a further CCl_2 group into the molecule and the subsequent isolation of the compound 293 in 25% yield.

In all the photochlorinations, NMR signals pertaining to CHCl groups were never observed in large intensity. It can therefore be concluded that a singly chlorinated CHCl group will continue to react faster than the initial chlorination rate of a new CH_2 group.

In order to prepare CHCl-containing derivatives of $(Cl_2Si-CH_2)_3$, the chlorination must be carried out with SO_2Cl_2 and benzoyl peroxide. In this way 50% of $(Cl_2Si-CH_2)_3$ could be chlorinated. The chlorination product 295 was obtained in 68% yield, together with compound 293 containing two CCl_2 groups in 32% yield.

2.1.2 C-Brominated Carbosilanes

The use of pure bromine under irradiation by UV light has been shown to be rather ineffective in the photobromination of $(Cl_2Si-CH_2)_3$. More effective is the use of

BrCl, or bromine activated by means of Cl_2 under UV irradiation. Under these conditions $(Cl_2Si-CH_2)_3$ reacts in CCl_4 solution to generate a CBr_2 group.

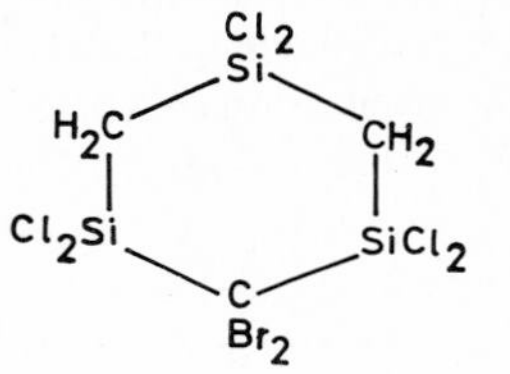

The bromination product is a white crystalline solid with mp. 86–97 °C which sublimes at 100 °C/10^{-3} mm Hg. To some extent it undergoes an exchange reaction, forming a CClBr group. As expected, the first stage in this reaction is the formation of a compound with a CHBr group [64].

While bromination proceeded without difficulty to the stage where one CHBr or CBr_2 group existed in the compound, any further bromination was retarded. Under harsher conditions, competitive chlorination played a predominant role in determining the reaction pathway. In an attempt to synthesize $(Cl_2Si-CBr_2)_3$ the intended reaction path was outweighed by the formation of a chlorinated product of composition $Si_3Cl_6C_3Cl_{5.5}Br_{0.5}$. An informative reaction was carried out, namely, further bromination of the compound with one CBr_2 group under reflux in CCl_4. After an irradiation time of 60 hours the following product distribution emerged:

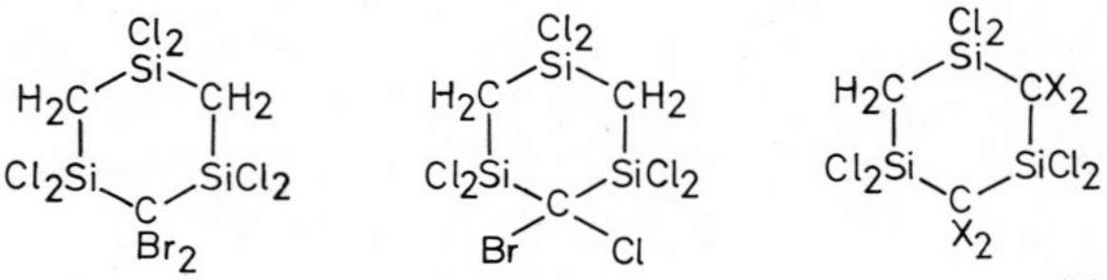

25%   28%   35% (about the same amounts of n = 1, 2, 3)

$Si_3Cl_6C_3H_2Br_{4-n}Cl_n$ (n = 1, 2, 3; X = Cl, Br)

The formation of more extensively brominated derivatives of $(Cl_2Si-CH_2)_3$ did not occur because incorporation of even one CBr_2 group impeded the reaction course and only resulted in a rechlorination reaction.

It appeared logical then to attempt the synthesis of a C-brominated 1,3,5-trisila-cyclohexane such as $(Br_2Si-CBr_2)_3$ through $(Br_2Si-CH_2)_3$. However, the bromination of $(Br_2Si-CH_2)_3$ took on another aspect, namely, the formation of compound 250. In a complicated and not yet fully elucidated manner, a SiBr/SiCl exchange occurred parallel to the bromination of a CH_2 group. The only isolated bromination product is compound 250, also accessible through one other independent reaction pathway already mentioned.

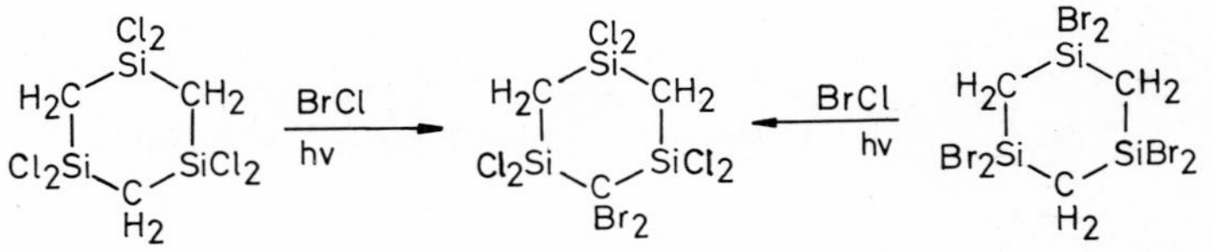

250

The bromination of $(Cl_3Si-CH_2)_2SiCl_2$ occurred in a fashion similar to that of $(Cl_2Si-CH_2)_3$. Here $Cl_3Si-CBr_2-SiCl_2-CH_2-SiCl_3$ will form by attack of BrCl under UV irradiation in refluxing CCl_4. During the initial period of the reaction, formation of the intermediate $Cl_3Si—CHBr—SiCl_2—CH_2—SiCl_3$ was indicated by a weak signal in the ^{1}H-NMR spectrum of the reaction mixture. The first major product was that containing a CBr_2 group. However, when half of the starting material is consumed and the reaction rate becomes slower, more and more of the compounds $Cl_3Si—CBrCl—SiCl_2—CH_2—SiCl_3$ and $Cl_3Si—CCl_2—SiCl_2—CH_2—SiCl_3$ appeared [64].

The formation of $(Cl_3Si)_2CBr_2$ via BrCl photobromination of $(Cl_3Si)_2CH_2$ in CCl_4 occurs without difficulty. The ^{1}H-NMR spectrum of the reaction mixture displays a signal pertaining to $(Cl_3Si)_2CHBr$. The rate at which the bromination proceeds is considerably less for this compound than for the bromination described above. The $(Cl_3Si)_2CBr_2$ is isolated in the form of white crystals of mp. 78 °C which can be re-crystallized from pentane, and which sublimes at 70 °C/10^{-3} mm Hg [64].

2.1.3 Photobromination of Si-Methylated Carbosilanes

Of special interest is the selective halogenation of the skeletal C atom in Si-methylated carbosilanes. A selective chlorination of the CH_2 group adjacent to an SiMe group, for example in Si—CH_2—Si—Me, is not possible due to the high and nonspecific reactivity of photochemically generated Cl radicals. On the other hand, the selective bromination of the CH_2 group in Si-methylated carbosilanes can be achieved. The bromination of $(Me_3Si—CH_2)_2SiMe_2$ in the absence of solvent is initiated immediately by adding a small amount of bromine during irradiation. This is recognized by the development of HBr. As soon as bromine is consumed in the reaction it must be supplemented by further small amounts.

After only 30 minutes ^{1}H-NMR signals pertaining to not only the starting material, but also to $Me_3Si—CBr_2—SiMe_2—CH_2—SiMe_3$ are observed. When about 70% of the starting material is consumed, signals relating to $(Me_3Si—CBr_2)_2SiMe_2$ appear in the ^{1}H-NMR spectrum. If the exposure time is extended, the percentage yield of this compound increases to 80%. Compounds containing a CHBr group are not observed in the ^{1}H-NMR spectrum, which means that any CHBr groups appearing will be converted rapidly to CBr_2 groups [77].

In an analogous way, $(Me_2Si—CH_2)_3$ is photobrominated. After approximately half of the starting material is converted to the 2,2-dibromo derivative 408, signals pertaining to compound 409 containing two CBr_2 groups begin to appear in the ^{1}H-NMR spectrum. Continued photobromination yields an 1:1 mixture of both compounds and proceeds eventually to 2,2,4,4-tetrabromo-hexamethyl-1,3,5-tri-silacyclohexane 409 in about 90% yield. In a side reaction, a methyl group instead of the second methylene group is brominated yielding a CBr_3 group (5–10%) [100].

408   409   410

An increasing degree of bromination decreases the reaction rate; increasing the reaction temperature favors the formation of unknown polymers which deposit on the walls of the reaction flask, preventing any further reaction [77].

The photobromination of Si-brominated carbosilanes occurs much slower. A requirement for a successful reaction is that the bromine used must be completely free of traces of moisture and chlorine. Reaction of $BrMe_2Si-CH_2-SiMe_2Br$ with bromine during irradiation in the absence of solvent produced HBr after 30 minutes. However, only after 100 hrs was the formation of $BrMe_2Si-CBr_2-SiMe_2Br$ recognized in the reaction mixture. An easier photobromination occurred with $BrMe_2Si-CH_2-SiMe_3$ producing $BrMe_2Si-CBr_2-SiMe_3$. The bromination of $(BrMe_2Si-CH_2)_2SiMe_2$ yielded $BrMe_2Si-CBr_2-SiMe_2-CH_2-SiMe_2Br$. It is clear that photobromination of the CH_2 group is hindered in those compounds by the presence of SiBr groups. This is also confirmed in the photobromination of $Me_3Si-CH_2-SiMe_2-CH_2-SiMe_2Br$. After a reaction time of 50 hrs this compound could not be seen any longer in the ^{1}H-NMR spectrum of the reaction mixture. A fraction of the reaction mixture obtained by distillation, representing 50% by volume of the original reaction mixture, consisted of $BrMe_2Si-CH_2-SiMe_2-CBr_2-SiMe_3$. Bromination occurs specifically on the CH_2 group not adjacent to the SiBr group.

1,1,3,3,5,5-Hexamethyl-2-trimethylsilyl-1,3,5-trisilacyclohexane was investigated as a representative of carbosilanes containing a tertiary CH group. The C atom of the $CH(SiMe_3)$ group is very quickly brominated; about 50% of the starting material is converted after 3 hrs. After 10 hrs the reaction was complete. Further exposure caused the formation of a CBr_2 group [77].

In the photochlorination of 1,3,5,7-tetramethyl-1,3,5,7-tetrasilaadamantane both the methyl and methylene groups are chlorinated. However, selective halogenation is possible by means of photobromination [80]. The possibility to form partially brominated Si-adamantanes arises from the results presented in Table 27. Here 0.25 molar solutions of Me_4Ad in CCl_4 were treated with varying amounts of bromine in mole ratios of $Me_4Ad:Br = 1:1$, $1:2$, $1:4$ and $1:8$ and irradiated with a high pressure Hg UV source.

From the NMR spectra of the product mixtures of reactions I to IV, the percentage of the major compounds present can be determined. Reaction I and II produced mainly Me_4AdBr_2. Increasing the bromine concentration in II doubled the amount of Me_4AdBr_2 produced, while increasing the irradiation time has only a small effect on product yield. Only by adding fresh amounts of Br_2 will the percentage of Me_4AdBr_2 increase again. Comparable amounts of Me_4AdBr_2 and Me_3BrAd were generated under the conditions of reaction III. Because no further bromine was added, a proportionate increase of yield was obtained, although slow in comparison to reaction II. Me_3BrAd was the main product present at the commencing stages of reaction IV,

Table 27. Photobromination of Me_4Ad in CCl_4 with varying amounts of Br_2 in various mole ratios of $Me_4Ad:Br_2$; in reaction I = 1:1; II = 1:2; III = 1:4; IV = 1:8. Compounds formed and their % content have been determined by ^{1}H-NMR spectroscopy

Reaction time	Compounds in product mixture		% Content in product mixture			
			I	II	III	IV
12 hrs.	Me_4Ad	39	95	90	88	78
	Me_4AdBr_2	39a	5	10	7	6
	Me_3BrAd	39b	—	—	5	16
36 hrs.		39	93	87	82	66
		39a	7	13	10	7
		39b	—	—	8	27
60 hrs.		39	85[a]	70[a]	72	—[b]
		39a	15	30	15	10
		39b	—	—	13	50
84 hrs.		39	83	66	69	[b]
		39a	17	34	17	
		39b	—	—	14	

[a] Br_2 added again
[b] Further higher brominated compounds are formed

before more extensively brominated compounds became evident. The formation of Me_3BrAd was generated through Si—C cleavage. If reactions are undertaken using a concentration range of 0.5 to 1.0 molar solutions of Me_4Ad in CCl_4, then Si—C cleavage can be even more selectively controlled.

Reaction II can be very easily controlled by observing the characteristic color of solutions containing low concentrations of bromine. If irradiation is ceased after 50% of Me_4Ad has converted, then after distilling off the solvent and removal of unreacted Me_4Ad by sublimation at 30 °C/10^{-1} mm Hg, Me_4AdBr_2 is separated at 60 °C/10^{-1} mm Hg as a white crystalline substance of mp. 114 °C.

2.2 Formation and Reactions of Si-Hydrogenated Carbosilanes

2.2.1 Hydrogenation of SiCl- and CH-Containing Carbosilanes

The Si—Cl groups in CH_2-containing carbosilanes are converted to Si—H groups by hydrogenation with $LiAlH_4$ without a change in the molecular skeleton [93, 94].

$$2 (Cl_3Si)_2CH_2 + 3 LiAlH_4 \rightarrow 2 (H_3Si)_2CH_2 + 3 AlCl_3 + 3 LiCl$$

$$(Cl_3Si—CH_2)_2SiCl_2 + 2 LiAlH_4 \rightarrow (H_3Si—CH_2)_2SiH_2 + 2 AlCl_3 + 2 LiCl$$

$$2 (Cl_2Si—CH_2)_3 + 3 LiAlH_4 \rightarrow 2 (H_2Si—CH_2)_3 + 3 AlCl_3 + 3 LiCl$$

Pure $(H_3Si)_2CH_2$ can be obtained in dibutylether in a yield higher than 80%. It has an extrapolated boiling point of 16.7 °C, is remarkably stable and is not self-igniting on contact with air. The same applies to $(H_3Si—CH_2)_2SiH_2$, formed as a colorless liquid (bp. 100 °C/760 mm Hg).

The related $(H_3Si-CH_2)_2SiCl_2$ is formed by reaction of $(Cl_3Si-CH_2)_2SiCl_2$ with a less than stoichiometric amount of $LiAlH_4$, along with the fully hydrogenated compound. The nonappearance of $(ClH_2Si-CH_2)_2SiH_2$ or $Cl_2HSi-CH_2-SiH_2-$ $-CH_2-SiH_3$ is indicative of less reactivity of the central $SiCl_2$ group compared to the terminal $SiCl_3$ group.

The compound $(H_2Si-CH_2)_3$ (bp. 142 °C/760 mm Hg) forms white crystals below 10 °C, with a yield greater than 80%. The formation of $(H_2Si-CH_2)_3$ occurs only after addition of an excess of $LiAlH_4$; otherwise a mixture of SiCl-containing compounds is obtained.

2.2.2 SiH-Bromination of Carbosilanes

The bromination of SiH_4 [95] and SiH-containing alkylsilanes is known [96] and can be used in the quantitative determination of Si—H groups present [97]. Partial bromination of carbosilanes, whereby Si—H groups are still contained in the compounds, is of interest in a number of aspects. Partial chlorination with Cl_2 could be achieved by careful manipulation of the reaction. Reaction with Br_2 offers more favorable possibilities [64].

2.2.2.1 $(H_3Si)_2CH_2$

Reaction of $(H_3Si)_2CH_2$ in the absence of solvent at −70 °C by careful dropwise addition of bromine, proceeds quantitatively. Depending upon which molar ratio is selected, a variety of Si-brominated derivatives is obtained. Reaction of the Si—H group with HBr can be excluded under the conditions just mentioned.

Starting with a stoichiometric ratio of 1:1, the reaction products isolated through fractional condensation were $H_3Si-CH_2-SiH_2Br$ in 13% yield and $H_2BrSi-CH_2-$ $-SiH_2Br$ in 25% yield. Some unreacted $(H_3Si)_2CH_2$ was left. When the reaction is carried out in pentane, the yield of $H_3Si-CH_2-SiH_2Br$ is increased to 40% [64].

Performing the reaction with a mole ratio of 1:2 in pentane at −70 °C yields $(H_2SiBr)_2CH_2$ as main product, in a 72% yield. The byproducts $H_3Si-CH_2-SiH_2Br$ and $H_2SiBr-CH_2-SiHBr_2$ were separated through fractional distillation.

Similarly, $(HSiBr_2)_2CH_2$ was obtained from the reaction of $(H_3Si)_2CH_2$ with bromine in a mole ratio of 1:6 in the absence of solvent. This was carried out by dropwise addition of bromine at −70 °C, followed by warming of the solution to 20 °C. When HBr evolution ceased, excess Br_2 was removed at 20 °C. The reaction products consisted essentially of $(HSiBr_2)_2CH_2$ together with a small amount of $HSiBr_2-CH_2-SiBr_3$ and $(Br_3Si)_2CH_2$. The $(HSiBr_2)_2CH_2$ was purified and separated through condensation. The yield was 88% [64].

A larger batch of $(Br_3Si)_2CH_2$ was formed by reaction of $(H_3Si)_2CH_2$ with bromine at −70 °C. By progressively brominating the Si atoms, the reaction progressed slower. After all the bromine had been added, the temperature was raised to 20 °C. After 4 days of continuous stirring the following mixture was encountered: $(HSiBr_2)_2CH_2$ (10%), $HSiBr_2-CH_2-SiBr_3$ (58%), $(Br_3Si)_2CH_2$ (32%). By using an excess of bromine for six days at 60 °C, pure $(Br_3Si)_2CH_2$ was obtained.

All brominated derivatives of $(H_3Si)_2CH_2$ form colorless liquids. The viscosity of the bromine derivatives increases as the number of bromine atoms in the compound increases; $(Br_3Si)_2CH_2$ itself is an oily liquid [64].

2.2.2.2 $(H_3Si-CH_2)_2SiH_2$

Reaction of $(H_3Si-CH_2)_2SiH_2$ with bromine in a mole ratio of 1:1 at $-50\ °C$ in methylcyclohexane produces a mono-brominated derivative with the bromine atom bound to the secondary central silicon atom. However, the obviously higher reactivity of the SiH_2 group is not sufficient to prevent entirely the reaction of terminal SiH_3 groups. Progressive bromination with another 2 moles of bromine yields $(H_2SiBr-$ $-CH_2)_2SiHBr$, which is obtained as a pure colorless oily liquid of bp. $74\ °C/10^{-2}$ mm Hg. Typical of all SiHBr-containing compounds, this compound smokes heavily when brought into contact with moist air [64].

To produce $(Br_3Si-CH_2)_2SiBr_2$, $(H_3Si-CH_2)_2SiH_2$ must react first with bromine in benzene solution, followed by 3 days of continual stirring at $20\ °C$. The solvent and remaining bromine are subsequently removed, and the reaction products are then maintained for 30 days at $60\ °C$ with a further excess of bromine until only a singlet appears in the ^{1}H-NMR spectrum pertaining to the CH_2 group. After removal of the excess bromine by evaporation and condensation, the reaction product maintains a brown color. After its vacuum distillation (bath temperature $170\ °C$), colorless $(Br_3Si-CH_2)_2SiBr_2$ is obtained in 60% yield [64].

2.2.2.3 $(H_2Si-CH_2)_3$

Reaction of $(H_2Si-CH_2)_3$ with bromine allows selective formation of partially brominated derivatives, because formation of an SiBr group reduces the reactivity of the remaining Si—H bond in the SiHBr group. If appropriate amounts of bromine, dissolved in methylcyclohexane are added dropwise at $-50\ °C$ to a 0.2 molar solution of $(H_2Si-CH_2)_3$ in methylcyclohexane, three bromination products are obtained:

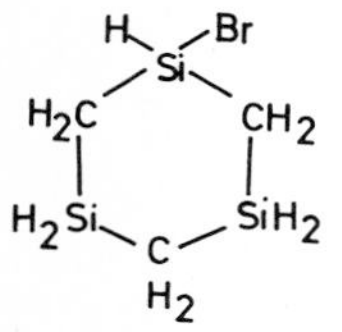

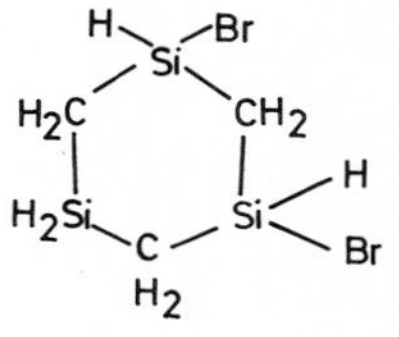

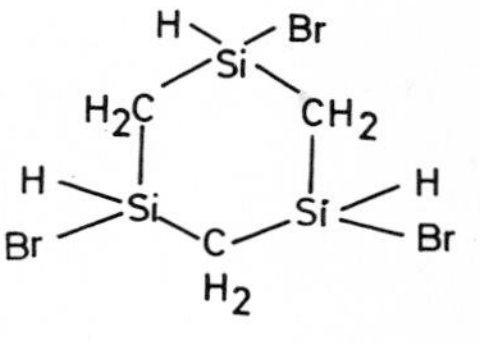

Using a mole ratio of 1:1 results in synthesis of 63% monobrominated plus 10% dibrominated compounds; a mole ratio of 1:2 yields 6.5% monobrominated, 65% dibrominated and 6.5% tribrominated compounds, a mole ratio of 1:3 produces the tribrominated compound in a 93% yield.

Further bromination of $(HBrSi-CH_2)_3$ is achieved in methylcyclohexane. However, the decreased reactivity means that the time required for complete reaction at room temperature is considerably longer. Partly brominated 1,3,5-trisilacyclohexanes exist at room temperature as colorless liquids which maintain an oily appearance as the extent of bromination is increased.

The compound $(Br_2Si-CH_2)_3$ is formed without difficulty by reaction of $(H_2Si-$ $-CH_2)_3$ with bromine in benzene solution. The reaction goes to completion in 4 days at $20\ °C$. After recrystallization from pentane and sublimation under vacuum, $(Br_2Si-CH_2)_3$ appears as a white crystalline substance of mp. $101\ °C$ [64].

2.2.3 C-Chlorinated, SiH-Containing Carbosilanes

In the previous Chapt. III.2.1, the formation of perchlorinated carbosilanes such as $(Cl_2Si-CCl_2)_3$ through photochlorination of $(Cl_2Si-CH_2)_3$ was reported. Chapter III.2.2.1 contained examples showing the conversion of SiCl groups to SiH groups by means of $LiAlH_4$, leading to the formation of $(H_2Si-CH_2)_3$. By combining these reactions, an approach to the synthesis of compounds such as $(H_3Si)_2CCl_2$ or $(H_2Si-CCl_2)_3$ is well within reason. The linear $H_3Si-CCl_2-SiH_3$ is formed according to [93, 105]:

$$2\,Cl_3Si-CCl_2-SiCl_3 + 3\,LiAlH_4 \rightarrow 2\,H_3Si-CCl_2-SiH_3 + 3\,LiCl + 3\,AlCl_3$$

The product is a limpid colorless liquid with an odor typical of SiH-containing carbosilanes, with bp. 83 °C (extrapolated from the vapor pressure measurement). It can be stored successfully in a sealed glass tube, but in contact with traces of H_2O or air, polymerization occurs. While pure $(H_3Si)_2CCl_2$ remains unchanged over a period of years, in the presence of $AlCl_3$ at 20 °C it degrades over a period of months; the starting material is completely consumed. Thirty percent of the Si in $(H_3Si)_2CCl_2$ is converted to SiH_4. The SiCl-containing compounds of higher boiling point which are present are $H_3Si-CH_2-SiH_2Cl$, $H_2ClSi-CH_2-SiH_2Cl$, $H_2SiCl-CH_2-SiHCl_2$, and $HSiCl_2-CH_2-SiH_3$. The last compound is the main degradation product.

Similarly, $(H_3Si-CCl_2)_2SiH_2$ is formed in a 62 % yield if the reaction

$$(Cl_3Si-CCl_2)_2SiCl_2 + 2\,LiAlH_4 \rightarrow (H_3Si-CCl_2)_2SiH_2 + 2\,AlCl_3 + 2\,LiCl$$

is carried out in diethyl ether at 0 °C with exact adherence to stoichiometric ratios. The product $(H_3Si-CCl_2)_2SiH_2$ is a colorless liquid which crystallizes at -18 °C. The compound is collected in a condenser cooled by liquid nitrogen. Its vapor pressure at 20 °C is 10^{-2} mm Hg.

The cyclic compound $(H_2Si-CCl_2)_3$ was isolated as a pure product in approximately 50 % yield by the reaction

$$2\,(Cl_2Si-CCl_2)_3 + 3\,LiAlH_4 \rightarrow 2\,(H_2Si-CCl_2)_3 + 3\,AlCl_3 + 3\,LiCl$$

Carrying out the reaction in diethyl ether, cannot be recommended because the $AlCl_3$ etherates sublime under the same conditions as the $(H_2Si-CCl_2)_3$. However, if dibutyl ether is used as solvent, then the less-volatile $AlCl_3$ etherate residue can be separated from $(H_2Si-CCl_2)_3$ via sublimation without difficulty. Pure $(H_2Si-CCl_2)_3$ is crystalline at 20 °C, forming long white needles which sublime readily in an evacuated ampoule by warming in the hand.

It must be emphasized that pure compounds of this class explode readily on warming [105], and this tendency is severely enhanced in the presence of unreacted $LiAlH_4$ and $AlCl_3$. Therefore, in the isolation of $(H_3Si)_2CCl_2$ the reaction mixture should be stirred continually in a mixture of half sulphuric acid and half ice. Volatile $H_3Si-CCl_2-SiH_3$ can then be separated easily from the ether extracts of the hydrolytically worked-up reaction mixture. The $(H_3Si-CCl_2)_2SiH_2$ is resistant to sudden movement and shaking, but explodes violently when exposed to a naked flame. Cyclic $(H_2Si-CCl_2)_3$ behaves in a similar manner. The extraordinary explosions caused by these substances in only small amounts, for instance one drop in a capillary, cannot be explained simply by oxidation in air, this has to do with transfer of chlorine.

The bond energy of Si—Cl is higher than that of C—Cl, while the bond energy of C—H is much higher than that of Si—H. Hence transfer of Cl from C to Si, with simultaneous transfer of H from Si to C, liberates much energy. Solutions of these compounds in ether, hydrocarbon solvents, or CCl_4 have been shown to be safe at room temperature. This explosive nature, which always appears when a pure sample of these compounds is heated, is attributed to a rearrangement, as illustrated by placing face to face the isomers of 1,3,5-trisilacyclohexane:

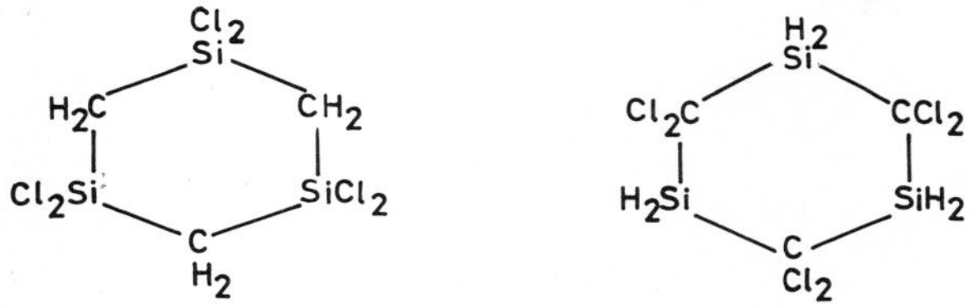

Notice that CH and SiCl groups exist in $(Cl_2Si—CH_2)_3$, which represents a thermodynamically favored arrangement of bound atoms in the molecule. In $(H_2Si—CCl_2)_3$, however, the SiH and CCl groups occupy the opposite positions. The course of such explosive reactions has not been investigated in detail, however.

The hydrogenation of SiCl groups with $LiAlH_4$ in carbosilanes containing CH_2 groups, for instance in the conversion of $(Cl_3Si)_2CH_2$ or $(Cl_2Si—CH_2)_3$ to yield $(H_3Si)_2CH_2$ or $(H_2Si—CH_2)_3$, respectively, proceeds quantitatively without cleavage of the Si—C-skeleton, but corresponding reactions of perchlorinated compounds are complicated heavily by side reactions. This becomes especially important in reactions where $LiAlH_4$ is used in excess. The following results illustrate this.

Reactions of $(Cl_3Si)_2CCl_2$ with $LiAlH_4$ produce the byproducts SiH_4 and $(H_3Si)_2CH_2$. The amount of SiH_4 and $(H_3Si)_2CH_2$ increases considerably with the concentration of $LiAlH_4$, the reaction temperature, and the reaction time, as Table 28 shows. Optimal conditions for the synthesis of $(H_3Si)_2CCl_2$ prevail in reaction 2, whereas the formation of $(H_3Si)_2CH_2$ dominates in reaction 4.

Regarding the synthesis of $(H_3Si—CCl_2)_2SiH_2$ from $(Cl_3Si—CCl_2)_2SiCl_2$, the formation of the byproducts SiH_4, $(H_3Si)_2CCl_2$, $H_3Si—CH_2—SiH_2—CCl_2—SiH_3$ and $H_3Si—CHCl—SiH_2—CCl_2—SiH_3$ could not be totally avoided. The amount of byproducts depended greatly upon the reaction conditions. Table 29 gives an account of the reaction products obtained as well as of the reaction conditions used.

The mole percentage is calculated with respect to the starting material $(Cl_3Si—CCl_2)_2SiCl_2$, so that the sum of the mole percentages given in the horizontal rows refers to the production of volatile and distillable compounds. The remaining reaction products are found in the non-volatile residue. The formation of $(H_3Si—CCl_2)_2SiH_2$ is optimal only if for every Si—Cl group present one Al—H group is employed.

The side reactions include a) the formation of CH-containing compounds, and b) the cleavage of the SiC bond.

a) Only a small amount of CH-containing compounds results, maintaining the complete skeleton of the starting substance. The end product $(H_3Si—CH_2)_2SiH_2$, obtained as a stable product from $(Cl_3Si—CH_2)_2SiCl_2$, is formed only in a small yield from its perchlorinated analogue $(Cl_3Si—CCl_2)_2SiCl_2$. The related $H_3Si—CH_2—SiH_2—CCl_2—SiH_3$ was produced in this reaction in a slightly larger yield. Small

Table 28. Influence of reaction conditions on the reaction between $(Cl_3Si)_2CCl_2$ and $LiAlH_4$ [93]

Reaction No.	Mole equiv. $LiAlH_4$ per SiCl group	Temperature during addition of $LiAlH_4$ °C	Reaction time hrs.	Reaction time °C	Reaction products in Mole % SiH_4	$(H_3Si)_2CH_2$	$(H_3Si)_2CCl_2$	Total reaction turnover %
1	0.27	−8	3	≦8	0.53	1.4	62	
2	0.42	0	2	20	3	2.0	76	62.4
3	0.72	30	3 / 6	40 / 50	n. m.[a]	33	23	
4	0.72	50	6½	56	30	70	<1	

[a] n. m., not measured

Table 29. Influence of reaction conditions on the reaction between $(Cl_3Si-CCl_2)_2SiCl_2$ and $LiAlH_4$ [93]

Reaction No.	Mole equiv. $LiAlH_4$ per SiCl group	Solvent	Temperature at start of reaction °C	Reaction time hrs.	Reaction time °C	395	396	397	398	399	400	401	261	Total reaction turnover %
1	0.78	Et_2O	34	24	34	not measured	not measured		2	<1	<1	0	0	
2	0.81	$n\text{-}Bu_2O$	0	12	60	not measured	32.4	<1	30	<2	not measured	not measured	0	62.4
3	0.82	$n\text{-}Bu_2O$	50	2.5	60	6.0	16.5	<1	50.5	<2	not measured	not measured	<1	73.0
4	0.80	Et_2O	0	{ 1 / 5	20 / 34	47.0	<1	0	36	<1	3.0	0	<1	86.0
5	0.28	Et_2O	−6	2.5 to 17	55	6.3	0	0	5.7	0	<1	<1	62	74
6	2.48	$n\text{-}Bu_2O$	22	{ 8 / 1	55 / 85	69	2	0	0	0	not measured	not measured		69

395 = SiH_4
396 = $(H_3Si)_2CH_2$
397 = $(H_3Si)_2CHCl$
398 = $(H_3Si)_2CCl_2$
399 = $(H_3Si-CH_2)_2SiH_2$
400 = $H_3Si-CH_2-SiH_2-CCl_2-SiH_3$
401 = $H_3Si-CHCl-SiH_2-CCl_2-SiH_3$
261 = $(H_3Si-CCl_2)_2SiH_2$

amounts of the CHCl-containing compound $H_3Si-CHCl-SiH_3-CCl_2-SiH_3$ could only be detected in the presence of $(H_3Si-CCl_2)_2SiH_2$. The difference in the ratios of CH_2- to CCl_2-containing compounds in reactions 2 and 3 is similarly observed in the reaction of $(Cl_3Si)_2CCl_2$ with $LiAlH_4$. By employing a longer reaction time in reactions 2 and 3, the CCl_2 group is extensively converted to the CH_2 group.

b) Because of the appearance of SiH_4, $(H_3Si)_2CCl_2$ and $(H_3Si)_2CH_2$ (Table 29), it is recognized that the cleavage reaction may predominate over all. Even utilizing the smoothest reaction conditions (No. 5, Table 29), more than 12% of $(Cl_3Si-CCl_2)_2SiCl_2$ is cleaved.

The interpretation of the cleavage reactions leading to the formation of SiH_4 and $(H_3Si)_2CCl_2$ presented some difficulties initially. This is mainly because of the presumption that in such cases where complete cleavage of SiC bonds occurs, methane or suitably-related derivatives appear in the reaction mixture. In our work, however, such compounds could not be observed even with the most sophisticated work-up procedure. The interpretation of this cleavage reaction comes from investigation of the residue left after reaction. Hydrolysis of the residue yields $MeSiH_3$ [93].

$$H_3Si-CH_2-AlR_2 \xrightarrow{H_2O} H_3Si-CH_3 + Al(OH)R_2$$

Reaction with $(Cl_3Si-CCl_2)_2SiCl_2$ yields not only substitution of H atoms for Cl atoms, but also cleavage of the Si—C bond

$$\ce{>Si-C<} + alH \rightarrow \ce{>SiH} + al-\ce{C<} \quad (alH = 1/4\ LiAlH_4)$$

The reaction proceeds according to:

$$H_3Si-CCl_2-SiH_2-CCl_2-SiH_3 + alH \rightarrow (H_3Si)_2CCl_2 + H_3Si-CH_2al$$

$$H_3Si-CCl_2-SiH_3 + alH \rightarrow SiH_4 + H_3Si-CH_2al$$

The last reaction coincides with the behavior of $(Cl_3Si)_2CCl_2$.

A thorough investigation of the reaction products shows that the reaction does not proceed according to

$$H_3Si-CCl_2-SiH_2-CCl_2-SiH_3 + alH$$

$$\nrightarrow H_3Si-CCl_2-SiH_2-CX_2al + SiH_4$$

$$H_3Si-CCl_2-SiH_2-CX_2al \nrightarrow (alCX_2)_2SiH_2 + SiH_4 \quad (X = Cl, H)$$

because neither $H_3Si-CX_2-SiH_2-CH_3$ (X = Cl, H) nor Me_2SiH_2 were detected as reaction products nor as hydrolysis products from the non-volatile reaction residue.

The cleavage of $(Cl_3Si-CCl_2)_2SiCl_2$ proceeds through $(H_3Si-CCl_2)_2SiH_2$ and precedes the formation of a CH_2 group. This conclusion is made on the grounds of the stability of $(H_3Si-CH_2)_2SiH_2$ to $LiAlH_4$.

The behavior of the cyclic compound $(Cl_2Si-CCl_2)_3$ can be explained by considering the reaction temperature and the amount of $LiAlH_4$ used. Table 30 gives an account of the reaction conditions, as well as an indication of the volatile and sublimable compounds produced.

Table 30. Influence of reaction conditions on the reaction between $(Cl_2Si-CCl_2)_3$ and $LiAlH_4$ [93]

Reaction No.	Mole equiv. $LiAlH_4$ per SiCl group	Solvent	Temperature at start of reaction °C	Reaction time		Reaction products in Mole %				
				hrs.	°C	395	396	397	292	291
1	0.27	Et_2O	0	1	9	3.6	0	0	0	20
2	0.27	Bu_2O	0	3	20	0	0	>2	0	48
3	1.15	Bu_2O	0	6	60	2.5	13	>2	>2	0
4	4.0	Bu_2O	20	7	60	16.5	9	0	0	0
5	1.1	Et_2O	20	20	34	not measured			0	0

395, 396, 397 see Table 29, 292 $(H_2Si-CH_2)_3$, 291 $(H_2Si-CCl_2)_3$.

The reaction products are divided into two categories: volatile products including those obtainable by condensation of the reaction mixture, and compounds accessible after hydrolysis of the residual mixture. Of these, $(H_2Si-CCl_2)_3$ is isolated only through reactions 1 and 2 (Table 30) on the condition that 1/4 mole of $LiAlH_4$ is employed for every SiCl group present in the molecule, and reaction temperature does not exceed 20 °C. If the amount of $LiAlH_4$ is increased, no more $(H_2Si-CCl_2)_3$ is produced. The product from the reduction of $(H_2Si-CCl_2)_3$, namely $(H_2Si-CH_2)_3$, appears only as a byproduct in reaction 3 (Table 30). Partially reduced rings containing CHCl or CH_2 groups adjacent to an SiH_2 group were observed as minor products only after meticulous separation. The amount of volatile compounds was negligible (Table 30). The mole percent quoted refers to 1 mole of SiH_4 or $(H_3Si)_2CH_2$ produced per mole of $(Cl_2Si-CCl_2)_3$. In reactions 3 and 4, only about 25% of the $(Cl_2Si-CCl_2)_3$ employed was converted to volatile products, in contrast to the considerably higher yield from $(Cl_2Si-CCl_2)_2SiCl_2$ conversion.

An explanation for this was gained after hydrolytic work-up of the product residue, which led eventually to the isolation of $H_3Si-CCl_2-SiH_2-Me$ and H_3Si-CH_2- $-SiH_2-CH_2-SiH_2-Me$ in a ratio of 4:1. To understand the reaction pathway, it is significant that both compounds are liberated only on hydrolytic work-up of the residue. This is seen also by comparing the ^1H-NMR spectra of the residue before and after hydrolytic work-up. These two compounds obviously arise through cleavage of the ring. However, because $(H_2Si-CH_2)_3$ is remarkably stable against $LiAlH_4$, cleavage must have occured by attack on $(H_2Si-CCl_2)_3$ or a CCl-containing derivative:

A further cleavage is necessary from this originally formed chain in order to achieve the formation of $H_3Si-CCl_2-SiH_2-CH_3$:

$$H_3Si-CCl_2-SiH_2-CCl_2-SiH_2-CCl_2-al + alH \rightarrow H_3Si-CX_2-al$$
$$+ H_3Si-CCl_2-SiH_2-CCl_2-al \qquad\qquad (X = Cl, H)$$

In this case also, the cleavage must occur before reduction of the CCl_2 group. Subsequent hydrolysis of the organoaluminium compounds leads to isolation of the respective carbosilanes [93].

2.2.4 Reaction of $(Cl_3Si)_2CBr_2$ with $LiAlH_4$

If the reaction is carried out in a fashion similar to that in Chapt. III.2.2.3, the main product is $(H_3Si)_2CH_2$, while SiH_4, $(H_3Si)_2CBr_2$, $(H_3Si)_2)CHBr$ and $(H_3Si)_2CHCl$ are formed as byproducts. Conversion of a CBr_2 group to a CH_2 group is a considerably easier process than conversion of a CCl_2 group. This is also confirmed by the corresponding reaction of $Cl_3Si-CBr_2-SiCl_2-CH_2-SiCl_3$ [64].

2.2.5 Partly C-Chlorinated, SiH-Containing Carbosilanes

Reduction of partly C-chlorinated SiCl-containing carbosilanes enables preparation of the corresponding SiH-containing carbosilanes with a partly chlorinated bridging C atom:

$$2(Cl_3Si)_2CHCl + 3\ LiAlH_4 \rightarrow 2(H_3Si)_2CHCl + 3\ LiCl + 3\ AlCl$$

The resulting $(H_3Si)_2CHCl$ is a colorless liquid of bp. 63 °C (extrapolated from the vapor pressure curve) and can be obtained in a 70% yield [93].
Also $H_3Si-CHCl-SiH_2-CH_2-SiH_3$ and $H_3Si-CCl_2-SiH_2-CH_2-SiH_3$ are accessible through reaction of the corresponding SiCl-containing carbosilanes with $LiAlH_4$. The approximate 85% yield obtained for the SiH-containing carbosilanes is calculated relative to the SiCl-containing starting compounds. These SiH-containing derivatives again are colorless liquids that can be distilled under reduced pressure at room temperature and condensed in a liquid-nitrogen-cooled condenser.

The same procedure applies also to the following partly chlorinated cyclic compounds:

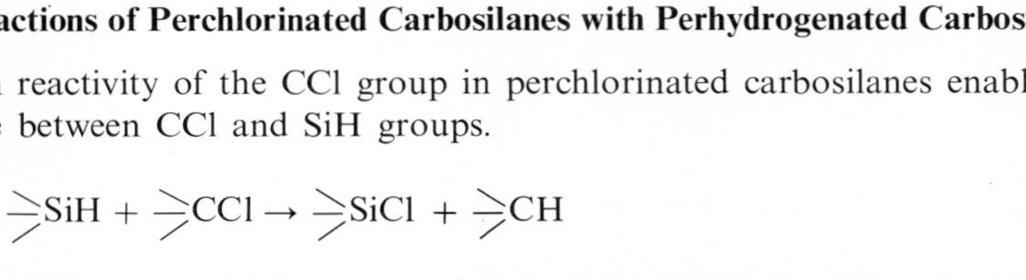

2.2.6 Reactions of Perchlorinated Carbosilanes with Perhydrogenated Carbosilanes

The high reactivity of the CCl group in perchlorinated carbosilanes enables gentle exchange between CCl and SiH groups.

$$\geq SiH + \geq CCl \rightarrow \geq SiCl + \geq CH$$

This reaction is useful for the synthesis of partly chlorinated carbosilanes. The exchange between SiH and CCl groups is also achieved in CCl_4 solutions of silanes, but is retarded by the solvent unless aided by a catalyst [106, 107] or through irradiation by UV light [108].

Reaction between $(Cl_3Si)_2CCl_2$ and $(H_3Si)_2CH_2$ in a mole ratio of 3:2 leads to chlorination of $(H_3Si)_2CH_2$ to 86% within 15 days at room temperature. No SiH-containing degradation products, such as SiH_4, appear, implying of course that no Si—C bonds are cleaved. Table 31 lists the products obtained in this reaction [92]

Table 31. Compounds from the reaction of $(Cl_3Si)_2CCl_2$ with $(H_3Si)_2CH_2$

$H_3Si-CH_2-SiH_2Cl$	$Cl_3Si-CHCl-SiCl_3$
$H_2ClSi-CH_2-SiClH_2$	$Cl_3Si-CH_2-SiCl_3$
$HCl_2Si-CH_2-SiH_3$	
$HCl_2Si-CH_2-SiClH_2$	
$HCl_2Si-CH_2-SiCl_2H$	

Complete exchange of all SiH groups was not observed. Some 56% of the $(H_3Si)_2CH_2$ used appeared as $H_3Si-CH_2-SiH_2Cl$, and 30% produced $H_2ClSi-CH_2-SiClH_2$. Further SiH-containing reaction products amounted to 14% of the total. On the other hand, 68% of the perchlorinated $(Cl_3Si)_2CCl_2$ was converted to $(Cl_3Si)_2CHCl$, and 7.5% reacted to form $(Cl_3Si)_2CH_2$.

Exchange occurring during reaction of $(Cl_3Si-CCl_2)_2SiCl_2$ with $(H_3Si-CH_2)_2SiH_2$ yields the respective partly chlorinated compounds. Carrying out the reaction in a 1:1 mole ratio for approximately 10 days at 95 °C yields the compounds listed in Table 32. No degradation of the carbosilane chain is observed. The total yield of the exchange reaction amounts to 80%.

Table 32. Reaction products from reaction of $(Cl_3Si-CCl_2)_2SiCl_2$ with $(H_3Si-CH_2)_2SiH_2$

$H_2ClSi-CH_2-SiH_2-CH_2-SiH_3$	$Cl_3Si-CHCl-SiCl_2-CCl_2-SiCl_3$
$H_3Si-CH_2-SiHCl-CH_2-SiH_3$	$Cl_3Si-CHCl-SiCl_2-CHCl-SiCl_3$
$H_3Si-CH_2-SiHCl-CH_2-SiH_2Cl$	$Cl_3Si-CH_2-SiCl_2-CCl_2-SiCl_3$
$H_2ClSi-CH_2-SiH_2-CH_2-SiH_2Cl$	$Cl_3Si-CHCl-SiCl_2-CH_2-SiCl_3$

Some 63% of the $(H_3Si-CH_2)_2SiH_2$ engaged in reaction is chlorinated; 70% of the chlorination product consists of monochlorinated $H_2ClSi-CH_2-SiH_2-CH_2-SiH_3$ and $H_3Si-CH_2-SiHCl-CH_2-SiH_3$ in a 3:2 ratio. The other 30% of the SiH-containing product consists of $ClH_2Si-CH_2-SiHCl-CH_2-SiH_3$ and ClH_2Si- $-CH_2-SiH_2-CH_2-SiH_2Cl$, whereby the unsymmetrical chlorinated compound is produced in greater yield. Chlorination of SiH groups coupled with the hydrogenation of CCl groups proceeded to produce most favorably $Cl_3Si-CHCl-SiCl_2-$ $-CCl_2-SiCl_3$, along with only a small amount of the other derivatives shown in Tab. 32. Complete conversion of CCl_2 to CH_2 groups was not observed. These SiCl-containing compounds (Tab. 32) react with $LiAlH_4$ producing $H_3Si-CHCl-SiH_2-$ $-CCl_2-SiH_3, H_3Si-CHCl-SiH_2-CHCl-SiH_3, H_3Si-CH_2-SiH_2-CCl_2-SiH_3$ and $H_3Si-CHCl-SiH_2-CH_2-SiH_3$.

The reaction between $(Cl_3Si-CCl_2)_2SiCl_2$ and $(H_3Si)_2CH_2$ proceeds, as far as C-chlorinated compounds are concerned, to the main product $Cl_3Si-CHCl-SiCl_2-$ $-CCl_2-SiCl_3$. The reaction of $(Cl_3Si-CCl_2)_2SiCl_2$ with $(H_3Si)_2CH_2$ in benzene in a 1:1.1 mole ratio was studied in order to establish the influence of solvent and also to determine whether stepwise addition of perhydrogenated $(H_3Si)_2CH_2$ increases the amount of partly C-chlorinated compounds. Apart from $H_2ClSi-CH_2-SiH_3$ no other Si-chlorinated compound was produced, and $Cl_3Si-CHCl-SiCl_2-CCl_2-$ $-SiCl_3$ remains the most abundant derivative of the perchlorinated starting material even if the perhydrogenated compound was used in excess with a mole ratio of 1:2. More than 58% of the $(Cl_3Si-CCl_2)_2SiCl_2$ was hydrogenated. The major reaction involving H-/Cl-exchange is described by the equation

$$(Cl_3Si-CCl_2)_2SiCl_2 + (H_3Si)_2CH_2 \rightarrow Cl_3Si-CHCl-SiCl_2-CCl_2-SiCl_3$$
$$+ H_2ClSi-CH_2-SiH_3$$

The same exchange reaction occurs also between linear and cyclic compounds, as the reaction between $(Cl_2Si-CCl_2)_3$ and $(H_3Si)_2CH_2$ indicates. The major SiCl-containing product arising from this reaction is $H_2ClSi-CH_2-SiH_3$, and the CHCl group forms in the cyclic compound. For experimental reasons it was decided to replace $(H_3Si)_2CH_2$ on account of its volatility by $(H_3Si-CH_2)_2SiH_2$. Reaction between these compounds in a 1:3.1 mole ratio proceeds in 90% conversion:

$$2(Cl_2Si-CCl_2)_3 + 6(H_3Si-CH_2)_2SiH_2 \rightarrow 2(Cl_2Si-CH_2)_3$$
$$+ 3 H_2ClSi-CH_2-SiHCl-CH_2-SiH_3 + 3 H_2ClSi-CH_2-SiH_2-CH_2-SiH_2Cl$$

The 1,3- and 1,5-dichloro-1,3,5-trisilapentanes arise as main products. The corresponding hydrogenation of CCl_2 groups in $(Cl_2Si-CCl_2)_3$ leads to $(Cl_2Si-CH_2)_3$.

The exchange reaction between $(Cl_2Si-CCl_2)_3$ and $(H_3Si-CH_2)_2SiH_2$ in a mole ratio of 1:1.5 proceeded to nearly complete consumption of the perhydrogenated chain, which incorporated three to five SiCl groups. The formation of HCl_2Si-CH_2- $-SiHCl-CH_2-SiH_2Cl$ and $(HCl_2Si-CH_2)_2SiHCl$ appeared to be favored, the former being the main product among the SiCl-containing derivatives. Because of very low yields, the mono- and dihalogenated compounds had to be neglected. The simultaneously occurring hydrogenation of $(Cl_2Si-CCl_2)_3$ produced CHCl-containing 1,1,2,2,3,3,4,5,5-nonachloro-1,3,5-trisilacyclohexane and $(Cl_2Si-CH_2)_3$, to a lesser degree also 1,1,2,2,3,3,5,5-octachloro-1,3,5-trisilacyclohexane.

An exchange between SiH and CCl_2 groups was also observed in the interaction of $(Cl_2Si-CCl_2)_3$ and Me_3SiH. Here Me_3SiCl was formed, as well as derivatives of 1,3,5-trisilacyclohexane containing CHCl groups, the latter in amounts too small to be isolated.

In connection with the conversion of partly chlorinated cyclic carbosilanes, the reaction between $(Cl_2Si-CCl_2)_3$ and $(H_2Si-CH_2)_3$ is of special interest. The reaction performed in a mole ratio of 1:1.1 in benzene proceeds to the compounds shown in Table 33. Of the 95% of $(H_2Si-CH_2)_3$ which was chlorinated, 63% underwent formation of one SiHCl group and 37% underwent formation of two SiHCl groups.

The reaction of $(Cl_2Si-CCl_2)_3$ with $(H_2Si-CH_2)_3$ in a mole ratio of 1:1.5 led to complete hydrogenation of the CCl_2 groups forming $(Cl_2Si-CH_2)_3$. No CHCl-

Table 33. Reaction products from $(Cl_2Si—CCl_2)_3$ and $(H_2Si—CH_2)_3$

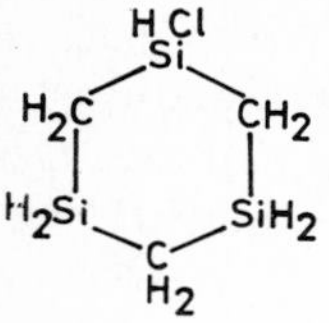

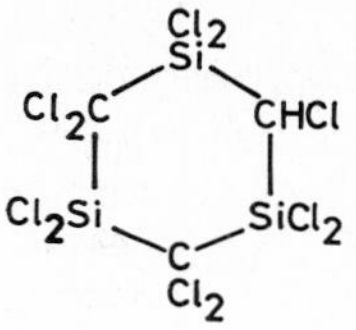

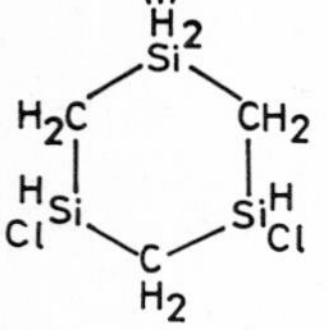

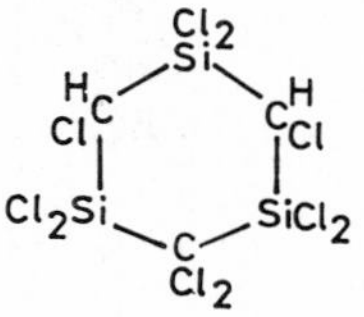

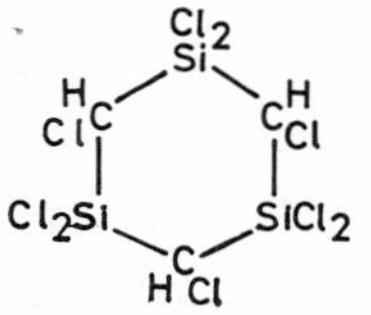

containing compounds were observed. Three to five SiCl groups were incorporated into the reaction partner $(H_2Si—CH_2)_3$:

$$(Cl_2Si—CCl_2)_3 + (H_2Si—CH_2)_3 \rightarrow (Cl_2Si—CH_2)_3 + (Cl_xH_ySi—CH_2)_3$$
$$x + y = 3$$

Taking $H_3Si—CHCl—SiH_3$ as an example, it was shown that such compounds gradually rearrange at 20 °C. Indeed, $(H_3Si)_2CH_2$, $H_3Si—CH_2—SiH_2Cl$, $H_3Si——CHCl—SiH_2Cl$, $H_2ClSi—CH_2—SiH_2Cl$ and $H_3Si—CH_2—SiHCl_2$ all form within 2 days. Warming the solution to 60 °C for 10 days accelerated the rearrangement considerably, in the course of which $H_3Si—CHCl—SiH_2Cl$ and $H_3Si—CH_2—SiH_2Cl$ resulted in approximately the same ratio [92].

2.3 Si-Fluorinated Carbosilanes

2.3.1 Fluorination of SiCl-Containing Carbosilanes

Fluorination of Si—Cl groups results when either ZnF_2, SbF_3 or anhydrous HF are used as fluorinating agents [87, 88]. Zinc fluoride was especially useful in the Si-fluorination of carbosilanes, because the cleavage of Si—C bonds to produce SiF_4 can thereby be suppressed [89]. Table 34 gives the experimental conditions under which SiCl-containing carbosilanes are converted to their SiF-, CH-containing analogues.

The compounds $(Cl_3Si—CH_2)_2SiCl_2$ and $(Cl_3Si—CCl_2)_2SiCl_2$ can be fluorinated in chloroform, as well as in acetonitrile and dichlorotrifluoroethane with ZnF_2. The separation of the fluorinated products $(F_3Si—CH_2)_2SiF_2$ and $(F_3Si—CCl_2)_2SiF_2$ from the solvent presented extreme difficulties. For the synthesis of Si-fluorinated 1,3,5-trisilapentanes on a preparative scale, hexachlorobutadiene was an especially suitable solvent. Zinc fluoride is effective as a fluorinating agent only if traces of water exist in the solid. Therefore the use of technical grade ZnF_2 is recommended, which must be partially dried at 120 °C for only one hour before use [90].

For the fluorination of $(Cl_3Si—CH_2)_2SiCl_2$, $N≡C—(CH_2)_4—C≡N$ can also be employed as a solvent. In this case the fluorination proceeds so quickly that ice-cold water must be used to control the temperature of reaction. The speed of the dropwise addition of the diluted starting material had to be adjusted so as to maintain the temperature of the reaction mixture [90].

Subsequent to the photochlorination of SiCl-containing carbosilanes, perchlorinated linear and cyclic carbosilanes were obtained [91, 92]. It was then of immense interest whether the formation of SiF-containing C-chlorinated compounds could be achieved through fluorination of perchlorinated compounds. It had already been shown that $(F_3Si)_2CCl_2$ can be obtained by reaction of SbF_3 with $(Cl_3Si)_2CCl_2$ [87]. With larger compounds, $Si—C$ cleavage occurs readily, so the use of ZnF_2 becomes more advisable. Table 35 presents the results obtained [89].

Investigations on $(Cl_3Si—CCl_2)_2SiCl_2$ show that reaction with ZnF_2 at 20 °C will produce $(F_3Si—CCl_2)_2SiF_2$ in high yield, while at 100 °C molecular cleavage occurs producing SiF_4, $F_3Si—CCl_2H$ and $F_3Si—CCl_2—SiF_3$. As Table 35 shows, the cyclic $(Cl_2Si—CCl_2)_3$ does not react with ZnF_2 to produce $(F_2Si—CCl_2)_3$. Furthermore, no reaction is evident under harsher reaction conditions, up to the point where complete cleavage of the molecule is observed. This applies also to other cyclic perchlorinated carbosilanes.

2.3.2 Cyclic Carbosilanes Containing SiF and CCl Groups

After it was discovered that fluorination of cyclic $(Cl_2Si—CCl_2)_3$ with ZnF_2 or other fluorination agents does not proceed, consideration was given to transfer of $Si—H$ to $Si—F$ groups with the help of PF_5. It is possible to convert $HSiMe_3$ and $HSiEt_3$ by reaction with PF_5 to $FSiMe_3$ and $FSiEt_3$, respectively. However, this reaction has little relevance to SiH-containing C-chlorinated carbosilanes, as the reaction of $(H_3Si—CCl_2)_2SiH_2$ with PF_5 or ZnF_2 shows.

The reaction of perchloro-1,3,5,7-tetrasilaadamantane with ZnF_2 does not proceed to its fluorinated derivative. At 110 °C in a sealed tube, cleavage occurs along with the formation of SiF_4 and an SiF-containing residue. No reaction occurs at lower temperatures.

The formation of C-chlorinated SiF-containing cyclic carbosilanes is possible through reversal of the reaction pathway, namely, via photochlorination of the corresponding SiF-, CH-containing compounds (see Table 36) [89].

Photochlorination occurs in CCl_4, in which compounds such as $(F_2Si—CH_2)_3$ are only sparingly soluble. At first, the partly C-chlorinated derivatives are formed

Table 34. Fluorination of Si—Cl groups in C—H containing carbosilanes

Starting material	g	Fluorination Conditions				Fluorinated Product	Yield %	Bp. °C	mp. °C in sealed glass-tubes	Sublimation temp. °C at 10^{-2} mm Hg
		Fluorinating agent	Solvent	Temp. °C	Time hrs.					
$Cl_3Si—CH_2—SiCl_3$	20	30 g ZnF_2 in 100 ml Hexachlorobutadiene-1,3	30 ml Hexachloro-butadiene-1,3	20	48	$F_3Si—CH_2—SiF_3$	80	35		
$(Cl_3Si—CH_2)_2SiCl_2$	10	10 g ZnF_2	Hexa-chloro-butadiene-1,3	20	24	$(F_3Si—CH_2)_2SiF_2$	40			
$(Cl_2Si—CH_2)_3$	60	70 g ZnF_2	300 ml CH_3CN	80	70	$(F_2Si—CH_2)_3$	75		143	~75
$(Cl_2Si—CH_2)_4$	1.8	10 g ZnF_2	CCl_4	80— 100	90		60		75	60
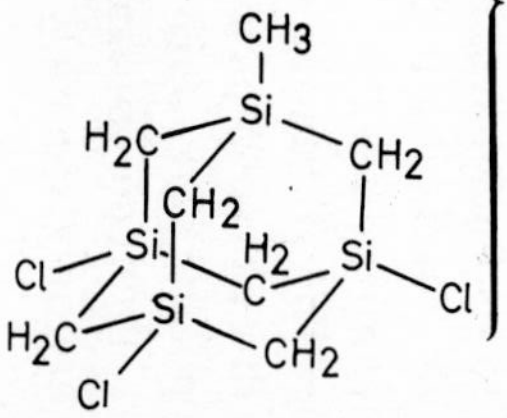 (cage carbosilane structures)	5	15 g ZnF_2		160	100	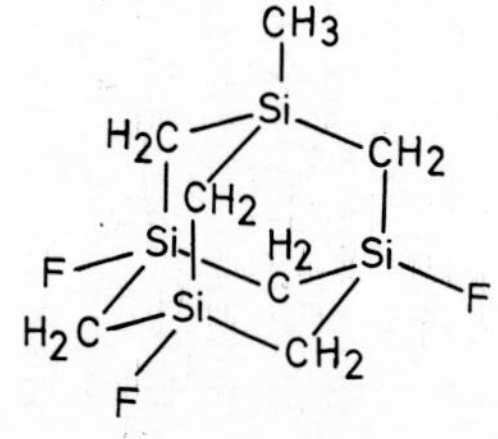 (fluorinated cage carbosilane structures)	82		260	60—70

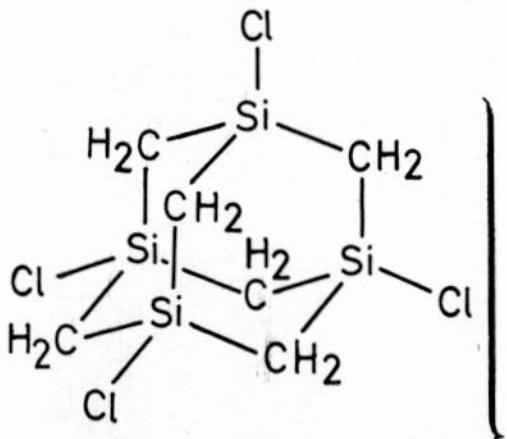

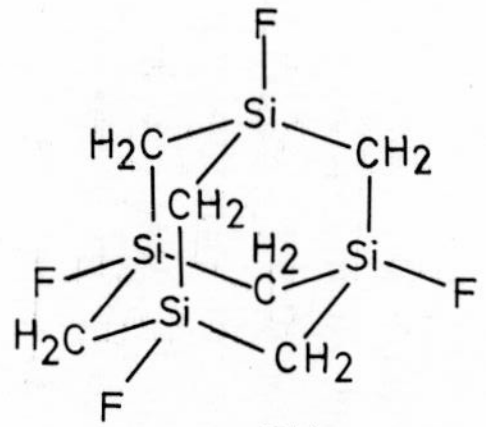

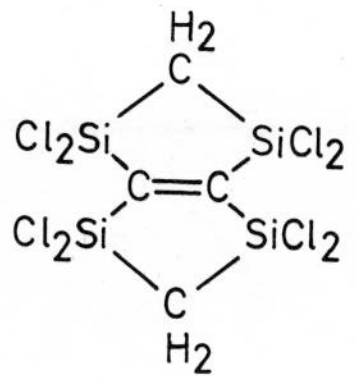

1.28

2.3 g ZnF$_2$ in 50 ml CCl$_4$

50 ml CCl$_4$

77

88

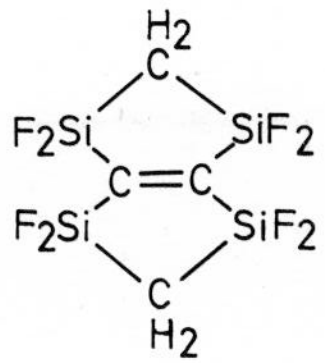

61

200–205   40

2

3.72 g ZnF$_2$

150 ml Toluene

111

22

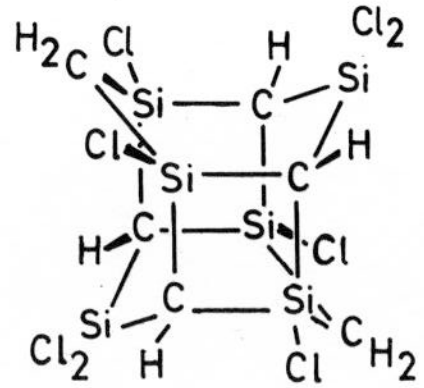

110

1.38

2.33 g ZnF$_2$

100 ml CCl$_4$

77

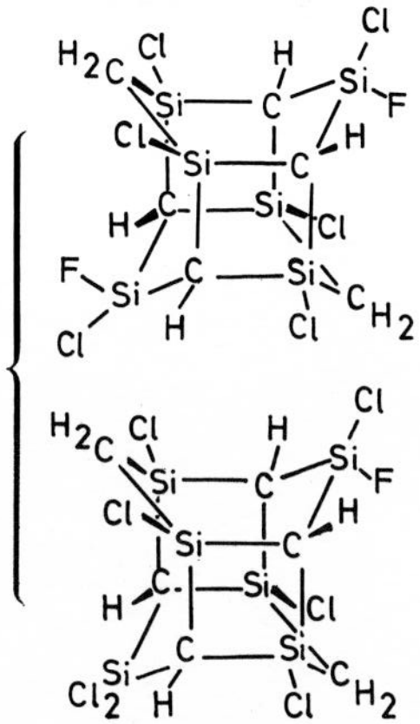

110

0.5

1.25 g ZnF$_2$

200 ml Hexa-chloro-butadiene-1,3

150

40

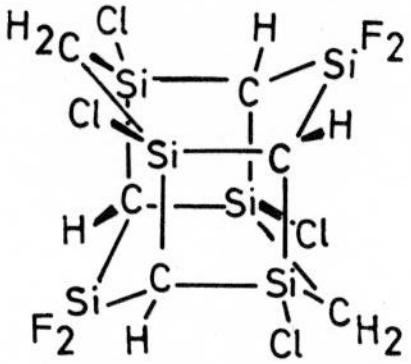

0.5

0.96 g SbF$_3$

200 ml CCl$_4$

77

40

Table 35. Fluorination of perchlorinated carbosilanes with ZnF₂

Starting material	g	Fluorinating agent	Solvent	Temp. °C	Time hrs.	Fluorinated product	Yield %
$Cl_3Si-CCl_2-SiCl_3$	105	200 g ZnF_2	700 ml Hexa-chlorobutadiene-1,3	20	100	$F_3Si-CCl_2-SiF_3$	95
$(Cl_3Si-CCl_2)_2SiCl_2$	35	50 g ZnF_2	250 ml Hexa-chlorobutadiene-1,3	20	96	$(F_3Si\text{-}CCl_2)_2SiF_2$ SiF_4 (0.5 g)	75
	5	6 g ZnF_2	without	100	48	0.7 g SiF_4 2.5 g of $F_3Si-CCl_2H$, $F_3Si-CCl_2-SiF_3$ $(F_3Si-CCl^2)_2SiF_2$ ratio 6:2:1	
$(Cl_2Si-CCl_2)_3$		ZnF_2	CCl_4	20	48		
		ZnF_2	Hexachloro-butadiene-1,3	90		some SiF_4	
		ZnF_2	Hexachloro-butadiene-1,3	150	48	SiF_4, $F_3Si-CCl_2H$	
		SbF_3	CCl_4	20	48	—	
		SbF_3	Hexachloro-butadiene-1,3	75	48	SiF_4 (15%) CCl_3F	

Table 36. Photochlorination of SiF-containing cyclic carbosilanes

Starting material	g	Solvent	Temp. °C	Irradiation time hrs.	Chlorination product	Yield	Method of isolation
$(F_2Si-CH_2)_2$	14 28	suspension in CCl_4	75	48	$(F_2Si-CCl_2)_3$	22 g 80% 50 g	Sublimation 50 °C/10⁻² mm Hg or recrystallization
$(F_2Si-CH_2)_4$	2.5	in 100 ml CCl_4		70	$(F_2Si-CCl_2)_4$		Solvent distilled off at 760 mm Hg
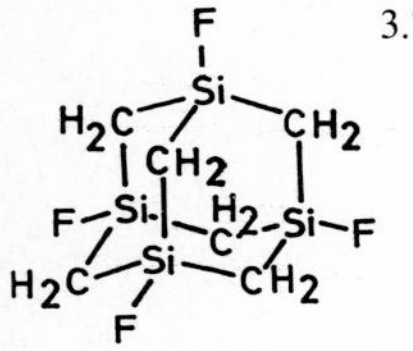 (tetrasilaadamantane structure, Si–F, CH₂ bridges)	3.7	in 200 ml CCl_4		200	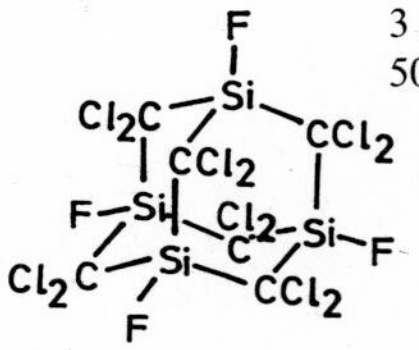 (tetrasilaadamantane structure, Si–F, CCl₂ bridges)	3 g 50%	pure through sublimation 180 °C/10⁻² mm Hg

followed by conversion to the perchlorinated compounds. With progressive increase in the degree of C-chlorination, one observes an increase in solubility so that near the end of the reaction a clear solution exists. The same behavior is observed for the formation of Si-fluorinated C-chlorinated 1,3,5,7-tetrasilaadamantanes.

Table 37. Fomation of party C-halogenated, SiF-containing cyclic carbosilanes

Starting material	Fluorination conditions					Fluorinated product	Yield %	mp. °C	Subl.
	g	Fluorinating agent	Solvent	Temp. °C	Time hrs.				
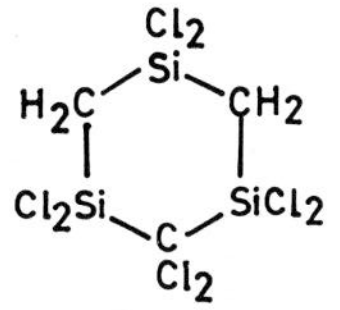 Cl_2Si–CH_2 / H_2C / Cl_2Si–C–$SiCl_2$ / Cl_2	4	6 g ZnF_2	50 ml benzene	60	70	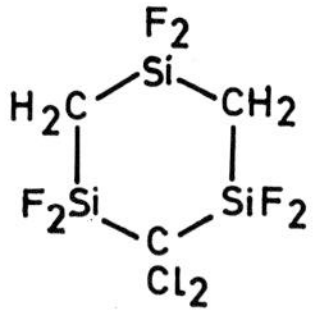 F_2Si–CH_2 / H_2C / F_2Si–C–SiF_2 / Cl_2	30	45	30 °C 0.01 mm Hg
Cl_2Si–CH_2 / H_2C / Cl_2Si–C–$SiCl_2$ / Br_2	6	9 g ZnF_2	60 ml CH_3CN	125	24	SiF_4, CH_3SiF_3, CH_2Cl_2			
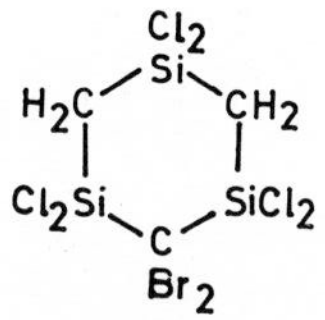	2.5	4 g ZnF_2	50 ml benzene	80	48	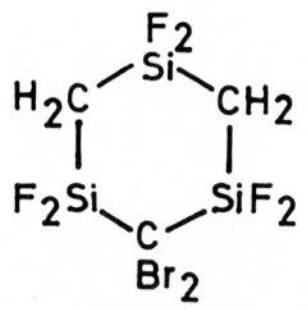 F_2Si–CH_2 / H_2C / F_2Si–C–SiF_2 / Br_2		66	50 °C 0.01 mm Hg
Cl_2Si–$SiMe_2$ / C=C / Cl, $SiMe_3$	1.9	2 g ZnF_2	60 ml CCl_4	60	72	F_2Si–$SiMe_2$ / C=C / Cl, $SiMe_3$			
Cl_2C–$SiMe_3$						Cl_2C–$SiMe_3$			
$MeClSi$–C≡C–$SiMe_3$						$MeFSi$–C≡C–$SiMe_3$			
Cl_2C–$SiMe_3$						Cl_2C–$SiMe_3$			
Cl_2Si–C≡C–$SiMe_3$						F_2Si–C≡C–$SiMe_3$			

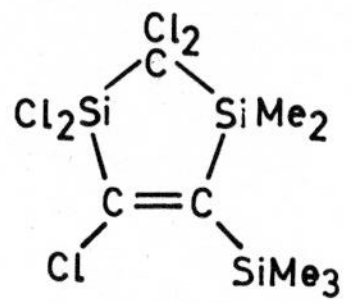

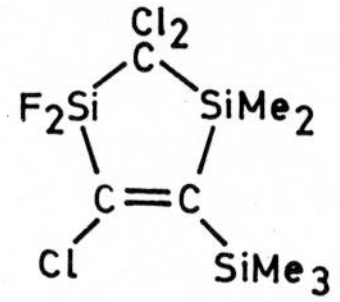

The photochlorination of SiF- and CH-containing carbosilanes to form $(F_3Si—CCl_2)_2SiF_2$, for example, is not a favorable reaction, because at temperatures as low as 60 °C, cleavage reactions occur.

2.3.3 Partly C-Halogenated, Si-Fluorinated Carbosilanes

The photochlorination of CH_2 groups in cyclic carbosilanes proceeds through intermediate stages. As monitored by NMR spectroscopy, 1,3,5-trisilacyclohexanes with 1, 2 or 3 CCl_2 groups have been identified, but compounds containing CHCl groups could not be isolated [89]. As shown in Table 37, Si-chlorinated and partly C-chlorinated cyclic carbosilanes can be fluorinated with ZnF_2 in acetonitrile or benzene between 60–80 °C without cleavage of the skeletal ring. The resulting compounds were purified through sublimation. Cleavage reactions begin appearing at around 125 °C. The following example illustrates how this ring cleavage occurs:

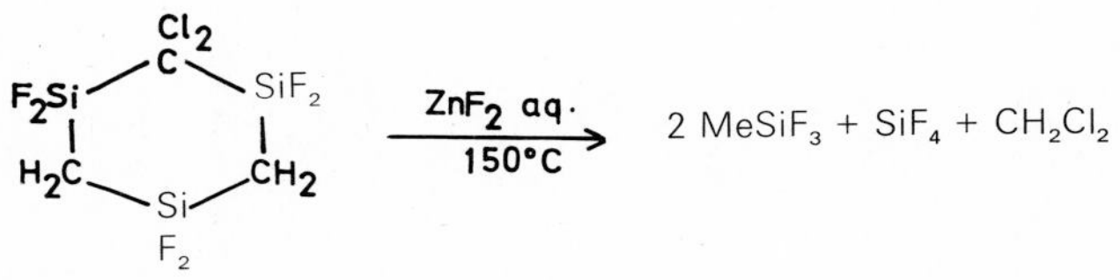

$$F_2Si \underset{H_2C}{\overset{\underset{C}{\overset{Cl_2}{|}}}{\diagdown}} \quad SiF_2 \quad \xrightarrow[150°C]{ZnF_2 \ aq.} \quad 2 \ MeSiF_3 + SiF_4 + CH_2Cl_2$$

Compounds containing a CBr_2 group will also undergo corresponding Si-fluorination.

3. Reactions with MeMgCl and MeLi

It is well established that the SiCl group determines the reactive behavior of Si-chlorinated carbosilanes containing only CH_2 or CH groups, whereas in reactions of SiCl-, CCl-substituted carbosilanes with organometallic reagents the CCl group is involved at the beginning of the reaction. The formation of SiH-, CCl-containing carbosilanes, as reported in the previous Chapt. III.2.2.3, gave rise to an investigation into the reactivity of such compounds towards MeMgCl compared to that of their SiH-, CH-substituted analogues.

3.1 SiH-Containing Carbosilanes

3.1.1 Reactions of $(H_3Si—CH_2)_2SiH_2$ and $(H_2Si—CH_2)_3$

SiH-containing compounds are looked upon as relatively inert towards Grignard reagents [109], excepted in a particular reactions [110]. Also in carbosilanes, the SiH group has been proved to be almost inert towards MeMgCl. For instance, in the reaction between $(H_3Si—CH_2)_2SiH_2$ and MeMgCl in Et_2O after 50 hours a product mixture was obtained containing 40% unreacted starting material, 46%

singly methylated and 14% doubly methylated derivative. Only traces of higher methylated derivatives were detected. The corresponding reaction in THF revealed complete reaction of the starting material $(H_3Si-CH_2)_2SiH_2$. The reaction products consisted of the trimethylated derivative in 64% yield, the dimethylated in 23% yield, and finally the tetramethylated 1,3,5-trisilapentane in 13% yield. The tri- and tetramethylated derivatives formed isomeric mixtures [94].

Similar results were obtained by analysing the products obtained in the reaction of $(H_2Si-CH_2)_3$ with MeMgCl in ether or THF. In no case was the molecular skeleton split. Even using a large excess of MeMgCl in ether caused only one SiH group to be methylated, while in the presence of THF, methylation was considerably enhanced, as is illustrated in Table 38 [94].

One observes from the NMR spectra of the reaction mixtures, that the multiply methylated 1,3,5-trisilacyclohexanes are cis-trans mixtures. This is also indicated during gas chromatographic separation by the largely differing melting points of the isomers observed in the product cooling traps [31]. As a result, by addition of excess

Table 38. Reaction products obtained from the reaction of $(H_2Si-CH_2)_3$ with MeMgCl

Mole equiv. of MeMgCl		18	0.5	0.8
Solvent		Et$_2$O	THF	THF (reflux)
Total yield of compounds shown		57%	80%	90%
Compounds formed	Number of Me groups	Percentage amount %	Percentage amount %	Percentage amount %
$(H_2Si-CH_2)_3$	0	87	1	—
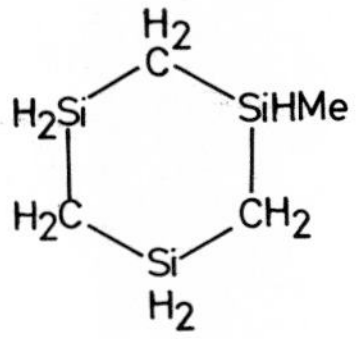	1	13	14.1	1
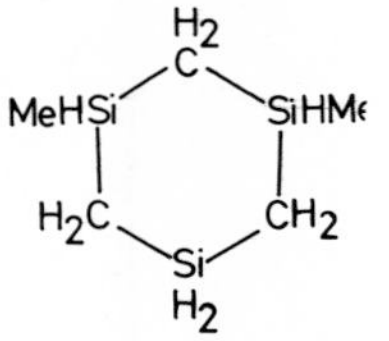	2	—	66.3	55
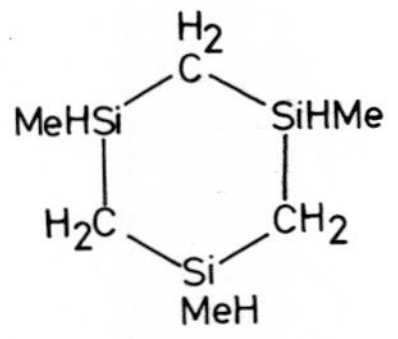	3		18.0	44

Table 39. Reaction of $(H_3Si)_2CCl_2$ with MeMgCl in Et_2O under varying reaction conditions

Mole equiv. of MeMgCl Total yield of compounds given below	0.6[a] 73%		3,7[a] 57%		12[a] 27%	
Compounds	Amount %	Amount of isomers %	Amount %	Amount of isomers %	Amount %	Amount of isomers %
$(H_3Si)_2CCl_2$	66					
$MeH_2Si-CCl_2-SiH_3$	30					
$MeH_2Si-CCl_2-SiH_2Me$	2		26.4			
$Me_2HSi-CCl_2-SiHMe_2$			61.0		3.9	
$Me_2HSi-CCl_2-SiH_2-CH_2-SiH_3$ }			3	1.5		
$MeH_2Si-CCl_2-SiHMe-CH_2-SiH_3$ }				1.5		
$Me_2HSi-CCl_2-SiHMe-CH_2-SiH_3$			7			
$Me_3Si-CHCl-SiH_2Me$					4.6	
$Me_3Si-CHCl-SiMe_3$					1.9	
$Me_3Si-CCl_2-SiHMe_2$					49.2	
$Me_3Si-CCl_2-SiMe_3$					26.1	
$\equiv Si-CHCl-\overset{\shortmid}{\underset{\shortmid}{Si}}-CH_3-Si\equiv$ [e]						0.5 (5 Me groups)
$Me_3Si-CCl_2-SiHMe-CH_2-SiH_3$						0.6
$Me_2HSi-CCl_2-SiMe_2-CH_2-SiH_3$ }					1.6	0.8
$Me_2HSi-CCl_2-SiHMe-CH_2-SiH_2Me$ }						0.2
$Me_2HSi-CCl_2-SiHMe-CH_2-SiH_3$						
$Me_3Si-CHCl-\overset{\shortmid}{\underset{\shortmid}{Si}}-CH_2-Si\equiv$ [e]						0.3 (3 Me groups)
$Me_3Si-CCl_2-SiHMe-CH_2-SiH_2Me$ }					8.0	5.3
$Me_3Si-CCl_2-SiMe_2-CH_2-SiH_3$ }						2.7
$Me_3Si-CCl_2-SiHMe-CH_2-SiHMe_2$					1.7	
$Me_2HSi-CHCl-SiH_2Me$						
$Me_2HSi-CHCl-SiHMe_2$						
$Me_3Si-CHCl-SiHMe_2$						
$\equiv Si-CHCl-\overset{\shortmid}{\underset{\shortmid}{Si}}-CH_2-Si\equiv$ [e]						
$Me_2HSi-CCl_2-SiHMe-CH_2-SiHMe_2$ }						
$Me_3Si-CCl_2-SiHMe-CH_2-SiH_2Me$ }						
$\equiv Si-CHCl-\overset{\shortmid}{\underset{\shortmid}{Si}}-CH_2-Si\equiv$ [e]						
not identified	2		2.6		2	
	(7 Compounds 1,3,5-Trisilapentanes)		(5 Compounds)		(4 Compounds)	

Table of compound types formed:

	3,7	12
Total amount of compounds containing the 1,3-disilapropane skeleton (%)	87.5	86
— containing a CCl_2-bridge	87.5	75
— containing a CHCl-bridge	—	11
Product with extended chain length	10	12

[a] MeMgCl in Et_2O added at -20 °C, reaction mixture kept at $+20$ °C for 20 hrs.

[b] similar to [a] but 40 or 240 hrs at 20 °C

[c] similar to [a] but 22 or 210 hrs at 20 °C

[d] MeMgCl added at -70 °C; reaction mixture gradually warmed to 20 °C over 16 hrs.; work up after another 2 hrs.

4.5 (40 hrs.)[b] 49%		7 (240 hrs.)[b] 39%		4.3[e] after 22 hrs.; 52%	after 210 hrs.	7[d] at −70 °C 67%	
Amount %	Amount of isomers %	Amount %	Amount of isomers %	Amount %	Amount %	Amount %	Amount of %
52.8		1.8		71.7	48.5		
				6.1	6.4 (Isomers)		
		5.4					
18.3		50.2		7.8	8.2	75	
		4.6				3.3	
	1.6						0.34
7.9	1.6					0.8	0.27
	4.7			4.4	5.5		0.19
6.9							
						1.3 (Isomers)	
						Traces	
1.5		5.0				4.1	
6.2				6.0	19.1	6.4	
1.4		23.0		1	4.3	5.8	
					2.9 (4 Me groups)		
				1	2.1		
						0.4 (5 Me groups)	
5 (12 Compounds)		10 (6.5% 1,3,5-Trisilapentanes)					
80				79.5		95	
71						85	
9				7	25	10	
15						3	

[e] no exact assignment of Me groups possible; remaining substituents are SiH groups

[f] yields with respect to employed $(H_3Si)_2CCl_2$. Percentage derived from GLC separation, or from NMR integration in the case of isomeric mixtures (}).

MeMgCl the methylation of the linear $(H_3Si-CH_2)_2SiH_2$ became a much easier process, in ether as well as in THF, than the methylation of the corresponding cyclic $(H_2Si-CH_2)_3$. In both cases the reaction progressed according to

$$\equiv Si-CH_2-\overset{|}{\underset{|}{Si}}-H + MeMgCl \rightarrow \equiv Si-CH_2-\overset{|}{\underset{|}{Si}}-Me + HMgCl$$

The increased reactivity of the linear compound is traced back to the primary terminal Si atoms. The increased reactivity in THF is attributed to the increased tendency of the solvent to coordinate, which causes a higher polarization of the Mg compound [94].

3.1.2 Linear CCl-, SiH-Containing Carbosilanes

Results obtained in reactions of appropriate carbosilanes with MeMgCl exist for $(H_3Si)_2CCl_2$, $(H_3Si)_2CHCl$, $H_3Si-CCl_2-SiH_2-CH_2-SiH_3$, $(H_3Si-CCl_2)_2SiH_2$ as well as for the simple silanes H_3Si-CH_2Cl, $H_3Si-CHCl_2$ and $H_3Si-CCl_3$ [110].

The results of the reaction between $(H_3Si)_2CCl_2$ and MeMgCl are shown in Tables 39 and 40. They indicate clearly that only methylation of SiH groups occurs and that the CCl_2 bridges are not methylated. The more hydrogen-rich Si atoms are always the sites for methylation, which progresses until complete substitution has been achieved. After a certain induction period, gradual hydrogenation of the CCl_2 bridges is observed, yielding a CHCl group. This tendency for gradual hydrogenation increases with respect to the methylation rate. Hydrogenation depends upon the amount of HMgCl, or on the mixture of MgH_2 and $MgCl_2$ already formed, and is obviously favored by longer reaction times. No hydrogenation occurs if low mole ratios of MeMgCl are used. With higher mole ratios, about 10% of the methylated compounds contain a CHCl bridge. On extreme extension of the reaction time, the amount of CHCl-bridged methylated derivatives rises to over 20%. Lowering the reaction temperature has no diminishing effect on the formation of CHCl-bridged methylated derivatives, if the reaction mixture is brought to room temperature before work-up.

The following equation presents the course of the reaction in a quite simplified manner:

$$H_3Si-CCl_2-SiH_3 + cMeMgCl \rightarrow H_{3-a}Me_aSi-CCl_2-SiMe_bH_{3-b} + cHMgCl$$
$$a + b = c \leqq 5 \qquad a\text{--}b = 0;\ 1$$

$$\equiv Si-CCl_2-Si\equiv\ +\ HMgCl \rightarrow\ \equiv Si-CHCl-Si\equiv\ +\ MgCl_2$$

Because formation of a CHCl bridge by reaction of 1,3-disilapropanes with a lower mole equivalence of MeMgCl fails to appear, the rearrangement

$$\equiv Si-CCl_2-Si\equiv\ +\ \equiv SiH \rightarrow\ \equiv Si-CHCl-Si\equiv\ +\ \equiv SiCl$$

has no real significance in the formation of a CHCl group under the applied reaction conditions.

Reaction of $(H_3Si)_2CCl_2$ with MeMgCl, preferably under conditions of high mole equivalence, also produces compounds which can be explained through Si—C cleavage in $(H_3Si)_2CCl_2$ and through molecular enlargement. Compounds arising through Si—C bond cleavage are SiH_4 and its methyl derivatives, excluding $SiMe_4$.

Compounds arising through molecular enlargement include most importantly SiH-containing methyl derivatives of 2,2-dichloro-1,3,5-trisilapentane. The proportion of these compounds will increase slightly with an increase in the mole equivalence of MeMgCl and with extended reaction time, but is suppressed by lowering the reaction temperature.

The side reactions occurring in Et_2O, namely, hydrogenation, elimination and molecular enlargement, become the main reactions if Et_2O is replaced by THF (Table 40) and similarly if MeMgCl is replaced by MeLi (Table 41). So, by the reaction of $(H_3Si)_2CCl_2$ in THF with an excess of MeMgCl, as shown in Tab. 40, the main reaction products obtained were cleavage and hydrogenation products with 5% of the CCl_2- and CHCl-containing compounds left [94].

If instead the reaction is conducted in the presence of MeLi (Table 41), methylation of the SiH groups is accompagnied by attack on the molecular skeleton. Apart from methylated silanes including a large amount of $SiMe_4$, carbosilanes containing more than two Si atoms were also produced, which comprised not only those compounds formed due to simple chain extension but also cyclic fully methylated 1,3,5-trisilacyclohexane. Even on addition of 12 mole equivalents of MeLi, the reaction ceases as soon as all products are Cl-free. Under no circumstances does methylation of bridging C atoms occur as is observed in reactions of perchlorinated carbosilanes with MeLi [122].

Table 40. Reaction of $(H_3Si)_2CCl_2$ with 12 mole equivalents of MeMgCl in THF

| Total percentage yield for given compounds | 20% |
Compounds formed	Yield %
$Me_3Si-CH_2-SiHMe_2$	95
$Me_3Si-CHCl-SiMe_3$	3
$Me_3Si-CCl_2-SiMe_3$	2

It has to be emphasized that the methylated derivatives of 2,2-dichloro-1,3,5-trisilapentane resulting from the reaction of $(H_3Si)_2CCl_2$ with MeMgCl are not generated by simple methylation of 2,2-dichloro-1,3,5-trisilapentane. In fact, the distribution of isomers in the product mixture is distinctly different from the distribution obtained from the reaction between 2,2-dichloro-1,3,5-trisilapentane and MeMgCl (given in Tab. 43). It resembles instead the isomeric distribution obtained from reactions of 2,2,4,4-tetrachloro-1,3,5-trisilapentane with MeMgCl. These observations are explained in the following manner [94]. The cleavage reactions

$$\equiv Si-\overset{|}{\underset{|}{C}}-Si\equiv \ + \ HMgCl \rightarrow \ \equiv SiH \ + \ \equiv Si-\overset{|}{\underset{|}{C}}-MgCl$$

and

$$\equiv Si-\overset{|}{\underset{|}{C}}-Si\equiv \ + \ MeMgCl \rightarrow \ \equiv SiMe \ + \ \equiv Si-\overset{|}{\underset{|}{C}}-MgCl$$

Table 41. Reaction of $(H_3Si)_2CCl_2$ with MeLi in Et_2O

Mole equiv. of MeLi	2[a]		12[a]
Total yield of compounds given below	27%		25%

Compounds formed	Percentage	Isomers	Percentage
$MeH_2Si-CCl_2-SiH_3$	20.0		
$MeH_2Si-CH_2-SiH_2-CH_2-SiH_2Me$	5.2		
$MeH_2Si-CCl_2-SiH_2Me$	12.0		
$H_3Si-CCl_2-SiH_2-CH_2-SiH_3$	17.0		
$H_3Si-CCl_2-SiH_2-CH_2-SiH_2Me$	21.6	17.1	
$MeH_2Si-CCl_2-SiH_2-CH_2-SiH_3$		4.5	
$MeH_2Si-CCl_2-SiH_2-CH_2-SiH_2Me$	5.1		
$MeH_2Si-CCl_2-SiH_2-CHCl-SiH_3$	2.0	1.75	
$H_3Si-CCl_2-SiH_2-CHCl-SiH_2Me$[b]		0.25	
$Me_2HSi-CCl_2-SiHMe-CH_2-SiH_3$		0.2	
$Me_2HSi-CCl_2-SiH_2-CH_2-SiH_2Me$	0.6	0.2	
$MeH_2Si-CCl_2-SiHMe-CH_2-SiH_2Me$		0.2	
$Me_3Si-CH_2-SiMe_3$			67
$(Me_3Si-CH_2)_2SiHMe$			13
$(Me_3Si-CH_2)_2SiMe_2$			10
$(Me_2Si-CH_2)_3$			7
unidentified	16.5		3

[a] MeLi in Et_2O added at $-20\ ^\circ C$; reaction mixture then kept at $20\ ^\circ C$.
[b] evidence from mass spectrometry, NMR spectra not fully assigned.
[c] yields with respect to employed $(H_3Si)_2CCl_2$. Percentages derived from GLC separation, or from NMR integration in the case of isomeric mixtures (}).

recognized by the formation of silanes or methylsilanes, lead to the formation of new organometallic compounds which readily undergo partial methylation on Si and react further:

$$\equiv Si-\overset{|}{\underset{|}{C}}-MgCl + H-\overset{|}{\underset{|}{Si}}-CCl_2-Si\equiv \rightarrow \equiv Si-\overset{|}{\underset{|}{C}}-\overset{|}{\underset{|}{Si}}-CCl_2-Si\equiv + HMgCl$$

This occurs via reactive SiH groups that exist in 2,2-dichloro-1,3-disilapropane or its corresponding methylated products, for instance, producing the next higher carbosilane series containing three Si atoms.

According to their reactivities, the carbosilanes formed will undergo further methylation with any available MeMgCl, or even undergo C-hydrogenation. The cleavage of $(H_3Si)_2CCl_2$, favored under conditions of increased mole equivalence of MeMgCl, or use of THF, or by reaction with MeLi, is chiefly responsible for the large decline in percent yield of reaction products shown in Tables 39, 40 and 41 [94]. This is explained by formation of volatile silanes and compounds of higher molecular weight, and is recognized by the appearance of products similar to siloxanes in the product residue. The effects of cleavage and hydrogenation attributed to HMgCl in reactions carried out on Si-hydrogenated, C-chlorinated carbosilanes with $LiAlH_4$ correspond closely to those discussed in Chapt. III.2.2.3.

3.1.3 (H₃Si)₂CHCl, H₃Si—CClH₂, H₃Si—CCl₂H and H₃Si—CCl₃

The behavior of $(H_3Si)_2CHCl$ was investigated in order to establish to what extent the CHCl bridge influences the course of reaction in comparison to the CCl_2 bridge in $(H_3Si)_2CCl_2$ when these compounds react with MeMgCl.

The reaction of $(H_3Si)_2CHCl$ with 11 mole equivalents of MeMgCl in Et_2O produced the only representative compound containing two Si atoms, $(Me_2HSi)_2CHCl$, in approximately 50% yield along with SiH_4 and methylated silanes. Experiments with the SiH-, CCl-containing silanes $H_3Si-CClH_2$, $H_3Si-CCl_2H$ and $H_3Si-CCl_3$ with MeMgCl were undertaken in order to determine what sort of influence increasing C-chlorination has on the reaction pathway. The results are shown in Table 42 [94].

Table 42. Reaction of $H_3Si-CClH_2$, $H_3Si-CCl_2H$ and $H_3Si-CCl_3$ with MeMgCl

Reaction of	$H_3Si-CClH_2$		$H_3Si-CCl_2H$		$H_3Si-CCl_3$
Mole equiv. of MeMgCl	3,5	6	2.8	6	3.5
Total yield of compounds given below	70	82	75		25
Compounds formed	Percentage	Percentage	Percentage	Percentage	Percentage
$Me_2HSi-CClH_2$	98	95			
$Me_3Si-CClH_2$	2	5			91
$Me_2HSi-CCl_2H$			48		
$Me_3Si-CCl_2H$			52	99	
$Me_3Si-CClH_2$					3
unidentified					6

The related compounds $H_3Si-CClH_2$ and $H_3Si-CCl_2H$ were shown to yield as reaction products SiH_4 and Me_2SiH_2, along with methane. The reaction proceeded mainly along the lines of methylation of the SiH groups. On the other hand, while cleavage of $H_3Si-CCl_3$ yielded mainly SiH_4, $MeSiH_3$, Me_2SiH_2 and Me_3SiH, no compounds appeared containing a CCl_3 group. Reaction products containing an uncleaved Si—C bond amounted to 25%, of which $Me_3Si-CCl_2H$ was the most important. This corresponds in principle to the behavior of the appropriate chloromethyl chlorosilanes towards Grignard reagents, in the course of which $Cl_3Si-CCl_3$ also undergoes cleavage and rearrangement.

3.1.4 H₃Si—CCl₂—SiH₂—CH₂—SiH₃

The CCl_2 group in 2,2-dichloro-1,3,5-trisilapentane considerably influences the reactivity of the different SiH groups in the molecule against MeMgCl. The reactions were carried out using 1.7, 4, 5, and 15 mole equivalents of MeMgCl. The results are shown in Table 43 [94].

Already in the reaction involving 1.7 mole equivalents of MeMgCl, volatile silicon compounds such as SiH_4, $MeSiH_3$, Me_2SiH_2, Me_3SiH and $(H_3Si)_2CH_2$ appeared, indicating cleavage of the starting material. The proportion of such compounds

Table 43. Reaction of $H_3Si-CCl_2-SiH_2-CH_2-SiH_3$ with MeMgCl

Mole equiv. of MeMgCl		1.7[a]		4[a]	
Total yield of compounds given below[e]		60%		58%	
Compounds formed	Number of Me groups	Percentage	Percent of isomers	Percentage	Percent of isomers
$MeH_2Si-CCl_2-SiH_2-CH_2-SiH_3$	1	36.7			
$Me_2HSi-CCl_2-SiH_2-CH_2-SiH_3$ } $MeH_2Si-CCl_2-SiHMe-CH_2-SiH_3$ $MeH_2Si-CCl_2-SiH_2-CH_2-SiH_2Me$ }	2	51.3	24.6 24.6 2.1		
$Me_2HSi-CCl_2-SiHMe-CH_2-SiH_3$ $Me_2HSi-CHCl-SiHMe-CH_2-SiH_3$[d]	3	3.9		60.4 12.8	
$\equiv Si-CHCl-\overset{\mid}{\underset{\mid}{Si}}-CH_2-Si\equiv$[c]	2	5.3			
$Me_3Si-CCl_2-SiHMe-CH_2-SiH_3$ } $Me_2HSi-CCl_2-SiMe_2-CH_2-SiH_3$ $Me_2HSi-CCl_2-SiHMe-CH_2-SiH_2Me$ }	4			21.6	8.1 7.3 6.2
$Me_3Si-CCl_2-SiHMe-CH_2-SiH_2Me$ } $Me_3Si-CCl_2-SiMe_2-CH_2-SiH_3$ $Me_2HSi-CCl_2-SiMe_2-CH_2-SiH_2Me$ }	5				
$Me_3Si-CCl_2-SiHMe-CH_2-SiHMe_2$ } $Me_3Si-CCl_2-SiMe_2-CH_2-SiH_2Me$ }	6				
$Me_3Si-CHCl-SiMe_2-CH_2-SiH_3$ } $Me_3Si-CHCl-SiHMe-CH_2-SiH_2Me$[d] $Me_2HSi-CHCl-SiMe_2-CH_2-SiH_2Me$[d] }	5				
$Me_3Si-CHCl-SiMe_2-CH_2-SiH_2Me$ } $Me_3Si-CHCl-SiHMe-CH_2-SiHMe_2$ $Me_2HSi-CHCl-SiMe_2-CH_2-SiHMe_2$ }	6				
unidentified		8.1 (7 Compounds)		5.2 (8 Compounds)	

[a] MeMgCl in Et$_2$O added at —20 °C; then 20 hrs. at 20 °C

[b] Same as [a] but in refluxing Et$_2$O

[c] no definite assignment of Me groups, remaining Si substituents are SiH groups

increases with increasing concentration of MeMgCl. Neither SiMe$_4$ nor unreacted starting material were observed. While reactions involving a lower mole equivalence of MeMgCl produced a 60% yield of less volatile condensable reaction products the yield using 15 mole equivalents of MeMgCl decreased to 46% and even to 34% if the reaction was performed in Et$_2$O under reflux. The composition of the product mixture was determined by the CCl$_2$ group in the molecule. This group exerts an activating influence on the neighboring SiH groups. However, this tendency decreases with progressive methylation in such a way that in the tetramethylated derivative all Si atoms possess a comparable reactivity towards MeMgCl. The next reaction step involves the methylation of the SiH groups adjacent to the bridging CH$_2$ group, resulting in a mixture of fivefold methylated derivatives. The decreased tendency of the SiH groups to react due to the elevated degrees of methylation is also obvious in the reaction residues, which contain unconsumed MeMgCl, while sufficient SiH groups still exist in the reaction mixture. Reactions involving an excess of MeMgCl

5[a] 60%		15[a] 46%		15[b] 34%	
Percentage amount %	Percent of isomers	Percentage amount %	Percent of isomers	Percentage amount %	Percent of isomers
31.5					
7.5		4.4		4.5	
	7 (4 Me groups)	8.9		18.3	
	24.1		10.1		6.6
46.4	19.5	20.9	10.8	13.2	6.6
	2.8				
	3		19.5		13.1
4.6	0.7	39.9	12.4	23.8	4.8
	0.9		8.0		5.9
		5.8	1.4	3.7	1.5
			4.4		2.2
			3.7		4.8
		11.4	5.8	23.8	13.4
			1.9		5.7
					0.5
		1.6		3.9	0.5
					2.8
3.0 (7 Compounds)		7.1 (12 Compounds)		8.8 (14 Compounds)	

[d] concluded from mass spectra

[e] yields calculated with respect to the starting amount of $H_3Si-CCl_2-SiH_2-CH_2-SiH_3$. Percentages are derived from glc separation, or from NMR integration in the case of isomeric mixtures (}).

revealed that trimethylated derivatives will in all cases react further, while 66% of the tetramethylated derivatives and only approximately 13% of the pentamethylated derivatives react further. This also corresponds to the hexamethylated derivatives, which react to a 7% degree. Table 44 illustrates the reaction course taken by the methylation of $H_3Si-CCl_2-SiH_2-CH_2-SiH_3$. The compounds containing CHCl groups are not considered here, on the grounds that reaction pathways become too complicated [94].

In each case, at the single and threefold methylation stage only one derivative is obtained, whereas in the remaining methylation stages isomeric mixtures are obtained without exception.

Apart from SiH-methylation, attention must be paid to C-hydrogenation (that is, formation of a CHCl group as a competing and paralleling reaction). The latter compounds will increase in yield as the degree of methylation increases. This yield will therefore be determined from the amounts of MeMgCl used. Moreover, the pro-

Table 44. Reaction pathway pertaining to the methylation of $H_3Si-CCl_2-SiH_2-CH_2-SiH_3$ without considering CHCl-containing products. Numbers give the fraction in the isomeric mixture

$$H_3Si-CCl_2-SiH_2-CH_2-SiH_3$$
$$1$$
$$\downarrow$$
$$Me_2HSi-CCl_2-SiH_2-CH_2-SiH_3$$
$$1$$

$Me_2HSi-CCl_2-SiH_2-CH_2-SiH_3 \qquad\qquad MeH_2Si-CCl_2-SiHMe-CH_2SiH_3$
$0,5 \qquad\qquad\qquad\qquad\qquad\qquad\qquad\qquad\qquad 0,5$

$$Me_2HSi-CCl_2-SiHMe-CH_2-SiH_3$$
$$1$$
$$\downarrow$$

$Me_3Si-CCl_2-SiHMe-CH_2-SiH_3 \quad Me_2HSi-CCl_2-SiHMe-CH_2-SiH_2Me \quad Me_2HSi-CCl_2-SiMe_2-CH_2-SiH_3$
$0,5 \qquad\qquad\qquad\qquad\qquad\qquad\qquad 0,1 \qquad\qquad\qquad\qquad\qquad\qquad\qquad\qquad 0,4$
$\downarrow \qquad\qquad\qquad\qquad\qquad\qquad\qquad\qquad\qquad\qquad\qquad\qquad\qquad\qquad\qquad\qquad\qquad \downarrow$

$Me_3Si-CCl_2-SiHMe-CH_2-SiH_2Me \quad Me_3Si-CCl_2-SiMe_2-CH_2-SiH_3 \qquad Me_2HSi-CCl_2-SiMe_2-CH_2-SiH_2Me$
$0,5 \qquad\qquad\qquad\qquad\qquad\qquad\qquad 0,3 \qquad\qquad\qquad\qquad\qquad\qquad\qquad\qquad 0,2$

$$Me_3Si-CCl_2-SiHMe-CH_2-SiHMe_2 \qquad Me_3Si-CCl_2-SiMe_2-CH_2-SiH_2Me$$
$$0,2 \qquad\qquad\qquad\qquad\qquad\qquad\qquad\qquad\qquad 0,8$$

$$Me_3Si-CCl_2-SiMe_2-CH_2-SiHMe_2{}^{a}$$

---

[a] This compound was not isolated

portion of such compounds containing a CHCl bridge depends largely on the reaction temperature, as the reaction in Et_2O under reflux shows (see Table 43). This temperature increase doubles the amount of products containing a CHCl bridge. Moreover, Table 45 gives an account of the percentage distribution of compounds produced at the same methylation stage with either hydrogenated (CHCl) or unhydrogenated (CCl_2) bridges, as well as a mean degree of methylation achieved by means of addition of specified amounts of MeMgCl. Numbers in brackets show the percentage of compounds from the individual methylation stages, which undergo hydrogenation, and as a consequence are excluded from further methylation. The higher methylated compounds are more easily hydrogenated. Consideration given to continued methylation of mono- or di-methylated compounds after introduction of a CHCl bridge can be excluded because reactions done with an excess of MeMgCl will always give tri-methylated 2-chloro-1,3,5-trisilapentane, inspite of the failure of the corresponding 2,2-dichloro-1,3,5-trisilapentane to appear. Similarly, only $(Me_2HSi)_2CHCl$ can be obtained from the reaction of $(H_3Si)_2CHCl$ with an excess of MeMgCl [94].

3.1.5 $(H_3Si-CCl_2)_2SiH_2$

Reactions were carried out in Et_2O in the presence of 1, 2, 6, 9, and 18 mole equivalents of MeMgCl under comparable conditions as before. The results are presented in Table 46.

From the reaction of $(H_3Si-CCl_2)_2SiH_2$ with 1 mole equivalent of MeMgCl, 46% of the reaction products contained the molecular skeleton of the starting material, but only 24% still contained both CCl_2 bridges. More than 50% of the reaction pro-

Table 45. Products obtained from the reaction of $H_3Si-CCl_2-SiH_2-CH_2-SiH_3$ 260 with varying mole equivalents of MeMgCl. Percentage distribution of varying methylated derivatives of 260 is shown in terms of compounds containing a CCl_2 or a CHCl bridge. Numbers in brackets indicate the percentage of CCl_2-containing compounds that undergo C-hydrogenation [110] with respect to the actual remaining amount of compounds available for either further methylation or C-hydrogenation

No. of Me groups	Mole equivalents of MeMgCl									
	1.7		4		5		15		15[a]	
	CCl_2	CHCl	CCl_2	CHCl	CCl_2	CHCl	CCl_2	CHCl	CCl_2	CHCl
1	36.7									
2	51.3 (9)	5.3								
3	3.9		60.4 (14)	12.8	3.15 (8)	7.5	(5)	4.4	(5)	4.5
4			21.6		46.4 (12)	7	20.9 (10)	8.9	13.2 (21)	18.3
5					4.6		39.9 (19)	11.4	23.8 (43)	23.8
6							5.8 (22)	1.6	3.7 (51)	3.9
Σ^b	91.9	5.3	82.0	12.8	82.5	14.5	66.6	26.3	40.7	50.5
mean methylation degree	1.6		3.1		3.5		4.3		4.2	

[a] Reaction performed in diethyl ether under reflux.
[b] The percentage difference to 100 % is attributed to unidentified products.

Table 46. Products from the reaction of $(H_3Si-CCl_2)_2SiH_2$ with MeMgCl in Et_2O

Mole equiv. of MeMgCl[a] Total yield of compounds given below[d]	No. of Me groups	1[a] 65% Percentage	Percent of isomers	2[a] 64% Percentage	Percent of isomers
$H_3Si-CCl_2-SiH_3$		9.4			
$MeH_2Si-CCl_2-SiH_3$		31.8		3.0	
$MeH_2Si-CCl_2-SiH_2Me$		9.8		59.3	
$H_3Si-CCl_2-SiH_2-CH_2-SiH_3$		1.9		1.8	
$H_3Si-CCl_2-SiH_2-CHCl-SiH_3$		4.9			
$H_3Si-CCl_2-SiH_2-CCl_2-SiH_3$		2.8			
$MeH_2Si-CCl_2-SiH_2-CHCl-SiH_3$ } 1		6.2	4.6		
$H_3Si-CCl_2-SiH_2-CHCl-SiH_2Me$[c] }			1.6		
$MeH_2Si-CCl_2-SiH_2-CHCl-SiH_2Me$	2	8.7			
$MeH_2Si-CCl_2-SiH_2-CCl_2-SiH_3$ } 1			2.4		
$H_3Si-CCl_2-SiHMe-CCl_2-SiH_3$ }		3.5	1.1		
$MeH_2Si-CCl_2-SiH_2-CCl_2-SiH_2Me$	2	18			
$Me_2HSi-CCl_2-SiH_2Me$				11.1	
$Me_2HSi-CCl_2-SiHMe_2$				0.5	
$MeH_2Si-CCl_2-SiH_2-CH_2-SiH_3$				5.0	
$MeH_2Si-CCl_2-SiH_2-CH_2-SH_2Me$ }					9.8
$MeH_2Si-CCl_2-SiHMe-CH_2-SiH_3$ } 2				11.3	1.2
$Me_2HSi-CCl_2-SiH_2-CH_2-SiH_3$ }					0.3
$Me_2HSi-CCl_2-SiH_2-CH_2-SiH_2Me$ } 3					1.2
$MeH_2Si-CCl_2-SiHMe-CH_2-SiH_2Me$ }				2	0.8
$Me_3Si-CCl_2H$					
$Me_2HSi-CHCl-SiH_2Me$					
$Me_2HSi-CHCl-SiHMe_2$					
$Me_3-CCl_2-SiHMe_2$					
$\equiv Si-CHCl-\overset{\mid}{\underset{\mid}{Si}}-CH_2-Si\equiv$[b]					
$Me_2HSi-CCl_2-SiHMe-CH_2-SiH_3$	4				
$Me_2HSi-CCl_2-SiHMe-CH_2-SiH_2Me$					
$Me_2HSi-CCl_2-SiHMe-CH_2-SiHMe_2$ } 5					
$Me_3Si-CCl_2-SiHMe-CH_2-SiH_2Me$ }					
$Me_3Si-CHCl-SiHMe_2$					
$Me_3-CHCl-SiMe_3$					
$Me_3Si-CCl_2-SiMe_3$					
$\equiv Si-CHCl-\overset{\mid}{\underset{\mid}{Si}}-CH_2-Si\equiv$[b]					
$Me_3Si-CCl_2-SiHMe-CH_2-SiH_3$					
$Me_2HSi-CCl_2-SiMe_2-CH_2-SiH_3$ } 4					
$Me_2HSi-CCl_2-SiHMe-CH_2-SiH_2Me$ }					
$\equiv Si-CHCl-\overset{\mid}{\underset{\mid}{Si}}-CH_2-Si\equiv$[b]					
$Me_3Si-CCl_2-SiHMe-CH_2-SiH_2Me$ } 5					
$Me_2HSi-CCl_2-SiMe_2-CH_2-SiH_2Me$ }					
$Me_3Si-CCl_2-SiMe_2-CH_2-SiH_3$					
$Me_3Si-CCl_2-SiHMe-CH_2-SiHMe_2$ } 6					
$Me_3Si-CCl_2-SiMe_2-CH_2-SiH_2Me$ }					
$Me_3Si-CHCl-SiMe_2-CH_2-SiH_3$					
$Me_3Si-CHCl-SiHMe-CH_2-SiH_2Me$[c] } 5					
$Me_2HSi-CHCl-SiMe_2-CH_2-SiH_2Me$[c] }					
unidentified		3 (10 Compounds)		6 (6 Compounds)	

[a] MeMgCl/Et_2O added at $-20\ ^\circ C$. Reaction mixture then stirred at $20\ ^\circ C$ for 20 hrs.
[b] No certain assignment of Me groups; remaining groups are SiH groups.
[c] Confirmed by mass spectrometry; NMR spectrum not fully assigned.

6[a] 56%		9[a] 45%		18[a] 61%	
Percentage	Percent of isomers	Percentage	Percent of isomers	Percentage	Percent of isomers
49.5		3.3		0.5	
2.8		6.3		2.3	
2.7					
7.0		6.5		10.6	
3.0		46.3		22.0	
3.0 (4 Me groups)		1.0		0.9	
7.5					
20					
	2.7				
3.0					
	0.3	4.8		3.1	
		0.3		0.3	
		4.4		4.4	
		1.8 (3 Me groups)			
			2.0		4.3
		5	2.0	9.1	4.3
			1.0		0.5
		0.3 (6 Me groups)		1.0	
			9.8		20.2
		12.4	2.2	29.2	6.7
			0.4		2.3
		3.9	3.9	9.2	8.3
			0		0.9
					0.3
				3.0	1.9
					0.8
1.5		3.7		4.4	
(7 Compounds)		(9 Compounds)		(8 Compounds)	

[a] Yields are calculated with respect to the starting amount of $(H_3Si-CCl_2)_2SiH_2$. Percent values given are derived from gas chromatographic separation; values for isomeric mixtures were determined by NMR integration.

ducts arise through cleavage of the starting compound. Reaction with 2 mole equivalents of MeMgCl resulted in complete absence of compounds containing two CCl_2 bridges; 73% of the reaction products arose through cleavage. Similarly, reaction with 6 mole equivalents of MeMgCl produced cleavage products as the main reaction products. Reactions involving 9 mole equivalents of MeMgCl yielded 3.7% of unidentified products containing double-bonded compounds. These compounds undergo β-elimination on the addition of Br_2, producing Me_3SiBr. Hydrolytic work-up of the residue obtained from reaction with 18 moles of MeMgCl leads to the formation

Table 47. Reaction of $(H_3Si-CCl_2)_2SiH_2$ with MeMgCl in cyclopentane[a]

		1.8[b]		5.5[b]	
Mole equiv. of MeMgCl[b]					
Total yield of compounds given below		95%		90%	
	No. of Me groups	Percentage	Percent of isomers	Percentage	Percent of isomers
$MeH_2Si-CCl_2-SiH_3$		2.5			
$MeH_2Si-CCl_2-SiH_2Me$		5.3		5.2	
$H_3Si-CCl_2-SiH_2-CH_2-SiH_3$		1.4			
$Me_2HSi-CCl_2-SiH_2Me$		0.6		9.2	
$MeH_2Si-CCl_2-SiH_2-CH_2-SiH_3$	1	8.9		5.4	
$Me_2HSi-CCl_2-SiHMe$		0.3		1.7	
$H_3Si-CCl_2-SiH_2-CHCl-SiH_3$	0	4.7			
$Me_2HSi-CCl_2-SiH_2-CH_2-SiH_3$ ⎫			1.1		
$MeH_2Si-CCl_2-SiH_2-CH_2-SiH_2Me$ ⎬ 2	2	3.3	1.1		
$MeH_2Si-CCl_2-SiHMe-CH_2-SiH_3$ ⎭			1.1		
$MeH_2Si-CCl_2-SiH_2-CHCl-SiH_3$ ⎫ 1	1	37.4	32.7		
$H_3Si-CCl_2-SiH_2-CHCl-SiH_2Me$[c] ⎭			4.7		
$MeH_2Si-CCl_2-SiH_2-CHCl-SiH_2Me$ ⎫ 2	2	14.3	13.2	8.4	0.1
$MeH_2Si-CCl_2-SiHMe-CHCl-SiH_3$ ⎭			1.1		8.3
$MeH_2Si-CCl_2-SiH_2-CCl_2-SiH_3$ ⎫ 1	1	7.7			
$MeH_2Si-CCl_2-SiH_2-CCl_2-SiH_2Me$ ⎭ 2	2	13.6			
$MeH_2Si-CCl_2-SiHMe-CH_2-SiH_3$ ⎫					7.3
$Me_2HSi-CCl_2-SiH_2-CH_2-SiH_3$ ⎬ 2	2			18.7	7.3
$MeH_2Si-CCl_2-SiH_2-CH_2-SiH_2Me$ ⎭					4.1
$Me_2HSi-CCl_2-SiHMe-CH_2-SiH_3$ ⎫					3.0
$MeH_2Si-CCl_2-SiHMe-CH_2-SiH_2Me$ ⎬ 3	3			10.9	3.0
$Me_2HSi-CCl_2-SiH_2-CH_2-SiH_2Me$ ⎭					4.9
$Me_2HSi-CCl_2-SiHMe-CH_2-SiH_2Me$	4			1.6	
$Me_2HSi-CCl_2-SiHMe-CHCl-SiH_3$ ⎫					10.1
$Me_2HSi-CCl_2-SiH_2-CHCl-SiH_2Me$ ⎬ 3	3			27.4	7.2
$MeH_2Si-CCl_2-SiHMe-CHCl-SiH_2Me$ ⎭					10.1
$Me_2HSi-CCl_2-SiHMe-CHCl-SiH_2Me$	4			8.6	
unidentified				2.9	

[a] After MeMgCl was prepared in Et_2O, the residue was dried under vacuum until a fine white powder resulted. This was suspended in cyclopentane.

[b] The cyclopentane suspension was added at $-20\ °C$ and the reaction mixture was maintained subsequently for 20 hrs at 20 °C.

[c] Determined through mass spectral analysis, NMR peaks not fully assigned.

[d] Yields calculated are based on starting amount of $(H_3Si-CCl_2)_2SiH_2$. Percentage yields are determined by gas chromatographic separation; in cases of isomeric mixtures (⎬), values were derived from NMR integration.

Table 48. Distribution of raction products (%) obtained from the reaction of $(H_3Si-CCl_2)_2SiH_2$ 261 with varying mole equivalents of MeMgCl. The product yields are arranged according to carbosilanes having the same molecular skeleton. The symbols Si3, Si2, Si1 indicate the number of Si atoms present in the carbosilane, and CCl_2/CCl_2, $CCl_2/CHCl$, CCl_2/CH_2, $CHCl/CH_2$, CCl_2 and $CHCl$ represent the appropriate number of bridging groups

Mole equiv. of MeMgCl	Si3	CCl_2/CCl_2	$CCl_2/CHCl$	CCl_2/CH_2	$CHCl/CH_2$	Si2	CCl_2	$CHCl$	Si1	Unidentified
<1	79.5	70.4[a]	8.6	0.5		17.5	17.5			3
1	46	24.3[b]	19.8	1.9		51	51			3
2	22.1		2[c]	20.1		73.9	73.9			4
6	33.5			30.5	3	62.2	52.5	9.7	2.7	1.5
9	24.4			21.3	3.1	65.6	54	11.6	6.3	3.7
18	52.4			47.5	4.9	43.2	26.9	14	2.3	4.4
1.8[d]	91.3	21.3	56.4	13.6		8.7	8.7			
5.5[d]	81		44.4	36.6		16.1	16.1			2.9

[a] still contains 50% of starting material
[b] still contains 2.8% of starting material
[c] assignment not certain
[d] reactions conducted in cyclopentane

of Me_2SiH_2. Unidentified substances (4.4%), also containing double-bonds, were detected by NMR spectroscopy, as well as through bromination to produce Me_3SiBr.

Table 47 is made up of results from the reaction of $(H_3Si—CCl_2)_2SiH_2$ with 1.8 and 5.2 mole equivalents of MeMgCl in cyclopentane, and can be compared with the results from Table 46. In the transition from $H_3Si—CCl_2—SiH_2—CH_2—SiH_3$ to $(H_3Si—CCl_2)_2SiH_2$ a change in reactivity towards MeMgCl is recognized, as shown by the composition of the reaction products presented in Tables 43 and 45 or 46 and 47 [94].

The results in Table 48 on the methylation of $(H_3Si—CCl_2)_2SiH_2$ are presented in such a way that the methyl derivatives having the same molecular skeleton, formed by reaction of $(H_3Si—CCl_2)_2SiH_2$ with varying mole equivalents of MeMgCl, are summed in their yield and presented together. Especially noteworthy is the cleavage of $(H_3Si—CCl_2)_2SiH_2$ by reaction in Et_2O, which occurs even with low mole equivalents of MeMgCl and determines the course of the reaction. With an excess of MeMgCl in Et_2O, more than 50% of the reaction products are traced back to cleavage of $(H_3Si—CCl_2)_2SiH_2$. However, if the reaction is performed in cyclopentane, Si-methylated 1,3,5-trisilapentanes containing either one CHCl or one CH_2 group are formed predominantly. In none of these reactions will Si-methylated 2,2,4,4-tetrachloro-1,3,5-trisilapentanes make up more than 25% of the reaction mixture; in Et_2O no higher than two-fold methylation occurs. Nevertheless, compounds derived through hydrogenation of their CCl_2 groups are obtained in over 70% yield.

If $(H_3Si—CCl_2)_2SiH_2$ reacts with two moles of MeMgCl in Et_2O, no more reaction products are found containing two CCl_2 groups, and only a very small amount containing one CCl_2 and one CHCl group. Using higher mole equivalents of MeMgCl, the formation of methylated 1,3,5-trisilapentanes with one CCl_2 group and one CH_2 group predominates, while compounds containing one CCl_2 and one CHCl bridging group are not found in the reaction mixture. One recognized from the ratio of isomers of 2,2-dichloro-1,3,5-trisilapentane formed from $(H_3Si-CCl_2)_2SiH_2$ that they are formed via a pathway different from that in which the derivatives of $H_3Si-CCl_2-$ $-SiH_2-CH_2-SiH_3$ are formed. In the latter case, the SiH groups are replaced by Si—Me groups according to their respective reactivities. On the other hand, in the reaction of $(H_3Si-CCl_2)_2SiH_2$ only those of 2,2-dichloro-1,3,5-trisilapentane are methylated derivatives obtained in which the least reactive positions on the Si atoms are substituted by methyl groups. This can only be partially explained by hydrogenation of their corresponding higher-chlorinated analogues. The largest amount of compounds identified resulted from cleavage of the molecular skeleton of $(H_3Si-$ $-CCl_2)_2SiH_2$. This refers mainly to 2,2-dichloro-1,3-disilapropane and its methyl derivatives. Analogous cleavage of 2,2,4,4-tetrachloro-1,3,5-trisilapentane by $LiAlH_4$ has already been observed [93].

Table 48 shows that the yield of cleavage products increased to 70% with a corresponding increase in mole equivalents of MeMgCl, but then decreased with a further increase in the mole equivalents of MeMgCl. This decrease resulted in the formation of methyl derivatives of 2,2-dichloro-1,3,5-trisilapentane. A similar dependance in these molecular enlargement reactions was observed, although not to the same extent, in reactions of 2,2-dichloro-1,3-disilapropane with increasing mole equivalents of MeMgCl. In both cases this molecular enlargement led to the formation of isomeric methyl derivatives of 2,2-dichloro-1,3,5-trisilapentane, which differ considerably in the ratio of isomers obtained from the reaction of 2,2-dichloro-1,3,5-trisilapentane with MeMgCl. By employing an excess of MeMgCl these 2,2-dichloro-1,3,5-trisilapentanes react according to their individual reactivities to produce more extensively-methylated derivatives. If the degree of methylation in these products, and in the cleavage products is sufficiently high, subsequent reactions occur converting CCl_2 groups to CHCl or CH_2 groups [94].

3.2 Si- and C-Chlorinated Carbosilanes

3.2.1 Reaction of $(Cl_3Si)_2CCl_2$

In Chapt. III.2.1 the formation of Si- and C-chlorinated compounds was discussed. The simplest representative of this class of compounds is $(Cl_3Si)_2CCl_2$. Reactions between $(Cl_3Si)_2CCl_2$ and an excess of MeMgCl, calculated with respect to all Cl atoms in the molecule, were undertaken in the initial stages of the investigations. This was mainly because the chemical behavior of this smallest functional carbosilane containing one CCl_2 group is obviously not straightforward, as was first thought, at least in comparison to $(Cl_3Si-CCl_2)_2SiCl_2$ and $(Cl_2Si-CCl_2)_3$. Reactions of this carbosilane led to the formation of the following compounds [111]:

$$(Cl_3Si)_2CCl_2 \xrightarrow[Et_2O]{MeMgCl} \begin{array}{l} (Me_3Si)_2C=CH_2 \ (84.6\%); \ (Me_3Si)_2CHMe \ (2.1\%) \\ (Me_3Si)_2CMe_2 \ (2.3\%); \ (Me_3Si)_2CHCl \ (2.2\%) \end{array}$$

Corresponding reactions with MeLi led to the same compounds, but in very different yields:

$$(Cl_3Si)_2CCl_2 \xrightarrow[Et_2O]{MeLi} \begin{array}{l} (Me_3Si)_2C=CH_2 \ (3.0\%); \ (Me_3Si)_2CH_2 \ (5.0\%) \\ (Me_3Si)_2CHMe \ (23\%); \ (Me_3Si)_2CMe_2 \ (25\%) \\ (Me_3Si)_2CHCl \ (28\%) \end{array}$$

To answer the question whether reaction between $(Cl_3Si)_2CCl_2$ and MeMgCl begins on the SiCl group or the CCl group, reactions had to be undertaken between $(Cl_3Si)_2CCl_2$ and differing mole ratios of MeMgCl. The yield of 1,3-disilapropanes still containing an unchanged CCl_2 group was 93% after reaction with 1 mole equivalent of MeMgCl, but decreased to about 75% with 2 mole equivalents and eventually to about 40% with 3 mole equivalents. This demonstrates clearly that the reaction begins with substitution of the SiCl group. Attack on the CCl_2 group becomes more significant with an increase in MeMgCl mole equivalence [112]. The yield of 1,3-disilapropanes having a bound CHMe group from the reaction with 3 mole equivalents of MeMgCl is approximately 18%, and with 4 mole equivalents of MeMgCl approximately 59%. However, this reduces to approximately 40% with 5 mole equivalents. At this stage of the reaction, methylation of the SiCl groups is not at all complete. The $>C=CH_2$ group appears for the first time in the product mixture from the reaction utilizing 4 mole equivalents of MeMgCl in 4% yield, increasing to 16% by reaction with 6 mole equivalents of MeMgCl. Using 7 mole equivalents, it rises to approximately 23%. Overall, 55 different 1,3-disilapropanes produced in such reactions were identified and could be classified into 8 different groups containing CCl_2, CClMe, $>C=CHMe$ CHCl. CMe_2, $>C=CH_2$, CH_2 and CHMe bridges.

The reaction of $(Cl_3Si)_2CCl_2$ with MeLi is now clearly understood. It begins with the methylation of the CCl group

$$(Cl_3Si)_2CCl_2 + MeLi \rightarrow (Cl_3Si)_2CMeCl + LiCl$$

so that by using 1 to 1.5 mole equivalents of MeLi, $(Cl_3Si)_2CMeCl$ in 93% yield could be isolated from the reaction mixture. Similarly, $(Cl_3Si)_2CMe_2$ was isolated in 90% yield using 2 mole equivalents of MeLi. Only after this reaction was completed did methylation of silicon begin. Thus, by reaction of $(Cl_3Si)_2CCl_2$ with 4 mole equivalents of MeMgCl, $(ClMe_2Si)_2CMe_2$ is found in the reaction mixture in 86% yield [112].

3.2.2 Mechanism of Formation of Methylidene Groups

In reactions with MeMgCl, the formation of methylidene groups has always been observed when the starting material contains an isolated $Si-CCl_2-Si$ group. For example, the methylidene group is formed in reactions of 2,2-dichloro-1,3,5-trisila-cyclohexane and $Me_3Si-CCl_2-SiCl_2-C\equiv C-SiMe_3$ [101]. The mechanism of formation of the methylidene group was thoroughly elucidated in reactions of $(Cl_3Si)_2CCl_2$ and its derivatives. Regarding the formation of intermediates in this mechanism, only compounds containing CCl_2 or CMeCl groups were considered, as the reactivity of other occurring intermediates was too low. In this investigation, reactions were induced using different Si-methylated and chlorinated 2,2-dichloro-1,3-disilapropanes, as well as varying Si-methylated and chlorinated 2-methyl-2-chloro-1,3-disilapropanes, which were especially prepared and purified for this purpose [113, 114].

3.2.2.1 $(Me_3Si)_2CMeCl$

In the initial stages of the investigation, $(Me_3Si)_2CMeCl$ was thought to be the immediate precursor of $(Me_3Si)_2C=CH_2$ because compounds containing the methylidene group were isolated only in cases where the silicon atom is fully methylated or contained at the most one SiCl group. Two mechanisms were considered regarding the formation of methylidene groups.

1. The organometallic reagent MeMgCl abstracts a proton from the C-methyl group and causes the elimination of the Cl substituent from the β-carbon, thus forming a double bond.

$$\equiv Si \quad \diagdown C \diagup \quad \overset{Me}{\underset{Cl}{}} + \overset{\ominus}{Me}\overset{\oplus}{Mg}Cl \rightarrow \equiv Si \quad \diagdown C = CH_2 + CH_4 + MgCl_2$$

2. A transmetallation occurs, followed by the elimination of HMgCl:

$$\equiv Si \diagdown C \diagup \overset{Me}{\underset{Cl}{}} \xrightarrow[-MeCl]{+MeMgCl} \left[\equiv Si \diagdown C \diagup \overset{Me}{\underset{MgCl}{}} \right] \rightarrow \equiv Si \diagdown C = CH_2 + HMgCl \, .$$

Metal transfer reactions have already been confirmed in association with carbosilanes [115], as well as the elimination of hydrides from Grignard compounds [116]. Attempts to bring about reaction of $(Me_3Si)_2CMeCl$ with MeMgCl in Et_2O and n-butyl ether proved unsuccessful. The reaction of $(Me_3Si)_2CMeCl$ with an excess of MeLi in Et_2O or THF produced only a small amount of $(Me_3Si)_2CMe_2$. This can be attributed to the failure of the organometallic reagents to cause deprotonation and metal transfer. This means that an elimination reaction necessary for the formation of a methylidene group in all probability will not occur, because the protons on the C-methyl group are not sufficiently acidic.

The metallation which did not succeed using MeMgCl or MeLi could in fact be achieved using lithium or magnesium. Reaction of $(Me_3Si)_2CMeCl$ with lithium in cyclohexane or benzene proceeds quantitatively, based on the amount of $(Me_3Si)_2CMeCl$ used, to produce $(Me_3Si)_2C=CH_2$ and $(Me_3Si)_2CHMe$ in a ratio of 55:45, which is explained by the formation of a lithiated intermediate:

$$\equiv Si \diagdown C \diagup \overset{Me}{\underset{Cl}{}} \xrightarrow[-LiCl]{+2Li} \left[\equiv Si \diagdown C \diagup \overset{Me}{\underset{Li}{}} \right] \rightarrow \equiv Si \diagdown C = CH_2 + LiH$$

$$\equiv Si \diagdown C \diagup \overset{Me}{\underset{Cl}{}} + LiH \rightarrow \equiv Si \diagdown C \diagup \overset{Me}{\underset{H}{}} + LiCl$$

Using magnesium in Et_2O, the reaction proceeds to $(Me_3Si)_2C=CH_2$ <u>262</u> and $(Me_3Si)_2CHMe$ <u>263</u> in a ratio of 43:57. This is explained by means of the reaction scheme:

$$\left[\equiv Si \diagdown C \diagup \overset{Me}{\underset{MgCl}{}} \right] \begin{cases} \xrightarrow{K_1} \equiv Si \diagdown C = CH_2 + HMgCl \quad &(a) \\ \xrightarrow{K_2} \equiv Si \diagdown C \diagup \overset{Me}{\underset{H}{}} \quad \text{ether cleavage} \end{cases}$$

$$\equiv Si \diagdown C \diagup \overset{Me}{\underset{Cl}{}} + HMgCl \rightarrow \equiv Si \diagdown C \diagup \overset{Me}{\underset{H}{}} + MgCl_2 \, . \quad (b)$$

Analogous to the reaction with lithium, the formation of an MgCl-containing intermediate occurs initially, followed by the elimination of HMgCl. This reduces the unreacted starting material $(Me_3Si)_2CMeCl$ to $(Me_3Si)_2CHMe$. Paralleling this is an ether cleavage resulting also in the formation of $(Me_3Si)_2CHMe$.

It can be concluded from these experimental results that reaction (a) described by the reaction rate K_1, together with reaction (b), must be three times as fast as ether cleavage which proceeds with a rate constant K_2. This emphasizes a large difference in comparison to reactions performed in aprotic solvents such as cyclohexane and benzene. The mechanism of Eqs. (a) and (b) is also supported by the reaction of $(Me_3Si)_2CCl(CD_3)$ with magnesium in Et_2O. The formation of $(Me_3Si)_2C=CD_2$, via DMgCl elimination, should occur analogous to Eq. (a). According to Eq. (b), reduction of the reactant $(Me_3Si)_2C(CD_3)Cl$ with DMgCl produces $(Me_3Si)_2CD(CD_3)$, and through ether cleavage of the intermediate $(Me_3Si)_2C(CD_3)(MgCl)$, $(Me_3Si)_2CH(CD_3)$ arises. The compounds $(Me_3Si)_2C=CD_2$, $(Me_3Si)_2CD(CD_3)$ and $(Me_3Si)_2CH(CD_3)$ are produced in the expected ratio of 3:3:1. The actual undertaking of this reaction, as well as the identification of the reaction products from the reaction of $(Me_3Si)_2CCl(CD_3)$, confirms this mechanism [113].

3.2.2.2 Varying Si-Chlorinated 2-Methyl-2-chloro-1,3-disilapropanes

While no reaction occurs between $(Me_3Si)_2CMeCl$ and MeMgCl, and formation of $(Me_3Si)_2C=CH_2$ 262 and $(Me_3Si)_2CMeH$ 263 is achieved only by using magnesium, Table 49 below shows that Si-chlorinated derivatives are capable of forming 262 and 263 on reaction with MeMgCl [113].

While $Me_3Si-CMeCl-SiMe_2Cl$ reacts to produce $(Me_3Si)_2C=CH_2$ and $(Me_3Si)_2CMeH$ in 10% yield, $(ClMe_2Si)_2CMeCl$ appears decidedly more reactive than this. After 40 hours reaction with an excess of MeMgCl (calculated with respect

Table 49. Reaction of 2-methyl-2-chloro-1,3-disilapropanes with MeMgCl

Starting material	Products from the reaction of:			
$Me_3Si-CMeCl-SiMe_2Cl$ 246	246: MeMgCl = 1:3 in Et_2O	%		
	246	91		
	$(Me_3Si)_2C=CH_2$	6		
	$(Me_3Si)CMeH$	3		
$ClMe_2Si-CMeCl-SiMe_2Cl$ 247	247 + MeMgCl (excess) in Et_2O	%	247 + MeMgCl (excess) in THF	%
	247	45		
	$(Me_3Si)_2C=CH_2$	12	$(Me_3Si)_2CMe_2$	25
	$(Me_3Si)_2CMeH$	13	$Me_3Si-CMe_2-SiMe_2Cl$	75
	$Me_3Si-CMeCl-SiMe_2Cl$	25		
$Cl_2MeSi-CMeCl-SiCl_3$ 248	248: MeMgCl = 1:2 in Et_2O	%	Prod. mixt. (reaction 1:2) + MeMgCl (excess) in Et_2O	%
	248	57	$(Me_3Si)_2C=CH_2$	68
	$(Me_3Si)_2C=CH_2$	28	$(Me_3Si)_2CHMe$	30
	$(\equiv Si)_2CHMe$	15		
	$(\equiv Si)_2CClMe$			

to all Cl atoms present), 262 and 263 were isolated in 12 and 13% yield respectively. Moreover, $Me_3Si-CMeCl-SiMe_2Cl$ is produced in 25% yield. This indicates that C-metallation and Si-methylation reactions still occur simultaneously. The compound $Cl_2MeSi-CMeCl-SiCl_3$ is still more reactive, and proceeds with an excess of MeMgCl exclusively to the formation of $(Me_3Si)_2C=CH_2$ and $(Me_3Si)_2CHMe$ in 68% and 30% yield respectively. This proves that formation of the methylidene group is favored in reactions of highly Si-chlorinated 2-methyl-2-chloro-1,3-disilapropanes. If this result is carried over to the mechanism pertaining to Eq. (a), a more effective formation of the Grignard compound $(\geqslant Si)_2CMe(MgCl)$ is observed if the negative partial charge on the C atom is stabilized even further by electronegative substituents on the Si atoms. Chloro substitution on silicon generally stabilizes the carbanionic intermediate through delocalization of negative charge over the molecule. Correspondingly, the amount of 262 produced increases from the reaction of $Me_3Si-CMeCl-SiMe_2Cl$ 246 with MeMgCl, through $ClMe_2Si-CMeCl-SiMe_2Cl$ 247 to $Cl_2MeSi-CMeCl-SiCl_3$ 248.

From the isolated products, one cannot determine which level of substitution on silicon is decisive for a transmetallation reaction to take place. However, if the formation of the methylidene group is completed by elimination of HMgCl, subsequent methylation of the remaining SiCl group will be so rapid that formation of SiH groups will not occur. Through the influence of the $C=C$ double bond on the Si atoms, an increased reactivity of the Si—Cl bond becomes apparent.

3.2.2.3 Reactions of 2,2-Dichloro-1,3-disilapropanes

For an accurate assessment of the reaction of $(Cl_3Si)_2CCl_2$ with MeMgCl, an investigation into the behavior of varying Si-methylated and Si-chlorinated derivatives is of significance, because this allows for a wider perspective to be gained over different reaction products. Table 50 presents the results of the reaction of $(Me_3Si)_2CCl_2$, which was chosen because no complications attributable to SiCl groups can here arise.

It follows that reaction of $(Me_3Si)_2CCl_2$ 264 in Et_2O under otherwise similar conditions is strongly influenced by reaction temperature. Mixing the reagents at

Table 50. Reaction of $(Me_3Si)_2CCl_2$ 264 with MeMgCl (four mole equivalents excess) in Et_2O at varying temperatures

	1	2	3
Conditions	264 dissolved in Et_2O, MeMgCl added at −20 °C, warmed to 20 °C, work-up after 20 hrs.	264 dissolved in Et_2O, MeMgCl added at 0 °C, warmed to 20 °C, work-up after 20 hrs.	264 dissolved in Et_2O, MeMgCl added at 0 °C, 36 hrs. reflux before work-up
		%	%	%
Products	264	88	66	0
	$(Me_3Si)_2C=CH_2$	8	24	75
	$(Me_3Si)_2CHCl$	4	8	17
	$(Me_3Si)_2CHMe$	—	—	3
	$(Me_3Si)_2CMeCl$	—	—	5

0 °C followed by heating under reflux of the solvent causes quantitative reaction of 264 to produce $(Me_3Si)_2C=CH_2$ as the main product [113]. Hence it was proved that the formation of the methylidene group is not restricted to the initial formation of a CMeCl group.

Starting from $(Me_3Si)_2CCl_2$ it is much easier to produce the intermediate $(Me_3Si)_2CMe(MgCl)$ than from $(Me_3Si)_2CMeCl$. In the latter case, no reaction with MeMgCl could be induced. Therefore, the reaction course from $(Me_3Si)_2CCl_2$ does not begin by methylation of the bridging carbon atom, succeeded by metallation of the bridging carbon atom. Rather it must occur in the opposite sequence:

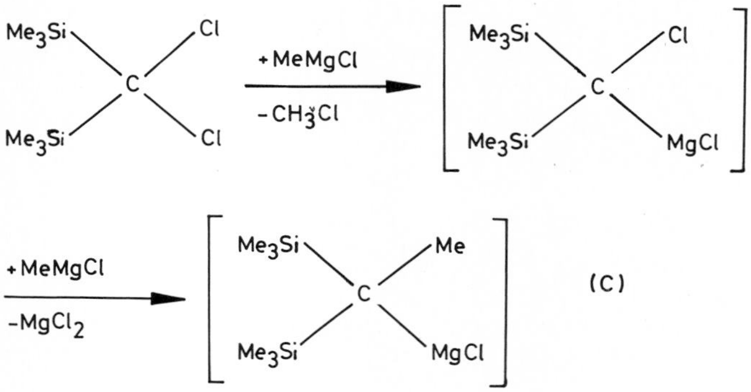

All of the present reaction products can be explained by considering this reaction pathway. The starting material $(Me_3Si)_2CCl_2$ will be methylated only to a totally subordinate extent to produce $(Me_3Si)_2CMeCl$. Nevertheless, the metallation reaction corresponding to Eq. (c) occurs much faster than C-methylation. A more effective stabilization of the carbanionic intermediate is responsible for an easier synthesis of $(Me_3Si)_2CMe(MgCl)$ from $(Me_3Si)_2CCl_2$ according to Eq. (c) than from $(Me_3Si)_2CMeCl$.

A number of reaction products arising from reaction of 2,2-dichloro-1,3-disilapropane with MeMgCl are found to contain a $(\equiv Si)_2CHCl$ group.

This can be obtained also through ether cleavage as well as through reduction of the CCl_2 group using HMgCl. That a CHCl group arises not only through ether cleavage but also through hydride transfer via HMgCl was demonstrated by reaction of $Me_3Si-CCl_2-SiMe_2Cl$ with MeMgCl in $(C_2D_5)_2O$. Here MeMgCl was produced in $(C_2D_5)_2O$ and added at 20 °C to a solution of $Me_3Si-CCl_2-SiMe_2Cl$. From the ^1H-NMR spectrum of the reaction products, the formation of $(Me_3Si)_2CHCl$ and $Me_3Si-CHCl-SiMe_2Cl$ was evident. In proton-free solvents the formation of this CHCl group can only occur through a reduction process involving HMgCl.

Table 51 presents the results obtained from the reaction of varying Si-methylated and Si-chlorinated 2,2-dichloro-1,3-disilapropanes with MeMgCl. The total reaction course is determined by two reactions, namely, through Si-methylation and C-metallation. The extent to which one process predominates over the other depends solely on the degree of Si-chlorination. When the 2,2-dichloro-1,3-disilapropanes are only slightly chlorinated on Si, C-metallation prevails, but when the degree of Si-chlorination is high, Si-methylation occurs. Hence all compounds derived from

$Me_3Si-CCl_2-SiMe_2Cl$ and $Me_3Si-CCl_2-SiMeCl_2$ arise through attack on the bridging $\overset{\cdot}{C}$ atom. On the other hand, the yield of Si-methylation products is only 20% from $Me_2ClSi-CCl_2-SiMe_2Cl$, 23% from $ClMe_2Si-CCl_2-SiMeCl_2$, and

Table 51. Reaction of 2,2-Dichloro-1,3-disilapropanes with MeMgCl in Et_2O

Starting material	Reaction products in % yield (starting material : MeMgCl)	1:1 %	1:3 %	1:5 %
$Me_3Si-CCl_2-SiMe_2Cl$	$(Me_3Si)_2C=CH_2$	11	25	66
	$Me_3Si-CHCl-SiMe_3$	1	12	33
	$Me_3Si-CHCl-SiMe_2Cl$	19	26	—
	$Me_3Si-CMeCl-SiMe_2Cl$	36	33	—
	Starting material	31	—	—
	unidentified	—	4	
$Me_3Si-CCl_2-SiMeCl_2$	$(Me_3Si)_2C=CH_2$	8	23	55
	$Me_3Si-CHCl-SiMe_3$	1	13	38
	$Me_3Si-CHCl-SiMeCl_2$	17	—	—
	$Me_3Si-CHCl-SiMe_2Cl$	—	33	—
	$Me_3Si-CMeCl-SiMeCl_2$	23	20	—
	Starting material	48	—	—
	unidentified	3	—	—
	$Me_3Si-CMe_2-SiMeCl_2$	—	8	
$ClMe_2Si-CCl_2-SiMe_2Cl$	$(Me_3Si)_2C=CH_2$	13	21[a]	
	$Me_3Si-CHCl-SiMe_3$	2	8	
	$Me_3Si-CHCl-SiMe_2Cl$	6	—	
	$Me_3Si-CCl_2-SiMe_2Cl$	20	—	
	Starting material	56	—	
	unidentified	3	—	
	$Me_3Si-CHMe-SiMe_3$	—	5	
	$Me_3Si-CHMe-SiMe_2Cl$	—	25	
	$Me_3Si-CMe_2-SiMe_3$	—	3	
	$ClMe_2Si-CMe_2-SiMe_2Cl$	—	32	
$ClMe_2Si-CCl_2-SiMeCl_2$	$(Me_3Si)_2C=CH_2$	13	54	
	$Me_3Si-CHCl-SiMe_3$	5	7	
	$ClMe_2Si-CHCl-SiMeCl_2$	7	—	
	$ClMe_2Si-CMeCl-SiMeCl_2$	1	—	
	$Me_3Si-CCl_2-SiMeCl_2$	17	—	
	$ClMe_2Si-CCl_2-SiMe_2Cl$	6	—	
	Starting material	47	—	
	unidentified	4	—	
	$Me_3Si-CHCl-SiMe_2Cl$	—	11	
	$Me_3Si-CHCl-SiMeCl_2$	—	14	
	$ClMe_2Si-CMeCl-SiMe_2Cl$	—	10	
	$Me_3Si-C(=CH_2)-SiMe_2Cl$	—	4	
$MeCl_2Si-CCl_2-SiCl_3$	$ClMe_2Si-CCl_2-SiCl_3$	20		
	$Cl_2MeSi-CCl_2-SiMeCl_2$	20		
	$ClMe_2Si-CHCl-SiCl_3$	5		
	$Cl_2MeSi-CMeCl-SiCl_3$	10		
	Starting material	45		

[a] Reaction time 60 hrs.

40% from $MeCl_2Si—CCl_2—SiCl_3$. The metallation and subsequent formation of methylidene groups, as well as the formation of CHCl groups, occurs considerably faster by incorporating more chlorine on silicon [113].

3.2.3 Reaction Pathways Taken by $(Cl_3Si)_2CCl_2$

Information on the influence of Si-substituents on the reaction course arises from the previously-mentioned reactions of varying Si-methylated 2,2-dichloro- and 2-methyl-2-chloro-1,3-disilapropanes. Because $(Cl_3Si)_2CCl_2$ contains functional groups which enable nucleophilic attack on silicon and carbon sites, as well as electrophilic attack on carbon, one has to expect Si- and C-methylation in the initial reaction stages, as well as C-metallation followed by a number of consecutive reactions. Because MeMgCl does not exist as a monomer, but rather as a dimer or even partly as an oligomer [117], the first reaction step in the Si-methylation to produce $MeCl_2Si—$ $—CCl_2—SiCl_3$ is kinetically controlled, and will be influenced by the steric requirements of the two reactants. One or two methyl groups are then attached to the Si atoms of $(Cl_3Si)_2CCl_2$, which makes further nucleophilic attack through RMgX unfavorable. This is attributed to the $+I$ effect of the methyl groups and not so much to its steric requirements [118]. For this reason it is possible to achieve consecutive C-metallation and C-methylation besides further Si-methylation. Compounds containing a CMeCl group, as well as those containing a CHCl group due to ether cleavage of a metallated intermediate, will occur as a result. By increasing the number of methyl groups on silicon, for example until $Me_3Si—CCl_2—SiMe_2Cl$ is achieved, one observes almost exclusively only attack on carbon, inspite of the remaining SiCl group. The electronic conditions on silicon are then so unfavorable that the steric disadvantages incurred in attacking carbon will be overcome, hence inducing attack of the CCl_2 group. That the C—Cl bond in these compounds is more reactive than the Si—Cl bond is best demonstrated by $(Cl_3Si)_2CCl_2$, which reacts with MeLi producing $(Cl_3Si)_2CMeCl$.

In an extension of the reaction pathway, intermediates containing a CMeCl or CCl_2 group can react further, producing a methylidene group as previously shown. This is preferred if the degree of chlorination on silicon is higher, mainly because this leads to a greater stabilization of the carbanions formed. Due to the elimination of HMgCl in the formation of a methylidene group, a reagent appears in the reaction mixture that causes reduction of CCl-containing compounds to form the corresponding CH-containing compounds. The formation of compounds containing CHCl and CHMe groups is explained equally as well through reduction with HMgCl as by ether cleavage of the corresponding Grignard compounds. The formation of compounds containing a CMe_2 group through a second methylation on carbon is also observed, but plays only a subordinate role in the reaction course.

On the other hand, the changed substitution on carbon has an influence on the substituents bound to silicon. For instance, only those methylidene-containing compounds were isolated that were fully methylated on Si or contained one Si—Cl group. The fact that $(Me_3Si)_2CCl_2$ reacts only slowly to form $(Me_3Si)_2C=CH_2$, and the complete failure of $(Me_3Si)_2CMeCl$ to react with MeMgCl to form $(Me_3Si)_2C=CH_2$, implies that double-bond formation occurs at the stage of higher Si-chlorination. The $>C=CH_2$ group then favors Si-methylation.

The reaction course cannot be explained by a statistical consideration. Attack on the Si—Cl group of $(Cl_3Si)_2CCl_2$ should in all probability be favored, while attack on the C atom of $Me_3Si—CCl_2—SiMe_2Cl$ would more commonly occur. All experimental findings rule out such a statistical interpretation of the reaction pathway: the exclusive C-methylation by MeLi in Et_2O, and also the reactions of MeMgCl and MeLi with the higher homologues $(Cl_3Si—CCl_2)_2SiCl_2$ and $(Cl_2Si—CCl_2)_3$. Similarly, the reactions of $Me_3Si—CCl_2—SiMe_2Cl$ and $Me_3Si—CCl_2—SiMe_3$ cannot be suitably explained.

3.2.4 Reactions of 1,3,5-Trisilapentanes

3.2.4.1 $(Cl_3Si—CCl_2)_2SiCl_2$ and MeMgCl

A considerable insight into the reaction between $(Cl_3Si—CCl_2)_2SiCl_2$ <u>265</u> and MeMgCl can be gained by carrying out this reaction in mole ratios of 1:18, 1:8, 1:4 and 1:1. The compounds formed and isolated are presented in Table 52 [121].

Table 52. Reaction products from the reaction of $(Cl_3Si—CCl_2)_2SiCl_2$ <u>265</u> with MeMgCl

Mole ratio	Compound No.		% in product mixture
1:18	<u>266</u>	$Me_3Si—C\equiv C—SiMe_3$	47
		$SiMe_4$	>30[a]
	<u>267</u>	$H_2C=C\begin{smallmatrix}CH(SiMe_3)_2\\SiMe_3\end{smallmatrix}$	24
	<u>268</u>	$\begin{smallmatrix}Me_3Si\\Me_3Si\end{smallmatrix}C=C\begin{smallmatrix}SiMe_3\\H\end{smallmatrix}$	6
	<u>269</u>	$Me_3Si—\underset{\overset{\|}{CH_2}}{C}—SiMe_2—\underset{\overset{\|}{CH_2}}{C}—SiMe_3$	5
1:8		$Me_3Si—C\equiv C—SiMe_3$	18
		$SiMe_4$	>10[a]
		<u>267</u>, <u>268</u> only traces	
1:4		$Me_3Si—C\equiv C—SiMe_3$	2
		$SiMe_4$ could not be proved	
		Me_3SiCl and Me_2SiCl_2 traces	
1:1		$Cl_3Si—CCl_2—SiCl_2—CHCl—SiCl_3$	

[a] Value too small due to work-up procedure

With an excess of MeMgCl, $Me_3Si—C\equiv C—SiMe_3$ appeared as the main product, along with $SiMe_4$. This Si—C bond cleavage, plainly evident from the formation of $Me_3Si—C\equiv C—SiMe_3$, appears diminished as the amount of MeMgCl used becomes less, and is eventually absent at a 1:1 molar ratio. In this cases the products recognized were $Cl_3Si—CCl_2—SiCl_2—CHCl—SiCl_3$ and Si-methylated derivatives of $(Cl_3Si—CCl_2)_2SiCl_2$ [121].

Table 53. Mechanism of the reaction of $(Cl_3Si-CCl_2)_2SiCl_2$ with MeMgCl

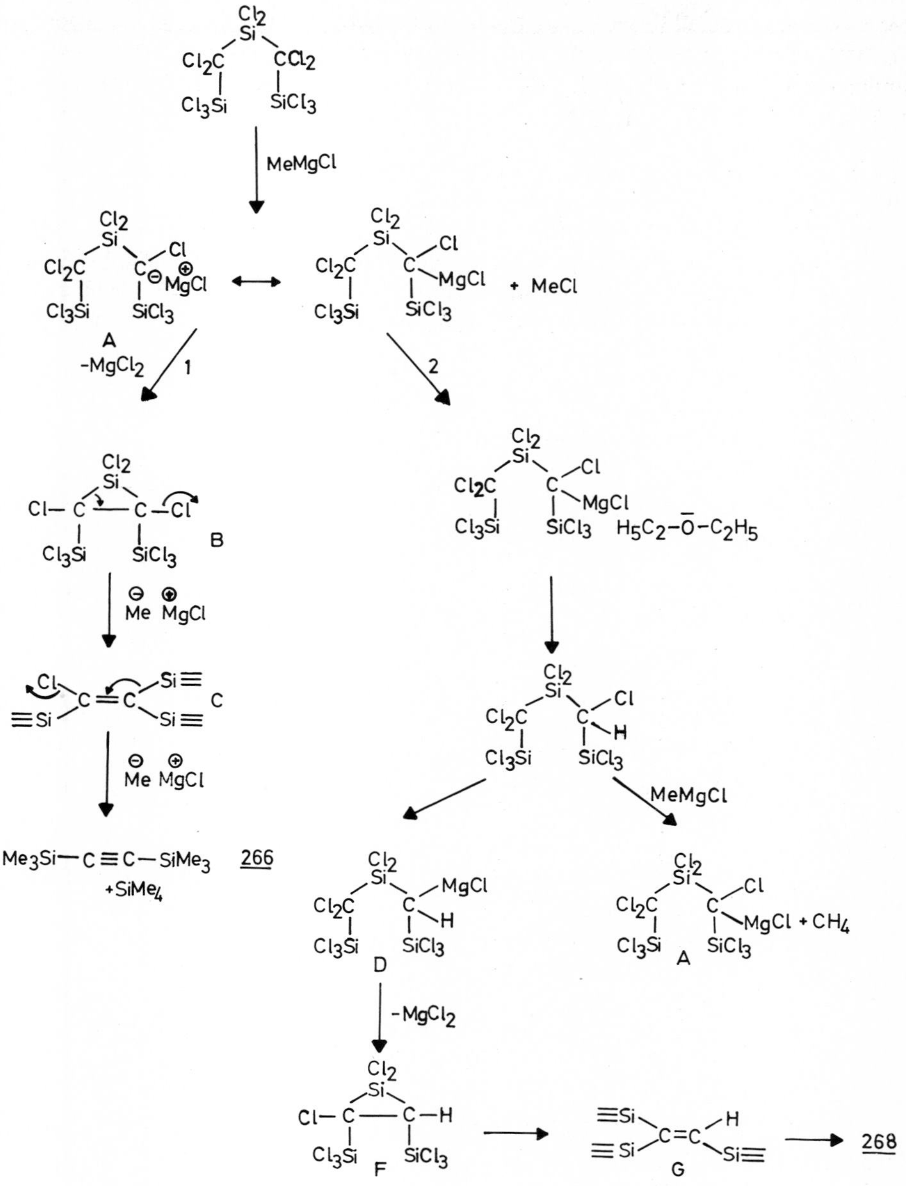

$$
\begin{array}{c}
Cl_2 \\
Si \\
Cl_2C \qquad CCl_2 \\
| \qquad\quad | \\
Cl_3Si \qquad SiCl_3
\end{array}
$$

$\downarrow$ MeMgCl

$$
\begin{array}{c}
Cl_2 \\
Si \qquad Cl \\
Cl_2C \qquad C\overset{\ominus}{}\overset{\oplus}{MgCl} \\
| \qquad\quad | \\
Cl_3Si \qquad SiCl_3 \\
A \\
-MgCl_2
\end{array}
\quad\longleftrightarrow\quad
\begin{array}{c}
Cl_2 \\
Si \qquad Cl \\
Cl_2C \qquad C\!-\!MgCl \\
| \qquad\quad | \\
Cl_3Si \qquad SiCl_3
\end{array}
\;+\; MeCl
$$

1 / 2

Route 1 (A, B, C):

$$
\begin{array}{c}
Cl_2 \\
Si \\
Cl\!-\!C\!-\!C\!-\!Cl \\
| \qquad | \\
Cl_3Si \qquad SiCl_3 \qquad B
\end{array}
$$

$\downarrow\; \overset{\ominus}{Me}\ \overset{\oplus}{MgCl}$

$$
\begin{array}{c}
Cl \\
\equiv Si \diagdown \quad \diagup Si\equiv \\
C=C \qquad C \\
\diagup \qquad \diagdown Si\equiv
\end{array}
$$

$\downarrow\; \overset{\ominus}{Me}\ \overset{\oplus}{MgCl}$

$$Me_3Si-C\equiv C-SiMe_3 \quad \underline{266}$$
$$+ SiMe_4$$

Route 2:

$$
\begin{array}{c}
Cl_2 \\
Si \qquad Cl \\
Cl_2C \qquad C\!-\!MgCl \\
| \qquad\quad | \\
Cl_3Si \qquad SiCl_3 \quad H_5C_2-\bar{O}-C_2H_5
\end{array}
$$

$\downarrow$

$$
\begin{array}{c}
Cl_2 \\
Si \qquad Cl \\
Cl_2C \qquad C\!-\!H \\
| \qquad\quad | \\
Cl_3Si \qquad SiCl_3
\end{array}
$$

$\swarrow \qquad\qquad \searrow$ MeMgCl

$$
\begin{array}{c}
Cl_2 \\
Si \qquad MgCl \\
Cl_2C \qquad C\!-\!H \\
| \qquad\quad | \\
Cl_3Si \qquad SiCl_3 \\
D
\end{array}
\qquad
\begin{array}{c}
Cl_2 \\
Si \qquad Cl \\
Cl_2C \qquad C\!-\!MgCl \;+\; CH_4 \\
| \qquad\quad | \\
Cl_3Si \qquad SiCl_3 \\
A
\end{array}
$$

$\downarrow\; -MgCl_2$

$$
\begin{array}{c}
Cl_2 \\
Si \\
Cl\!-\!C\!-\!C\!-\!H \\
| \qquad | \\
Cl_3Si \qquad SiCl_3 \\
F
\end{array}
\quad\longrightarrow\quad
\begin{array}{c}
\equiv Si \diagdown \quad \diagup H \\
C=C \\
\equiv Si \diagup \quad \diagdown Si\equiv \\
G
\end{array}
\quad\longrightarrow\quad \underline{268}
$$

In order to establish what influence the degree of C-chlorination of various 1,3,5-trisilapentanes has on the reaction pathway, the reactions of $Cl_3Si-CCl_2-$ $-SiCl_2-CH_2-SiCl_3$ and of $Cl_3Si-CCl_2-SiCl_2-CHCl-SiCl_3$ with an excess of MeMgCl were investigated. The amount of MeMgCl used was determined with respect to the total amount of Cl present. The reaction of $Cl_3Si-CCl_2-SiCl_2-$ $-CH_2-SiCl_3$ led to the development of CH_4 and C_2H_6 and to the formation of $Me_3Si-C(=CH_2)-SiMe_2-CH_2-SiMe_3$ in 70% yield. The reaction of Cl_3Si- $-CCl_2-SiCl_2-CHCl-SiCl_3$ with MeMgCl formed as main product $(Me_3Si)_2C=$ $=C(H)SiMe_3$, as well as $Me_3Si-CH_2-SiMe_2-CHCl-SiMe_3$. The number of CCl_2 groups in the 1,3,5-trisilapentane determines the course of the reaction. Table 53 illustrates the mechanism of this reaction [121].

The first reaction stage consists of a metallation. The resulting intermediate product A can lead to the formation of the compounds 266, $SiMe_4$, 267 and 268 in two different ways. Way 1 proceeds via abstraction of $MgCl_2$, through intermediate B to compound C. Under the influence of MeMgCl, this compound reacts by means of abstraction of a vinylic Cl atom to form $Me_3Si-C\equiv C-SiMe_3$ and $SiMe_4$. Reaction pathway 2 proceeds via intermediate product A through etheral coordination followed by cleavage to produce $Cl_3Si-CCl_2-SiCl_2-CHCl-SiCl_3$. This compound reacts further to produce the intermediate A and CH_4, or in another way through metallation to D. Compound D also reacts further via $MgCl_2$ abstraction and rearrangement through F to produce G. This compound reacts through methylation of Si—Cl groups to produce 268, which was isolated.

It appeared at first unusual to obtain $H_2C=C(SiMe_3)CH(SiMe_3)_2$ 267 (24% yield). Its formation is explained at the outset by the formation of compound C, which is converted to its methylated analogue E. This rearranges to yield 267 via a 1,3-isomerization.

$$\underset{Me_3Si}{\overset{Me_3Si}{>}}C=C\underset{Cl}{\overset{SiMe_3}{<}} + MeMgCl \rightarrow \underset{Me_3Si}{\overset{Me_3Si}{>}}C=C\underset{\underset{E}{Me}}{\overset{SiMe_3}{<}} + MgCl_2$$

There is evidence that E appears in the reaction mixture of $(Cl_3Si-CCl_2)_2SiCl_2$ with MeMgCl.

Decreasing the concentration of MeMgCl causes a corresponding decrease in cleavage reaction 1 and an increase in reaction 2 to form $Cl_3Si-CCl_2-SiCl_2-CHCl-$ $-SiCl_3$.

Reaction of $Cl_3Si-CCl_2-SiCl_2-CH_2-SiCl_3$ with MeMgCl leads to Me_3Si- $-C(=CH_2)-SiMe_2-CH_2-SiMe_3$, hence displaying a characteristic reaction of the CCl_2 group between two Si atoms, as already known from the reactions of $(Cl_3Si)_2CCl_2$. The reaction of $Cl_3Si-CCl_2-SiCl_2-CHCl-SiCl_3$ proceeds via rearrangement to produce $(Me_3Si)_2C=C(H)(SiMe_3)$. The reaction of $(Cl_3Si-CCl_2)_2SiCl_2$ with MeMgCl, producing $Me_3Si-C\equiv C-SiMe_3$ through cleavage, fails to appear. The reason lies in intermediate C, where no vinylic Cl atom is present, meaning the corresponding ethylene derivative must be stable.

3.2.4.2 $(Cl_3Si-CCl_2)_2SiCl_2$ and MeLi

Reaction of $(Cl_3Si-CCl_2)_2SiCl_2$ with an excess of MeLi produces the following compounds in considerable yields: $SiMe_4$, 267, 268, 270 and 271 [122].

SiMe$_4$ H$_2$C=C$\begin{smallmatrix}\diagup CH(SiMe_3)_2 \\ \diagdown SiMe_3\end{smallmatrix}$ $\begin{smallmatrix}Me_3Si \\ \diagup \\ Me_3Si\end{smallmatrix}$C=C$\begin{smallmatrix}\diagup SiMe_3 \\ \diagdown CH_3\end{smallmatrix}$

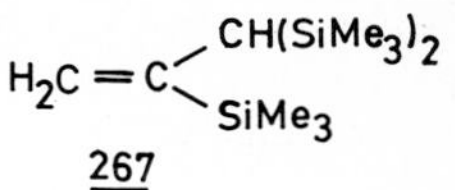

267 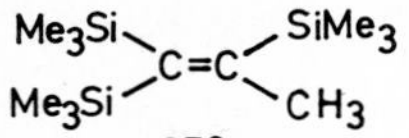   **270**

$\begin{smallmatrix}Me_3Si \\ \diagup \\ Me_3Si\end{smallmatrix}$C=C$\begin{smallmatrix}\diagup SiMe_3 \\ \diagdown H\end{smallmatrix}$

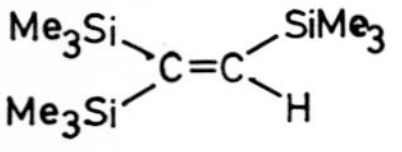

268 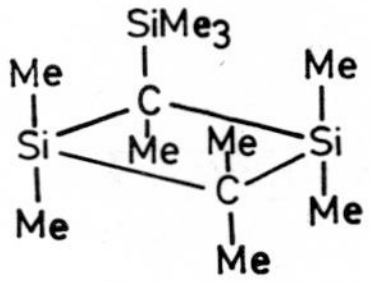   **271**

The byproducts of this reaction are <u>272</u>, <u>263</u> and <u>406</u>.

$$Me_3Si-CMe_2-SiMe_2-\underset{\underset{CH_2}{\parallel}}{C}-SiMe_3 \qquad Me_3Si-CHMe-SiMe_3$$

<u>272</u>   <u>263</u>

$$Me_3Si-CMe_2-SiMe_3$$

<u>406</u>

Compounds <u>263</u> and <u>406</u> are present as a result of chain degradation. Moreover, a certain amount of material of higher molecular weight is formed, but this was not identified. The formation of SiMe$_4$ and <u>270</u> is explained by consideration of the following reaction pathway, which corresponds to the reaction of $(Cl_3Si-CCl_2)_2SiCl_2$ with MeMgCl:

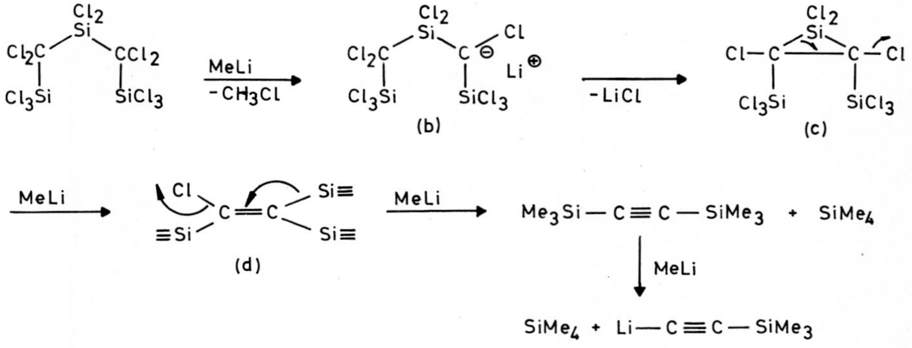

(b)   (c)

(d)

$$Me_3Si-C\equiv C-SiMe_3 + SiMe_4$$

$$\downarrow MeLi$$

$$SiMe_4 + Li-C\equiv C-SiMe_3$$

In the first reaction stage, a metal transfer occurs on the bridging C atom to (b). An intermolecular elimination of LiCl occurs, which proceeds to the transition stage (c), and subsequently rearranges; (d) reacts via β-elimination and Si-methylation yielding $Me_3Si-C\equiv C-SiMe_3$ and SiMe$_4$. Hence $Me_3Si-C\equiv C-SiMe_3$ reacts with an excess of MeLi according to the overall equation

$$Me_3Si-C\equiv C-SiMe_3 + 2\ MeLi \rightarrow 2\ SiMe_4 + LiC\equiv CLi$$

as the formation of HC≡CH through hydrolysis of the reaction products shows. The formation of <u>270</u> is also understandable according to this mechanism. The vinylic chlorine in the unisolated intermediate (d) can be methylated with MeLi to produce

$(Me_3Si)_2C=C(SiMe_3)Me$ 270. The formation of 268 is attributed to the initial hydrogenation of $(Cl_3Si—CCl_2)_2SiCl_2$. The reaction of $Cl_3Si—CHCl—SiCl_2—CCl_2—$ $—SiCl_3$ with MeMgCl proceeds exclusively to 268. An important difference in the chemical behavior of MeLi as opposed to MeMgCl towards $(Cl_3Si—CCl_2)_2SiCl_2$ lies in the formation of 271. The formation of this compound arises from stage (b) via the following reaction pathway:

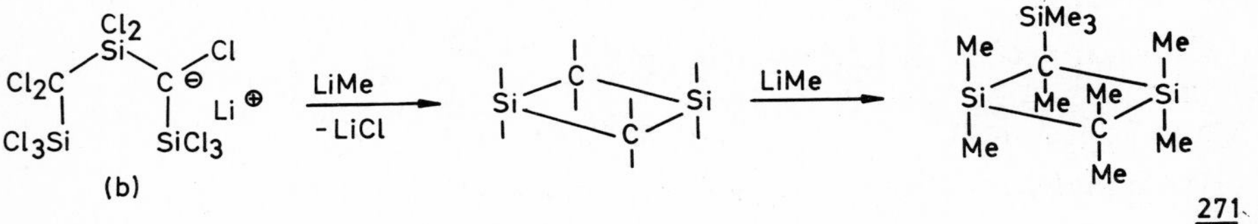

271

The carbanion of the lithium compound is obviously more reactive than the corresponding magnesium intermediate and will not only attack bridging C atoms, but also terminal Si atoms. The Si-methylation then produces 271. Compound 270 is rearranged in the presence of traces of water or acid, and will undergo 1,3-isomerization producing 267:

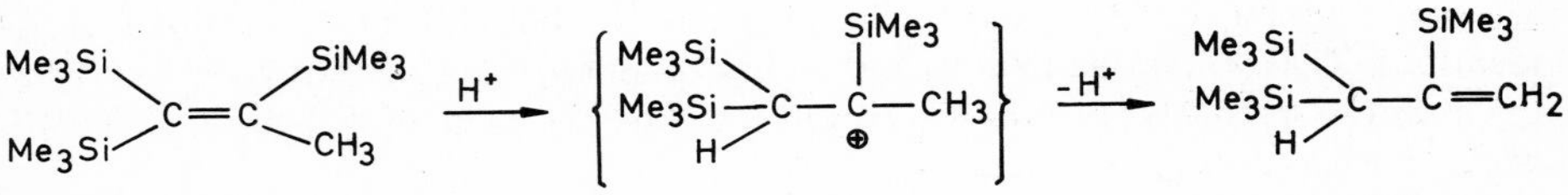

270   267

The proportions of the individual compounds in the reaction mixture are very dependant on the actual concentration of MeLi present during the course of reaction, as shown by Table 54 [122].

Regarding the reaction in which $(Cl_3Si—CCl_2)_2SiCl_2$ is added gradually to MeLi (Reaction A), reaction occurs instantaneously on mixing the starting materials. In the reverse reaction (B), Si-methylated compounds are also obtained but the major product present in an 80% yield is a viscous, brown polymeric mass. As a result of these observations it becomes apparent that by using lower concentrations of MeLi in the first stage of the reaction course, a condensation occurs after initial metallation:

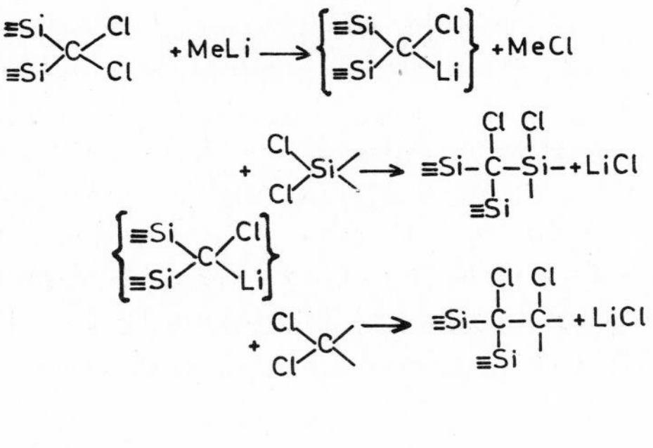

The behavior in Reaction B is such that a reduction in temperature causes a sharp reduction in the extent of polymer formation. In all reactions of $(Cl_3Si-CCl_2)_2SiCl_2$ where an excess of MeLi exists, $Me_3Si-C\equiv C-SiMe_3$ does not appear. On the contrary, by reaction of $(Cl_3Si-CCl_2)_2SiCl_2$ with MeLi in a mole ratio of 1:8, $Me_3Si-C\equiv C-SiMe_3$ is obtained in a yield of 11% of the total product. Under these conditions, the cleavage of $Me_3Si-C\equiv C-SiMe_3$ by MeLi fails to emerge [122].

3.3 Reactions of 1,3,5-Trisilacyclohexanes

3.3.1 Reactions of $(Cl_2Si-CCl_2)_3$ with MeMgCl

The fully chlorinated $(Cl_2Si-CCl_2)_3$ reacts with MeMgCl under ring contraction and ring cleavage to produce the following compounds [101, 115]:

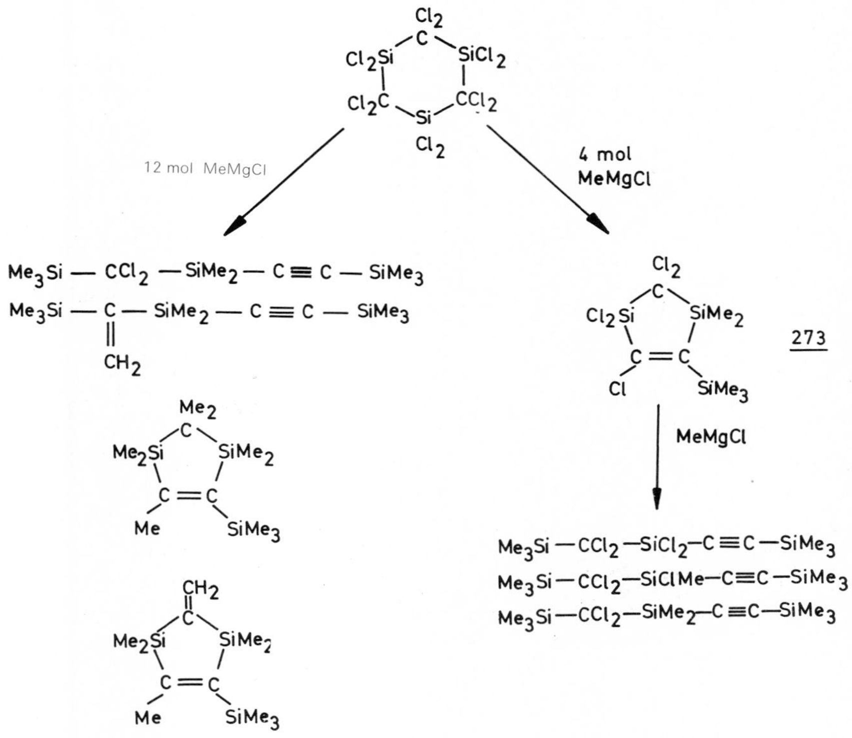

An excess of MeMgCl caused a preferred development of linear derivatives. The formation of 1,3-disilacyclopentene, 273, is obviously a deciding stage in the overall reaction path. It was shown that compounds with this molecular skeleton are formed in a 98% yield in the reaction of $(Cl_2Si-CCl_2)_3$ with 4 mole equivalents MeMgCl. Table 56 presents the reaction products obtained from the reaction of 1,3-disila-4-trimethylsilyl-cyclopentene 273 with MeMgCl.

Table 54. Distribution of isolated compounds among the reaction products from the reaction of $(Cl_3Si—CCl_2)_2SiCl_2$ 265 with MeLi (mole ratio 1:18)

Products	Reaction	A		B	
	Temperature	−20 °C	−78 °C	−20 °C	−78 °C
		Weight % of total products			
a) SiMe$_4$		23.6	11.5	14	14.7
b) Products condensable up to 70 °C/0.1 mm Hg		59	62.5	5.8	44.1
c) Residue		17.2	26.0	80.0	41.2

A: 265 added to MeLi;   B: MeLi added to 265

Table 55. Distribution of isolated compounds in the condensable fraction from the reaction of $(Cl_3Si—CCl_2)_2SiCl_2$ 265 with MeLi (cf. Table 54)

Compound	Reaction	A		B	
	Temperature	−20 °C	−78 °C	−20 °C	−78 °C
		Weight % of condensable fraction			
268		10	18	6	20
267		40	22	50	27
270 + 271		—	18	—	33
272		—	8	—	15
263 + 406		<10			
unidentified		40	24	34	5

A: 265 added to MeLi;   B: MeLi added to 265

The 1,3-disilacyclopentene 273 reacts further with MeMgCl to yield its acetylene derivatives, which are obtained directly from reaction of $(Cl_2Si—CCl_2)_3$ with excess MeMgCl, as shown in the mechanism

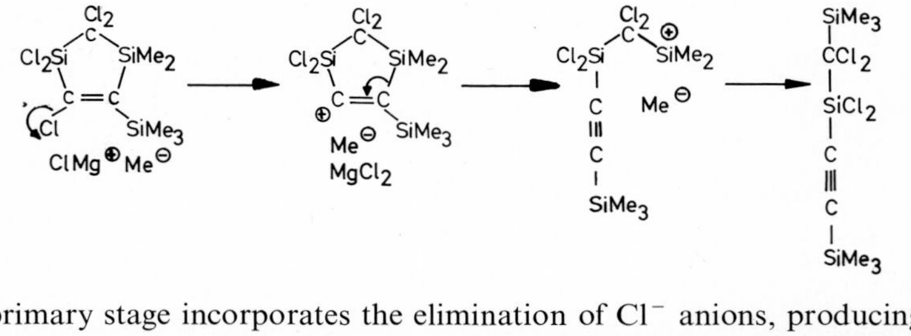

The primary stage incorporates the elimination of Cl^- anions, producing $MgCl_2$. The resulting cation undergoes rearrangement followed by nucleophilic addition by the methyl anion to produce the favored $—C\equiv C—$ group as well as Si-methylation. The cleavage reaction must originate from the vinylic chlorine substituent, since cleavage is not observed and only further methylation occurs on the C=C group of the methylated derivative 275. All further linear compounds produced are derived by reactions involving the functional groups of 274 [123].

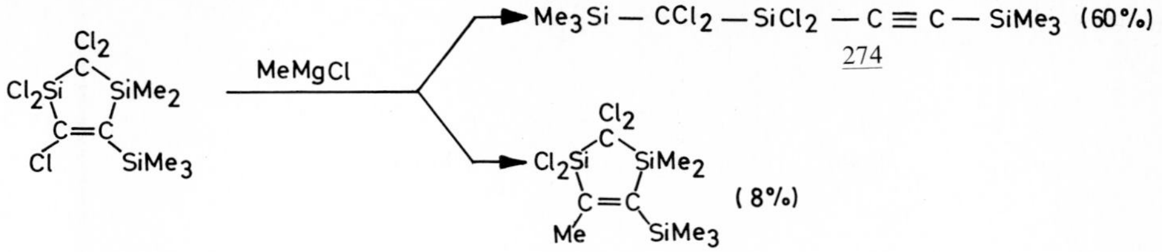

$Me_3Si - CCl_2 - SiCl_2 - C \equiv C - SiMe_3$ (60%)

274

(8%)

Table 56. Distribution of Compounds formed from the reaction of 273 with MeMgCl (1 and 3 mole equivalents)

No.	Compound formed	+1 mole equiv. of MeMgCl % yield[a]	+3 mole equiv. of MeMgCl % yield[a]
274	$Me_3Si - CCl_2 - SiCl_2 - C \equiv C - SiMe_3$	60	—
276	$Me_3Si - CCl_2 - SiClMe - C \equiv C - SiMe_3$	15	33
277	$Me_3Si - CCl_2 - SiMe_2 - C \equiv C - SiMe_3$	3	9
278	$Me_3Si - CHCl - SiMe_2 - C \equiv C - SiMe_3$	1	30
279	$Me_3Si - \overset{\parallel}{\underset{CH_2}{C}} - SiMe_2 - C \equiv C - SiMe_3$	1	10.5
275	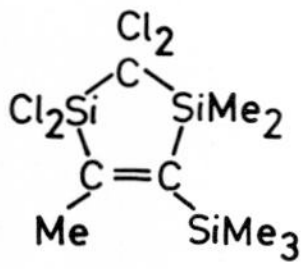	8	—
280	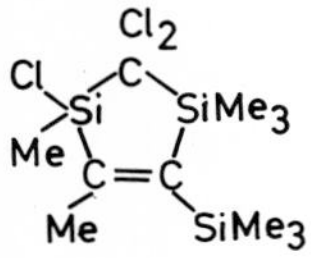	3.5	—
281	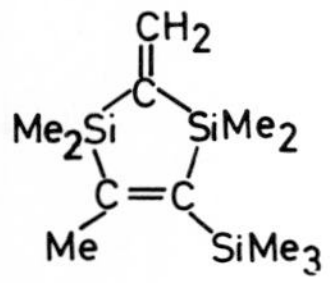	1	3

[a] Percentage yields refer only to distillation products obtained. Residue represents approximately 10%.

Depending on the amount of MeMgCl used in the reactions with $(Cl_2Si-CCl_2)_3$, the following compounds 282, 283, 284 and 285 were isolated after the initial stages of reaction.

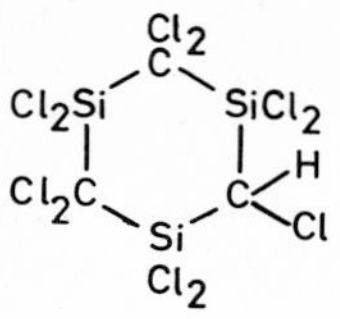

Structures 282, 283, 284, 285

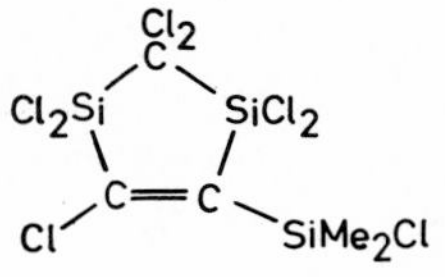

$$\underline{282} \qquad \underline{283}$$

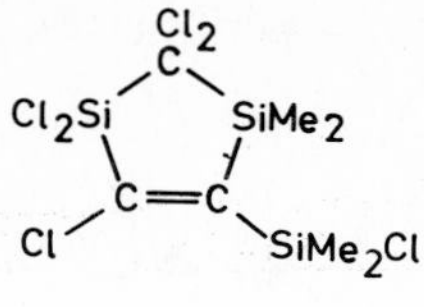

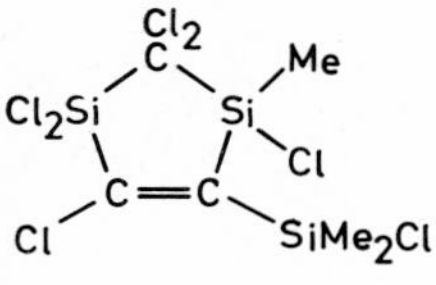

$$\underline{284} \qquad \underline{285}$$

The overall formation mechanism of these respective compounds is presented in Fig. 15.

An understanding of the overall reaction of $(Cl_2Si—CCl_2)_3$ with MeMgCl was gained by inspection of the pathway leading to the disilacyclopentene $\underline{273}$ as well as of the ring contraction mechanism.

The initial reaction stage involves metallation of $(Cl_2Si—CCl_2)_3$ with MeMgCl producing carbanion A. This can eventuate by two possible ways. Reaction pathway 1 involves first the formation of a bicyclic intermediate which is rearranged under the influence of MeMgCl, yielding the cyclopentene system $\underline{285}$. This in turn reacts further with MeMgCl through $\underline{284}$ and $\underline{283}$ to produce $\underline{273}$. Reaction pathway 2 starts out from carbanion A by reaction with ether, forming a CHCl group and resulting in $\underline{282}$. The proton of the CHCl group is acidic in character and as a result can react further with MeMgCl donating its proton and thereupon forming CH_4 and carbanion A. The reaction pathway from then on follows the lines of pathway 1. In addition, from the CHCl-containing compound $\underline{282}$ a renewed metal transfer can occur on a CCl_2 group causing subsequent rearrangement yielding B. This intermediate, unisolated to date, could explain the formation of $\underline{281}$ and $\underline{287}$, which indeed have also been observed as products in the reaction of $(Cl_2Si—CCl_2)_3$ with an excess of MeMgCl. Similarly, $\underline{273}$ could give rise to $\underline{280}$.

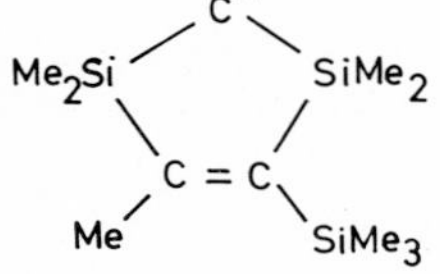

Structures 287, 280, 281

$$\underline{287} \qquad \underline{280} \qquad \underline{281}$$

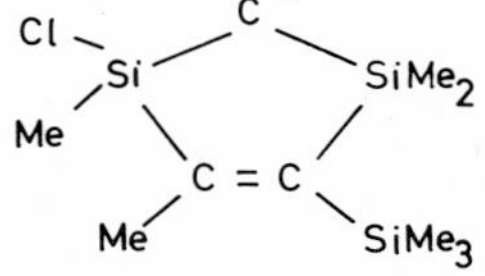

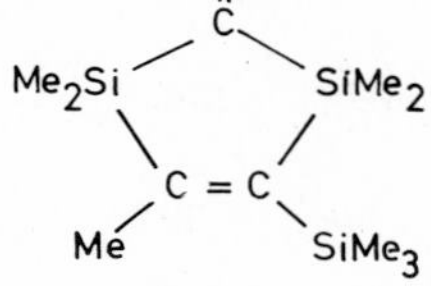

The choice of one reaction pathway over another depends on the concentration of MeMgCl used initially. If the amount of MeMgCl used is small, the reaction proceeds to the CHCl-containing compound $\underline{282}$, while an initial high concentration of MeMgCl directs the reaction via rearrangement into the 1,3-disilacyclopentene system. Also of note is the solvent effect on MeMgCl. If MeMgCl is prepared originally

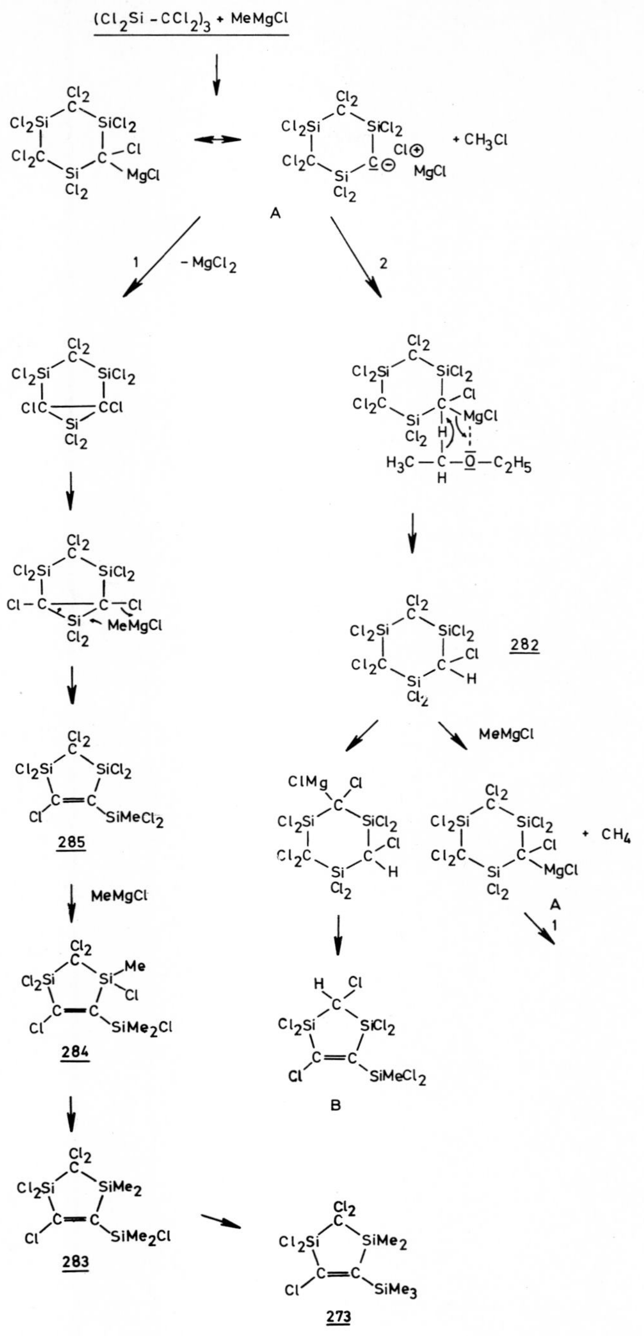

$(Cl_2Si-CCl_2)_3$ + MeMgCl

A

1 $-MgCl_2$

2

282

MeMgCl

285

MeMgCl

284

+ CH$_4$

A
1

B

283

273

Fig. 15. Formation mechanism pertaining to the reaction of $(Cl_2Si-CCl_2)_3$ with MeMgCl

in ether but is subsequently freed of ether by distillation under vacuum followed by reaction with $(Cl_2Si—CCl_2)_3$ in pentane or cyclopentane, then this reaction will produce the same types of compound as reaction in ether, but the reaction rate is much slower [115].

3.3.2 Reactions of Partly Chlorinated 1,3,5-Trisilacyclohexanes with MeMgCl

The previously-discussed reaction of $(Cl_2Si—CCl_2)_3$ with MeMgCl to yield the 1,3-disilacyclopentene skeleton requires at least two CCl_2 groups to be present for ring contraction. The reaction of the 1,3,5-trisilacyclohexane containing one CCl_2 group with an excess of MeMgCl proceeds in the following fashion:

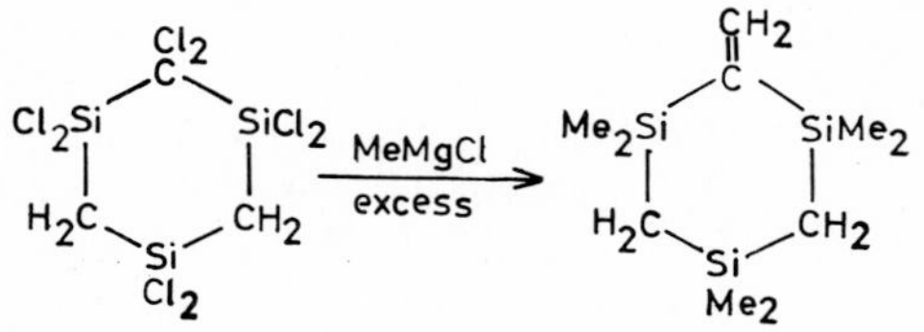

This corresponds closely to the reaction of one CCl_2 group between two Si-atoms in linear carbosilanes such as $(Cl_3Si)_2CCl_2$.

However, the 1,3,5-trisilacyclohexane 293 containing two CCl_2 groups reacts with an excess of MeMgCl via ring contraction, whereby cyclopentene 287 is produced in 80% yield.

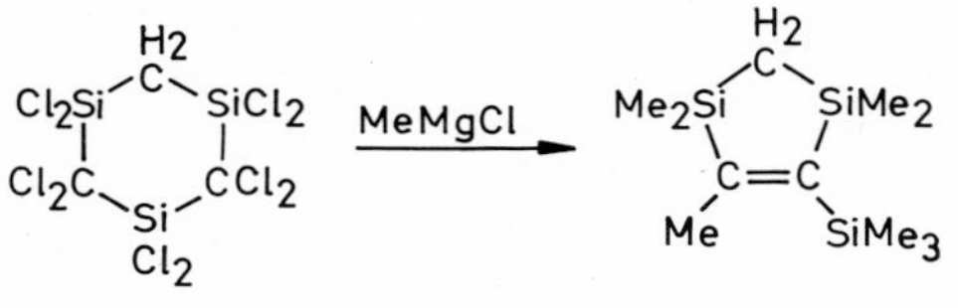

The first step involves metallation, followed by a rearrangement yielding a 1,3-disilacyclopentene. Finally a series of methylation reactions occurs:

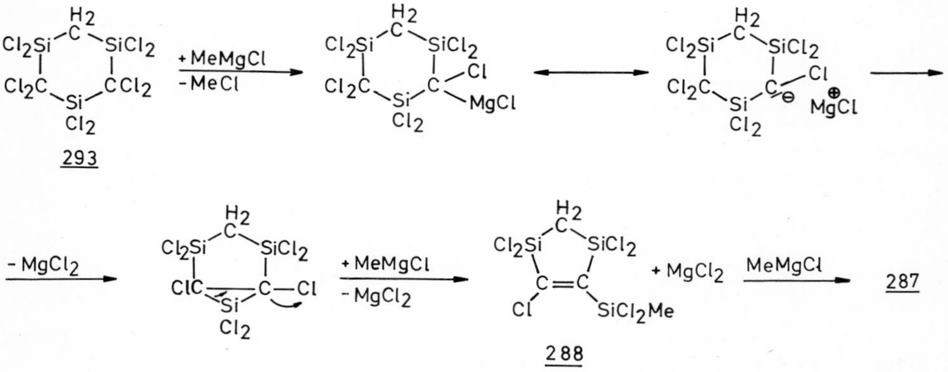

Further rearrangement producing linear products was not detected. In comparison to $(Cl_2Si—CCl_2)_3$, where ring opening under suitable conditions is the predominating reaction, reaction of 293 with MeMgCl is very specific. The failure of ring cleavage to occur is explained by the methylation of a vinylic Cl atom, which is also observed to a limited extent (approx. 10%) in reactions of $(Cl_2Si—CCl_2)_3$. Obviously, the methylation of vinyl Cl atoms and cleavage of the 1,3-disilacyclopentene system are competing reactions. In the case just described, the methylation is preferred to such an extent that ring cleavage is completely absent.

3.3.3 Si-Hydrogenated 1,3,5-Trisilacyclohexanes

Access to SiH-containing, C-chlorinated 1,3,5-trisilacyclohexanes enabled an investigation into the influence of the SiH group in these compounds on the reaction with MeMgCl in comparison to the SiCl-containing derivatives. In extending these investigations, the fully-hydrogenated ring $(H_2Si—CH_2)_3$ did not react with an excess of MeMgCl (mole ratio 1:15) within 48 hours at 20 °C. Compound 289 containing one CCl_2-group reacted under analogous conditions with MeMgCl via Si-methylation to produce 297 in 70% yield. The byproducts contained SiHMe groups.

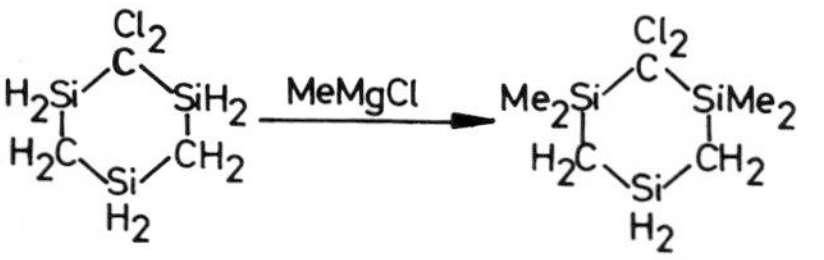

289   297

The reaction products gave no indication of the formation of a vinyl group.

Reaction of 1,3,5-trisilacyclohexane 290 containing two CCl_2 groups with MeMgCl generated

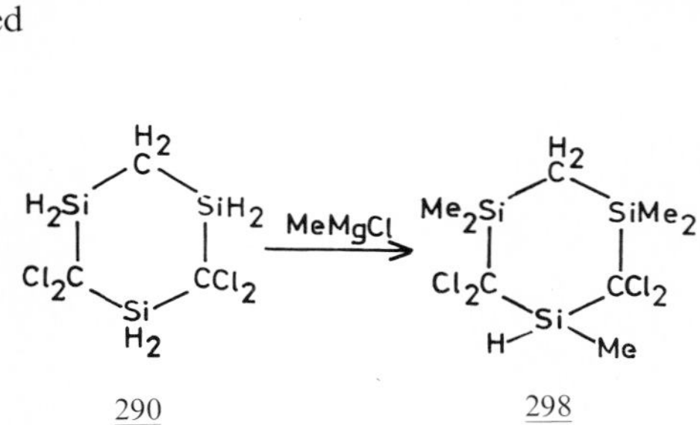

290   298

Compounds with a molecular weight of between 190–210 were detected, which were formed via cleavage reactions observed also in analogous reactions between similar compounds and $LiAlH_4$ [93, 94].

3.3.4 Reactions of Si-Methylated, Partly C-Brominated 1,3,5-Trisilacyclohexanes with BuLi and EtMgBr

In reactions with organometallic agents, the Si-methylated partly C-brominated 1,3,5-trisilacyclohexanes exhibit some similarities — and a number of differences — to their Si-chlorinated C-chlorinated analogues. For example, the 2,2,4,4-tetrabromo-hexamethyl-1,3,5-trisilacyclohexane yields with n-BuLi the dialkylated derivative.

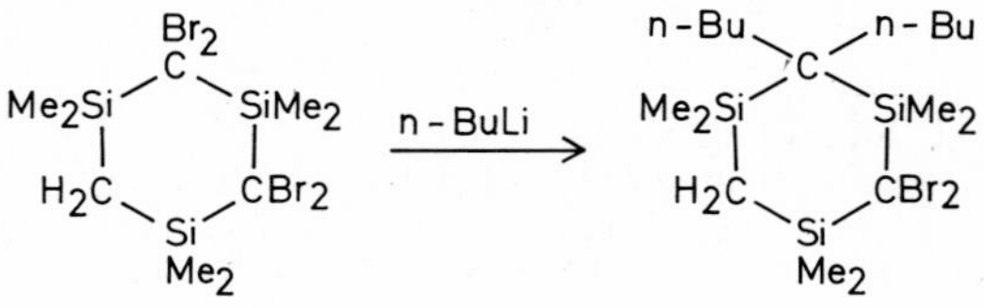

Exclusively one CBr_2 group is alkylated [100]; no formation of a $>C=CHR$ group occurs.

The reactions of 2,2,4,4-tetrabromo-hexamethyl-1,3,5-trisilacyclohexane 409 with t-BuLi or EtMgBr yield ring contraction and ring cleavage. This is evident from the compounds in the following scheme which were isolated after hydrogenation of the SiBr groups by means of $LiAlH_4$.

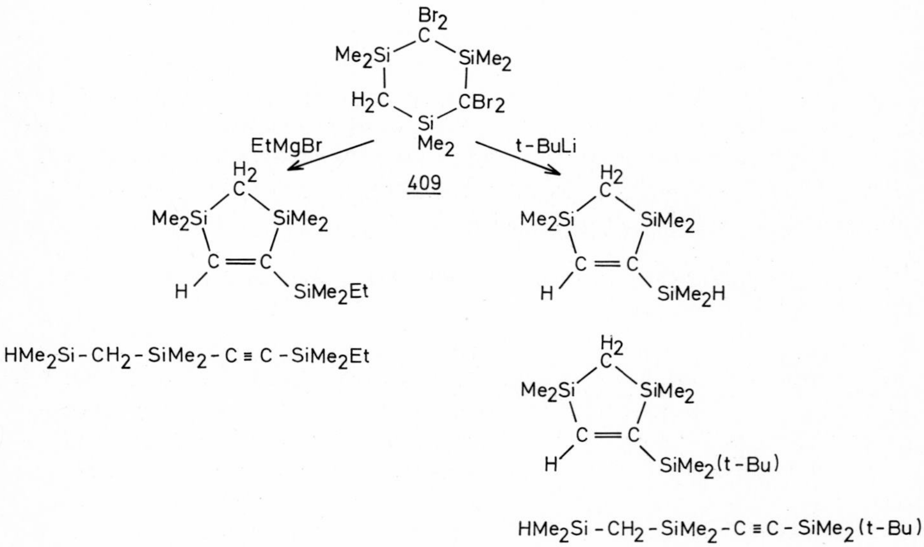

This behavior corresponds closely to that of the Si-chlorinated 1,3,5-trisilacyclohexane 293 containing two CCl_2 groups with the exception that in the latter case under comparable reaction conditions no cleavage of the five-membered ring occured [115].

When treated with t-BuLi and subsequently with MeI, the pentabromo derivative 410 of $(Me_2Si—CH_2)_3$ yields two products which show the dominating reactivity of the CBr_3 group:

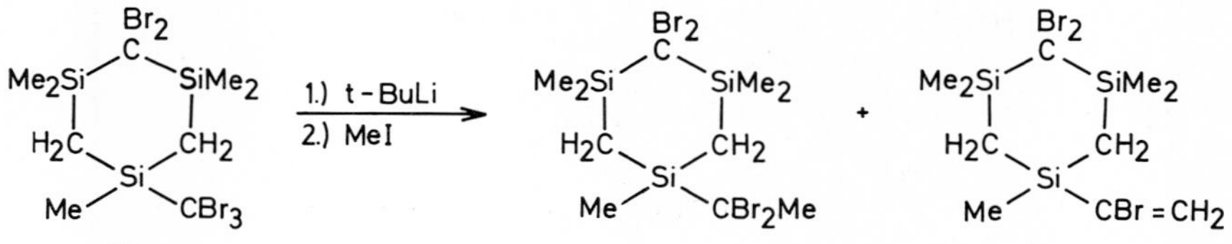

When a mixture of compounds 409 and 410 was reacted with n-BuLi at $-100\ °C$, white crystals of compound 411 could be isolated in minor yield from a number of not yet fully identified products.

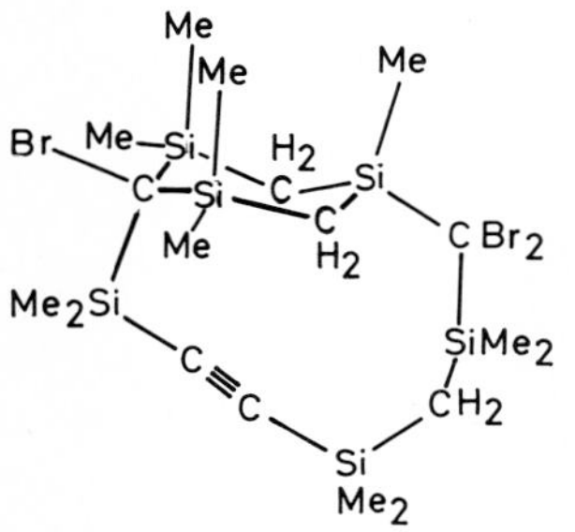

The completely unexpected structure of this molecule was ascertained by an X-ray structure determination using a single crystal of compound 411 [98]. It enables to propose a reaction path for the formation of 411 which includes first the ring cleavage of the 1,3,5-trisilacyclohexane 409 yielding the acyclic compound $BrMe_2Si—$ $—CH_2—SiMe_2—C≡C—SiMe_2Br$, and then coupling of this intermediate with the CBr_3 and CBr_2 groups of the 1,3,5-trisilacyclohexane 410.

3.3.5 Consideration of Chemical Behavior and NMR Chemical Shift of 1,3,5-Trisilacyclohexanes

From the just reported results, the following generalized view can be derived: $(Cl_2Si—$ $—CH_2)_3$ will react with MeMgCl to produce $(Me_2Si—CH_2)_3$, while $(H_2Si—CH_2)_3$ under these conditions does not react. By introducing a CCl_2 group into the skeleton, the neighboring SiH_2 group will undergo methylation, while the CCl_2 group remains unaffected. The CCl_2 group has an activating effect on the SiH_2 group, but does not participate in this reaction. Only in SiCl-, CCl-containing compounds is the CCl_2 group included in the reaction. However, the intended substitution of $(H_2Si—CCl_2)_3$ with MeMgCl to produce Si-methylated compounds such as $(Me_2Si—CCl_2)_3$ must fail because in the course of continual replacement of CH_2 by CCl_2 groups, Si—C ring cleavage by means of organometallic reagents becomes increasingly favorable. This means that a continual increase in the Si—C polarization results, which in turn seems to be reflected to some extent in terms of the ^{29}Si-NMR chemical shift.

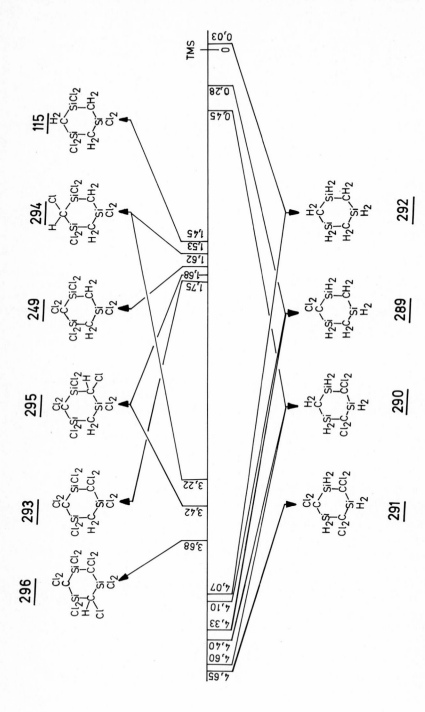

Fig. 16. Schematic representation of ^{1}H-NMR resonances (ppm) of the 1,3,5-trisilacyclohexanes 115, 249 and 289 to 296

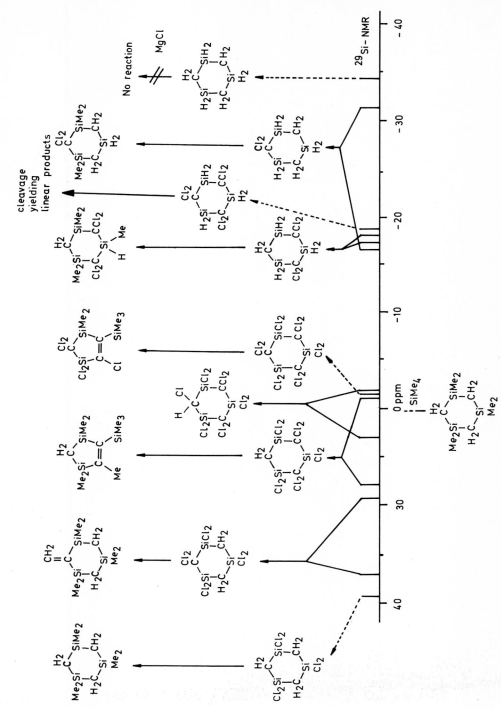

Fig. 17. ^{29}Si-NMR shift values of 1,3,5-trisilacyclohexanes and a schematic representation of their respective products from reaction with MeMgCl

Figure 16 presents a schematic representation of the ^{1}H-NMR chemical shift of various chlorinated 1,3,5-trisilacyclohexanes.

The extent of the shift in the ^{1}H-NMR signals of CH_2 and $CHCl$ groups to lower field strengths depends on the degree of chlorination of the C atom. For one Cl atom the shift difference measures ca. 0.08 ppm, or ca. 0.17 ppm per CCl_2 group. This range coincides closely with the measurements made on various derivatives of $(Cl_3Si)_2CCl_2$ and is caused by the high electronegativity of chlorine. In a similar way, the ^{1}H-NMR spectra for compounds hydrogenated on silicon show a continual down-field shift for the persisting CH_2 resonances with a stepwise increase in C-chlorination. The shift difference measured ca 0.27 ppm per CCl_2 group. The SiH signals show an analogous shift to low field as a consequence of the chlorination of carbon. The change in chemical shift by introducing one CCl_2 group is 0.27 ppm, and for the second 0.40 ppm. The shift is of the same order of magnitude as for the CH_2 protons, although these are separated from the CCl_2 group by an SiH_2 group.

Figure 17 presents the ^{29}Si-NMR spectral information from the investigated 1,3,5-trisilacyclohexanes, as well as their reaction products from reaction with MeMgCl.

The schematic representation of Fig. 17 shows, that progressing from $(H_2Si—CH_2)_3$ through $(Me_2Si—CH_2)_3$ and eventually to $(Cl_2Si—CH_2)_3$, increasingly electronegative substituents are bound to Si which causes absorption at a lower field strength. This is synonymous with a reduced electron density on the nucleus; anisotropic effects are not considered here. Regarding the $SiCl_2$-containing compounds, transformation of CH_2 groups to CCl_2 groups results in a chemical shift to higher field, synonymous with increased electron density on the silicon nucleus. This is concurrent with a decrease in Si—C bond polarization. Correspondingly, in compounds containing SiH_2 and CCl_2 groups the low field ^{29}Si-NMR chemical shift increases with a higher degree of chlorine substitution of C atoms. The Si—H bond naturally becomes more polar, which enables reaction with MeMgCl, while the chlorine-free $(H_2Si—CH_2)_3$ will not react with MeMgCl.

Over the whole area investigated, a surprisingly clear relationship exists between the substitution pattern in 1,3,5-trisilacyclohexanes and the chemical shift values in ^{29}Si-NMR spectra. Furthermore, it appears even possible to arrange the chemical shift scale into regions corresponding to varying reactivities of different 1,3,5-trisilacyclohexanes, mainly because chemical shifts result from variation in the electron density on the ^{29}Si nucleus, which in turn depends on the actual extent of Si—C bond polarization. This Si—C bond polarization also determines the reactive behavior of these compounds towards MeMgCl [115].

3.4 SiF-, CCl-Containing Carbosilanes

3.4.1 $(F_3Si)_2CCl_2$

As the results of the previous section show, Si- and C-chlorinated carbosilanes initiate their reactions with MeMgCl by attack on a CCl group. In Si-fluorinated, C-chlorinated carbosilanes however, the overall reaction path is governed by the reactivity of the SiF group. Hence, $(F_3Si)_2CCl_2$ reacts with MeLi to methylate the silicon.

$$(F_3Si)_2CCl_2 \xrightarrow[-LiF]{+MeLi} (Me_3Si)_2CCl_2$$

The byproducts of this reaction are $F_3Si-CCl_2-SiFMe_2$, $Me_2FSi-CCl_2-$ $-SiFMe_2$ and other partly-methylated derivatives of $(F_3Si)_2CCl_2$ [112].

3.4.2 $F_3Si-CCl_2-SiF_2-CH_2-SiF_3$ and $F_3Si-CCl_2-SiF_2-CHCl-SiF_3$

The reactions of MeMgCl with Si-fluorinated 1,3,5-trisilapentanes containing CCl_2 groups, such as $F_3Si-CCl_2-SiF_2-CH_2-SiF_3$ and $F_3Si-CCl_2-SiF_2-CHCl-$ $-SiF_3$, are distinct examples which show that methylation of the SiF group determines the course of the reaction. The results are presented in Tables 57 and 58 [90].

The stepwise Si-methylation of $F_3Si-CCl_2-SiF_2-CH_2-SiF_3$ proceeded to the symmetrically substituted $MeF_2Si-CCl_2-SiF_2-CH_2-SiF_2Me$. The methylation starts on the Si atom neighboring the CCl_2 group yielding at first $MeF_2Si-CCl_2-$ $-SiF_2-CH_2-SiF_3$. Apart from the Si-methylation of $F_3Si-CCl_2-SiF_2-CHCl-$

Table 57. Products from the reaction of $F_3Si-CCl_2-SiF_2-CH_2-SiF_3$ 300 with MeMgCl in diethylether (mole ratio = 1:12). A and B indicate experimental method. A: MeMgCl added dropwise (-10 °C) to 300; B: 300 added dropwise to MeMgCl. After addition of the solutions, the reaction mixture was brought to room temperature

Compounds formed	Proportion of reaction products, mole %	
	A	B
$Me_3Si-CCl_2-SiMe_2-CH_2-SiMe_3$	85	29
$Me_3Si-CHCl-SiMe_2-CH_2-SiMe_3$	5	—
$Me_3Si-C-SiMe_2-CH_2-SiMe_3$ $\parallel$ CH_2	3.5	—
$Me_3Si-CHCl-SiMeF-CH_2-SiMe_3$	1.5	3
$Me_3Si-CCl_2-SiMeF-CH_2-SiMe_3$	1.5	—
$Me_3Si-CCl_2H$	1.0	20
$Me_3Si-CH_2-SiMe_3$	Trace	18
$Me_3Si-CH_2-SiMeF-CH_2-SiMe_3$	—	28

Table 58. Products from the reaction of $F_3Si-CCl_2-SiF_2-CHCl-SiF_3$ 301 with MeMgCl (mole ratio = 1:12) in diethylether. Experimental method: Dropwise addition of MeMgCl solution to a solution of 301 at -10 °C followed by warming to 20 °C

Compounds formed	Proportion of reaction products, mole %
$Me_3Si-CCl_2-SiMe_2-CH_2-SiMe_3$	53
$Me_3Si-CCl_2-SiMe_2-CHCl-SiMe_3$	20
$Me_3Si-CHCl-SiMe_3$	7
$Me_3Si-CHCl-SiMe_2-CH_2-SiMe_3$	5
$Me_3Si-CCl_2-SiMe_3$	4
$Me_3Si-C-SiMe_2-CH_2-SiMe_3$ $\parallel$ CH_2	3
$Me_3Si-CH_2-SiMe_2-CH_2-SiMe_3$	1

—SiF$_3$, the transition of CCl to CH groups is also of major significance. This transition is explained by C-metallation followed by ether cleavage. It is not yet known at which Si-methylation stage reduction occurs. The formation of a $>$C=CH$_2$ group on the carbosilane skeleton is a minor reaction, although it appears as a main reaction in SiCl- and CCl-containing compounds.

3.4.3 (F$_3$Si—CCl$_2$)$_2$SiF$_2$

Distinct differences occur in the behavior of (F$_3$Si—CCl$_2$)$_2$SiF$_2$ (Table 59). For this compound the cleavage of the molecular skeleton is the determining reaction, followed by Si-methylation. To illustrate this, one should consider the reaction of (F$_3$Si—CCl$_2$)$_2$SiF$_2$ with 1 mole equivalent of MeMgCl in Et$_2$O to form the compounds F$_3$Si—CCl$_2$H, MeF$_2$Si—CCl$_2$—SiMeF$_2$ and Me$_2$FSi—CCl$_2$—SiMeF$_2$ [90].

Table 59. Compounds produced from the reaction of (F$_3$Si—CCl$_2$)$_2$SiF$_2$ with MeMgCl (mole ratio 1:12), A and B as in Table 57. C with hexachlorobutadiene as solvent

Compounds formed	Amount of reaction products, mole %		
	A	B	C
Me$_3$Si—CCl$_2$—SiMe$_3$	75	7	7
Me$_3$Si—CHCl—SiMe$_3$	5	—	—
Me$_3$Si—CCl$_2$—SiMe$_2$—CH$_2$—SiMe$_3$	3	—	—
Me$_3$Si—CCl$_2$H	2	2	Trace
Me$_3$Si—CMeCl—SiMe$_3$	2	—	—
H$_2$C=C(SiMe$_3$)[CH(SiMe$_3$)$_2$]	2	—	—
Me$_3$Si—CCl$_2$—SiMe$_2$—CHCl—SiMe$_3$	1	—	—
SiMe$_4$	in solvent fraction		
Me$_3$Si—CCl$_2$—SiMeF—CCl$_2$—SiMe$_3$	—	44	30
Me$_3$Si—CH$_2$—SiMe$_2$—CCl$_2$H	—	17	35
Me$_3$Si—CCl$_2$—SiMeF—CH$_2$—SiMe$_3$	—	7	—
Me$_3$Si—CCl$_2$—SiMe$_2$—CCl$_2$—SiMe$_3$	—	—	12

Table 60. Compounds from reaction of (F$_3$Si—CCl$_2$)$_2$SiF$_2$ with MeLi (mole ratio 1:10)

Compounds formed	Proportion of reaction products, mole %
Me$_3$Si—C≡C—SiMe$_3$	35
Me$_3$Si—CCl$_2$—SiMe$_3$	30
Me$_3$Si—C≡C—SiMe$_2$—C≡C—SiMe$_3$	8
Me$_3$Si—CCl$_2$—SiMe$_2$—CH$_2$—SiMe$_3$	5
Me$_3$Si—CCl$_2$—SiMe$_2$—C≡C—SiMe$_3$	3

Changes in the reaction pathway appear in the reaction of (F$_3$Si—CCl$_2$)$_2$SiF$_2$ with MeLi to form Me$_3$Si—C≡C—SiMe$_3$, as pointed out in Table 60. The formation of this compound through initial interaction of the starting material with MeLi emphasizes that the effect caused by this stronger metallation source acting on C-chlorinated, SiF-containing 1,3,5-trisilapentanes generates C-metallation as well as skeletal changes.

3.4.4 Reactions of Si-Fluorinated 1,3,5-Trisilacyclohexanes

3.4.4.1 Reactions with MeMgCl and MeLi

Si-fluorinated derivatives of 1,3,5-trisilacyclohexane behave in a totally similar fashion [124]. The presence of a CCl_2 group causes reaction according to

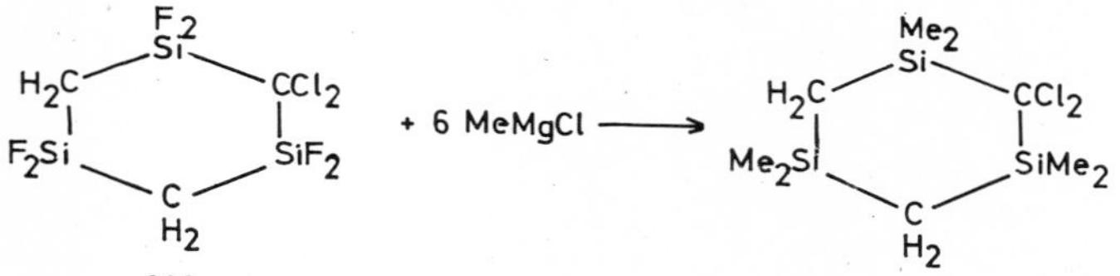

302

Reaction of 3 mole equivalents of MeMgCl with 302 produces 303, which in turn reacts with $LiAlH_4$ yielding 304.

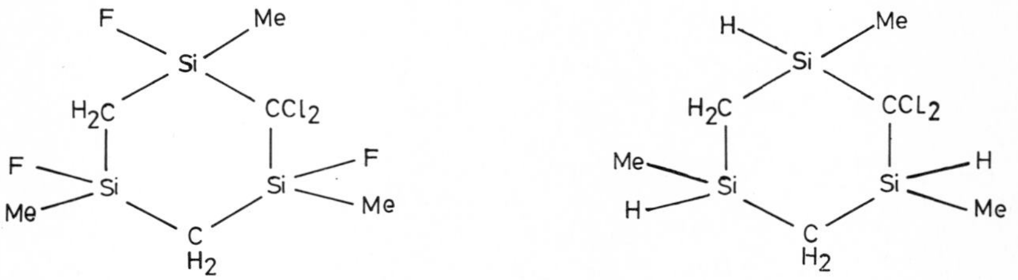

303

304

Also, Si-fluorinated 1,3,5-trisilacyclohexanes containing two CCl_2 groups will undergo Si-methylation:

305

306

Cleavage of the six membered ring is favored through the use of more polar solvents or through reaction with MeLi. For instance, 305 reacts with 6 mole equivalents of MeMgCl at $-70\ °C$ in THF causing complete cleavage. Reaction with six mole equivalents of MeLi in Et_2O at $-78\ °C$ also proceeds to complete cleavage of the ring. A further insight into the course of these reactions can be procured by considering the reaction of 307 with an excess of MeMgCl according to the following equation

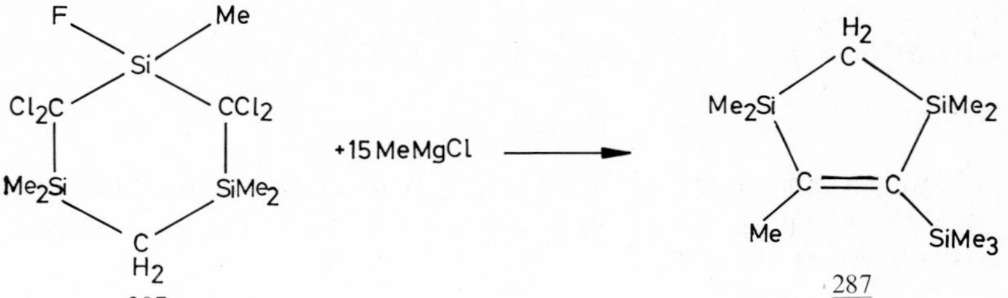

307

287

+

$Me_3Si—CH_2—SiMe_2—C{\equiv}C—SiMe_3$

By using an excess of MeLi, nearly quantitative cleavage resulted producing $Me_3Si—CH_2—SiMe_2—C{\equiv}C—SiMe_3$. Similarly, $(F_2Si—CCl_2)_3$ is initially methylated, allowing $(MeFSi—CCl_2)_3$ to be isolated in an 85% yield. Further reaction with MeMgCl or MeLi led to the splitting of the ring skeleton [124].

From these results it can be recognized that reactions of SiF- and CCl-containing 1,3,5-trisilacyclohexanes begin by normal SiF-methylation. After a particular stage of methylation is achieved, CCl_2 groups enter into reaction according to the mechanism known for the SiCl-containing derivatives undergoing ring contraction. Compound 306, for instance, represents the exact stage of methylation in which the CCl_2 group becomes active, as schematically depicted below.

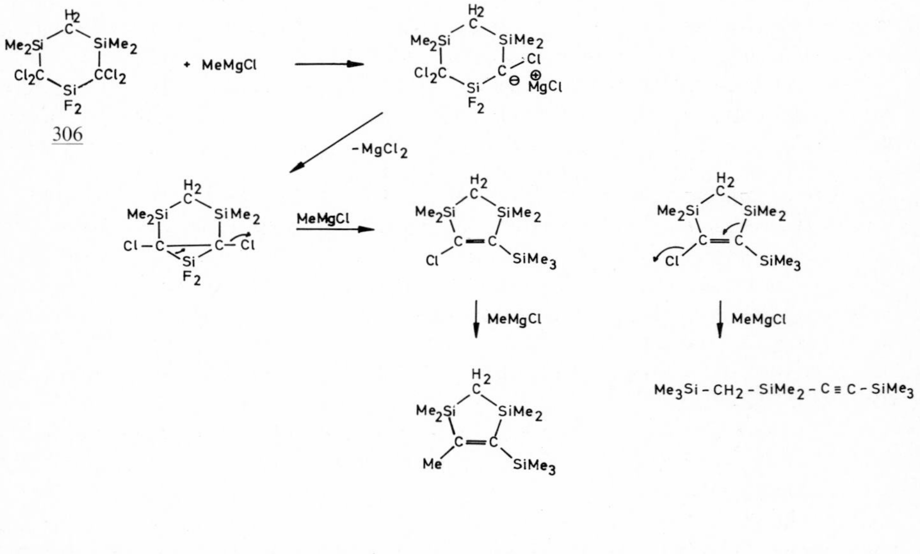

306

3.4.4.2 Reactions with Phenyllithium and PhMgBr

Si-fluorinated C-chlorinated 1,3,5-trisilacyclohexanes react with PhMgBr making the following compounds accessible [125].

$$\text{F}_2\text{Si} \underset{\text{H}_2\text{C} \underset{\text{Si}_{\text{F}_2}}{\diagup}}{\overset{\overset{\text{Cl}_2}{\text{C}}}{\diagdown}} \text{SiF}_2$$

$$\text{Ph}-\text{Si}\underset{\text{H}_2\text{C}}{\overset{\overset{\text{Cl}_2}{\text{C}}}{}}\text{Si}-\text{Ph}$$

$$\text{Ph}_2\text{Si}\overset{\overset{\text{Cl}_2}{\text{C}}}{}\text{SiPh}_2$$

$$\text{Ph}_2\text{Si}\overset{\text{Cl}\diagdown\diagup\text{H}}{}\text{SiPh}_2$$

308

However, compound 308 results only from reaction with PhLi. Attempts to use the CCl_2 group in these compounds in the construction of C-bridged carbosilanes have only been brought to the stage of pilot reactions.

3.4.5 Summary of the Behavior of SiH- and SiF-Containing C-Chlorinated Carbosilanes

Only one or two Si groups are methylated on reaction of $(H_2Si-CH_2)_3$ and $(H_3Si-CH_2)_2SiH_2$ with MeMgCl in Et_2O, while in THF the extent of methylation increases to about 3 or 4 Si groups. Chlorination of a C atom increases the reactivity of the neighboring SiH groups. Its reactivity decreases by increasing the degree of methylation in the compound. Hence, beginning with $H_3Si-CCl_2-SiH_2-CH_2-SiH_3$, the compounds $MeH_2Si-CCl_2-SiH_2-CH_2-SiH_3$ and $H_3Si-CCl_2-SiHMe-CH_2-SiH_3$ are formed. Other C-chlorinated derivatives, for example $(H_3Si)_2CCl_2$ or $(H_3Si-CCl_2)_2SiH_2$, react with MeMgCl in Et_2O via cleavage of the Si—C bond followed to a certain extent by molecular enlargement, as when $Me_2HSi-CCl_2-SiHMe-CH_2-SiH_3$ forms (among other products) from $(H_3Si)_2CCl_2$ in the described way. Besides Si—C bond cleavage, a transfer of CCl_2 to CHCl or CH_2 bridging groups occurs. This reaction is favored if cyclopentane is used as solvent. Under these conditions, Si—C bond cleavage is inhibited [94].

The enhancement of Si—H substitution and Si—C cleavage due to the use of more polar solvents confirms that changes of polarization in the molecule determine the course of reaction. In carbosilanes such as $(H_3Si-CH_2)_2SiH_2$, partial charge distribution can be expected $(\overset{\delta+}{\text{Si}}-\overset{\delta-}{\text{H}} \quad \overset{\delta-}{\text{C}}-\overset{\delta+}{\text{H}} \quad \overset{\delta+}{\text{Si}}-\overset{\delta-}{\text{C}})$ due to the different electronegativities of the elements. C-chlorination will also cause an increased polarization of the Si—H bond, generated by the high electronegativity of chlorine. This explains the increased reactivity of the SiH groups, as well as the highly cleavable nature of the Si—C bond in such compounds. In a similar manner Si-methylation is understood to decrease the reactivity of the remaining SiH groups on the partly methylated Si atom, as well as decreasing the tendency towards cleavage of the Si—C bond in the molecular skeleton.

The Si- and C-chlorinated derivatives differ from SiH-containing C-chlorinated compounds only through the Si—Cl group. Because the Si—Cl bond has a large electron density available for shielding of the Si atom, due to the increased C-chlorination a reduced polarization and reactivity of the Si—Cl bond is observed. Therefore the different behavior of SiH- and CCl-containing carbosilanes, as opposed to SiCl- and CCl-containing carbosilanes, can be understood.

The SiF- and CCl-containing carbosilanes such as $(F_3Si-CCl_2)_2SiF_2$ behave more like the SiH- and CCl-containing carbosilanes, such as $(H_3Si-CCl_2)_2SiH_2$, than like SiCl- and CCl-containing derivatives such as $(Cl_3Si-CCl_2)_2SiCl_2$. Reaction of $F_3Si-CCl_2-SiF_2-CH_2-SiF_3$ with MeMgCl begins by Si-methylation and not by C-metallation. A reverse situation occurs with Si- and C-chlorinated compounds. Moreover, CCl-hydrogenation occurs, meaning that the reaction proceeds from $F_3Si-CCl_2-SiF_2-CHCl-SiF_3$ to $Me_3Si-CCl_2-SiMe_2-CH_2-SiMe_3$. The related compound $(F_3Si-CCl_2)_2SiF_2$ reacts with MeMgCl via Si—C cleavage to produce mainly 1,3-disilapropanes [90].

3.4.6 CF$_2$-Containing 1,3-Disilapropanes

3.4.6.1 Formation of Si—CF$_2$—Si Groups by Insertion of a CF$_2$ Carbene into the Si—Si Bond

In connection with previously described investigations concerning C-chlorinated carbosilanes, the synthesis and subsequent reactions of C-fluorinated derivatives were of interest. These compounds became accessible by insertion of a CF_2 carbene into the Si—Si bond by the process

$$FMe_2Si-SiMe_2F + \overset{|}{C}F_2 \rightarrow FMe_2Si-CF_2-SiMe_2F$$

The carbene CF_2 is obtained by thermolysis of Me_3Sn-CF_3.

From the disilanes $XMe_2Si-SiMe_3$ (X = F, Cl, Br, OMe) and $FMe_2Si-SiMe_2F$, compounds such as $FMe_2Si-CF_2-SiMe_3$ and $FMe_2Si-CF_2-SiMe_2F$ were accessible. The trisilane $FMe_2Si-SiMe_2-SiMe_2F$ produced $FMe_2Si-CF_2-SiMe_2-SiMe_2F$ and $(FMe_2Si-CF_2)_2SiMe_2$. Insertion reactions of this type involving di- and trisilanes were observed, but only if the terminal Si atoms bear one or more —I substituents (F, Cl, Br, OMe). These substituents weaken the Si—Si bond and favor insertion of a CF_2 group. Consequently insertion fails to proceed with Si_2Me_6. On the other hand, higher fluorinated disilanes such as $F_2MeSi-SiMeF_2$ or Si_2F_6 decompose at temperatures which are not yet sufficient to generate CF_2 carbene from Me_3Sn-CF_3, so these are ruled out as starting materials.

3.4.6.2 Reactions of CF$_2$-Containing Carbosilanes

In contrast to comparable C-chlorinated carbosilanes, the reaction of $(FMe_2Si)_2CF_2$ with MeMgCl or MeLi at low temperatures results in an alkylation of silicon rather than in metallation of carbon, yielding $Me_3Si-CF_2-SiMe_2F$ and $(Me_3Si)_2CF_2$. With PhLi or PhMgBr the analogous products $PhMe_2Si-CF_2-SiMe_2F$ and $(PhMe_2Si)_2CF_2$ are obtained.

The related $(FMe_2Si)_2CF_2$ is converted by $LiAlH_4$ to $(HMe_2Si)_2CF_2$, and other derivatives are similarly hydrogenated on silicon. Reaction of $LiPMe_2$ with $(FMe_2Si)_2CF_2$ produces $Me_2P-Me_2Si-CF_2-SiMe_2F$ as well as $(Me_2P-SiMe_2)_2CF_2$ [131].

The CF_2 group in $(FMe_2Si)_2CF_2$ or $(Me_3Si)_2CF_2$ is capable of cleaving the Si—P bond in $Me_3Si-PMe_2$, forming ylides such as $(FMe_2Si)_2C=PMe_2-PMe_2$ or $(Me_3Si)_2C=PMe_2-PMe_2$, respectively [131].

4. Metallation of Carbosilanes

In contrast to the diverse reactions apparent on the Si atom, the CH_2 group appeared at first relatively unreactive. In order to extend the possibilities of carbosilane syntheses, the CH_2 group had to be incorporated into the preparative scheme. The formation of the perchlorinated carbosilanes and their reactions with MeMgCl and MeLi did not lead to a solution of this problem, because the CCl_2 group situated between two $SiCl_2$ groups displays an elevated reactivity, and therefore reactions with MeMgCl generate skeletal changes [101].

The following preparative alternatives were proposed for the formation of compounds containing the $Me_2Si-CCl_2-SiMe_2$ grouping:

a) Conversion of SiCl groups to SiF groups in perchlorinated carbosilanes, followed by their methylation (Chapt. III.3.4.3).
b) Organometallic synthesis via carbenoid intermediates (Chapt. II.4.5).

Starting with the $Me_2Si-CCl_2-SiMe_2$ grouping, further synthetic possibilities open up via its lithiation, producing a $Me_2Si-CClLi-SiMe_2$ grouping along with subsequent silylation of the skeletal C atom. So far, this had been proved only in simple cases [102].

4.1 Metallation of Skeletal C-Atoms in Si-Methylated Carbosilanes

A very promising means for incorporating CH_2 groups into the preparative scheme is via a selective acid-base reaction between $-Me_2Si-CH_2-SiMe_2-$ and LiR to produce $-Me_2Si-CHLi-SiMe_2- + RH$ [103]. The solution to this task demands a reagent which will selectively lithiate $\geq Si-CH_2-Si\leq$ groups and does not attack $(\geq Si)_3CH$, $\geq SiMe$, $>SiMe_2$ nor $-SiMe$ groups. This condition is fulfilled by using the n-BuLi/N,N,N',N'-tetramethylethylenediamine complex (TMEDA) [104].

4.1.1 $(Me_2Si-CH_2)_3$

For this investigation, 1,1,3,3,5,5-hexamethyl-1,3,5-trisilacyclohexane was chosen as model substance, because of its ability to undergo multiple lithiation and silylation reactions. Reaction of $(Me_2Si-CH_2)_3$ with n-BuLi/TMEDA in hexane in 2–3 hrs at 30 °C or 24 hrs at 20 °C quantitatively produced the monolithiated compound:

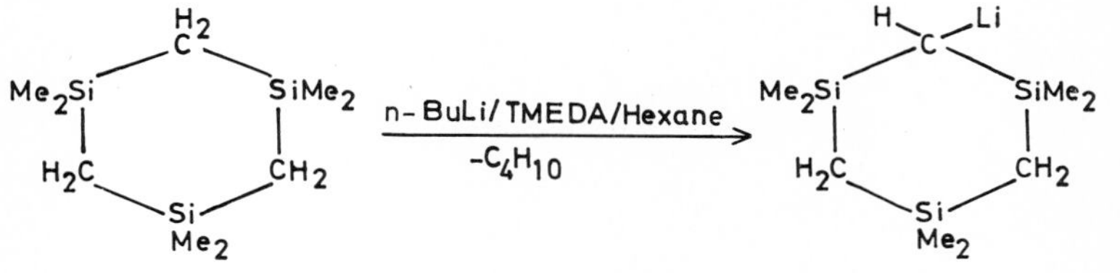

The reaction is thermodynamically controlled and the carbanion is stabilised by two adjacent $SiMe_2-CH_2$ groups. A kinetically controlled metallation of the SiMe group producing a Si—CH$_2$Li group was not observed. Even by employing an excess of n-BuLi no further metallation occurs at 20 °C. The monolithiated compound is stable for weeks in a sealed NMR tube.

Quantitative silylation occurs on reaction with trimethylchlorosilane, in which no byproducts were observed

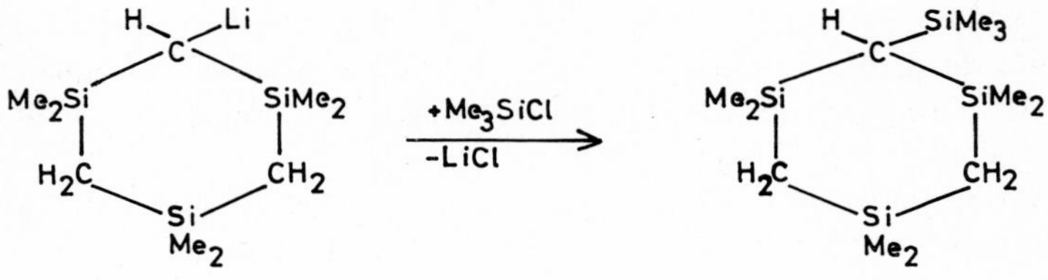

251

Two possible reaction products are conceivable by reaction of 251 with n-BuLi:

a) The thermodynamically preferred carbanion (≡Si)$_3$CLi, stabilized by delocalization of the lone pair of electrons in the d-orbitals of each of the three Si atoms.

b) A compound containing the kinetically favored sequence ≡Si—CHLi—Si≤.

By using n-BuLi/TMEDA as reagent, the compound containing a CHLi group arose as a result of the increased space requirement of the lithiating complex as well as through a high screening effect of the tertiary (≡Si)$_3$CH group. This reaction was quantitative; metallation of SiMe$_3$ and SiMe$_2$ groups did not occur. Further reaction with Me$_3$SiCl produced 252:

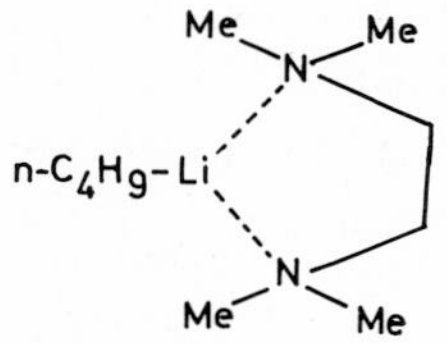

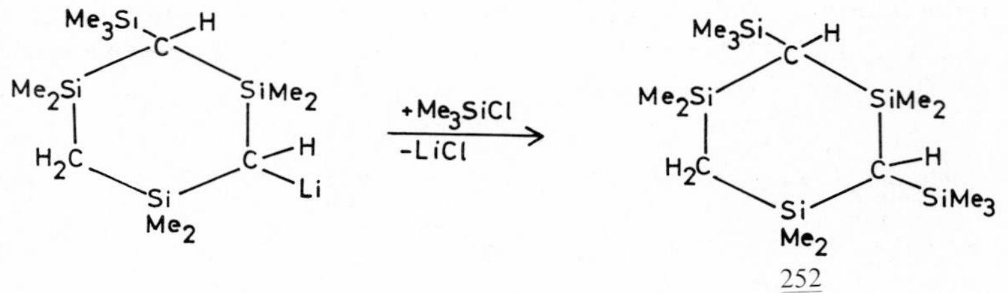

252

In this case both isomers, cis-2,4-bis(trimethylsilyl)-1,3,5-trisilacyclohexane in e,e- and trans-2,4-bis(trimethylsilyl)-1,3,5-trisilacyclohexane in e,a-configuration were formed in a ratio of 1 : 2 [103].

Further reaction of 252, consisting of 80 % trans and 20 % cis isomer, with n-BuLi/ TMEDA at 20 °C confirmed the selective lithiation:

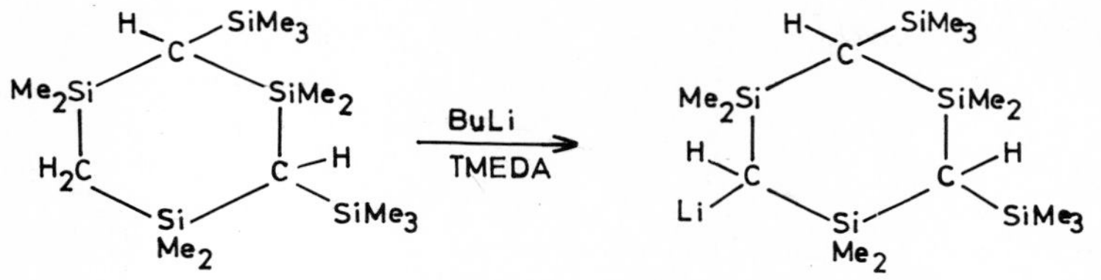

In the ^{1}H-NMR spectrum of the reaction mixture, signals appeared pertaining to CH:CHLi in a ratio of 2:1. The reaction was quantitative, as shown by the absence of byproducts. Further silylation yielded 253:

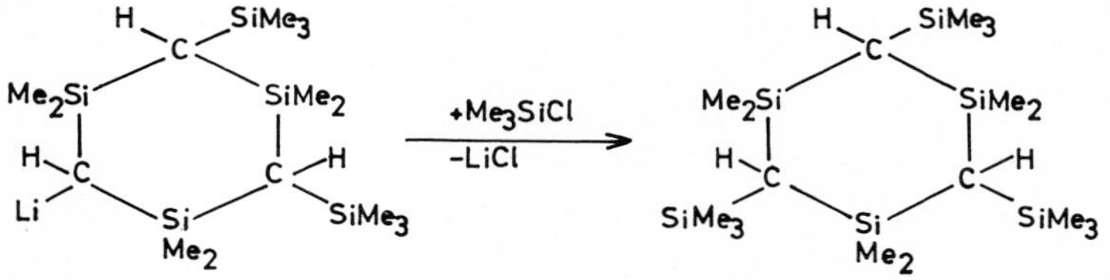

253

The isomeric mixture of $(Me_2Si—CHSiMe_3)_3$, as indicated by gas chromatography, consisted of 94 % cis-trans isomer and 6 % cis-cis isomer [103].

The carbanions below are the kinetically-controlled products of an initially

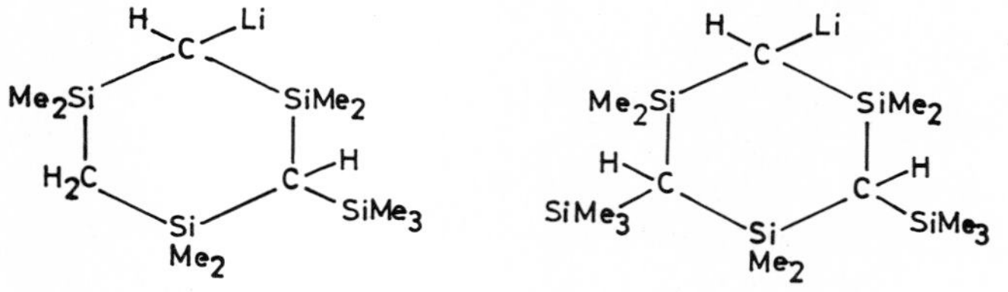

quantitative reaction. However, trapping these carbanions with Me_3SiCl yields not only the expected compounds 252 and 253, but also the derivatives 254 and 255 in 20 % and 30 % yield, both containing a $C(SiMe_3)_2$ group.

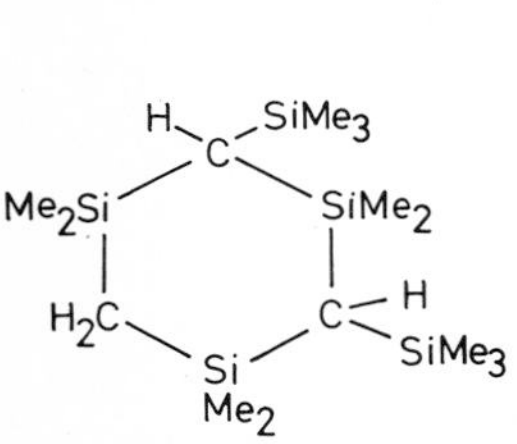

252 cis/trans

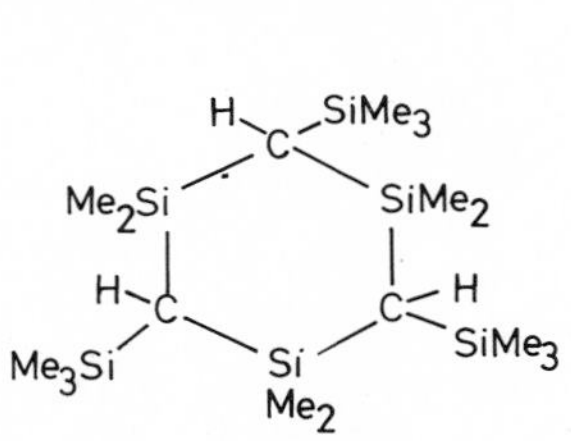

253 cis-cis/cis-trans

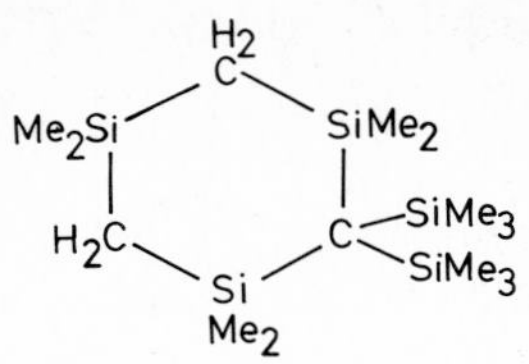

254 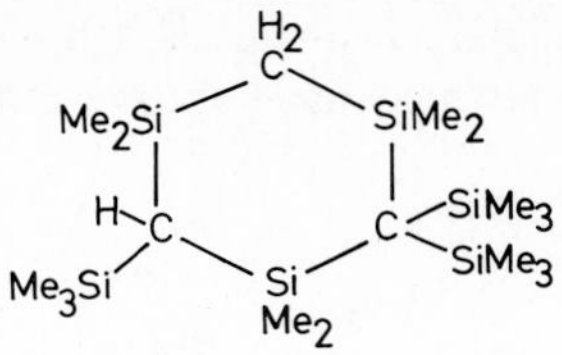   255

This is explained by a metal transfer reaction:

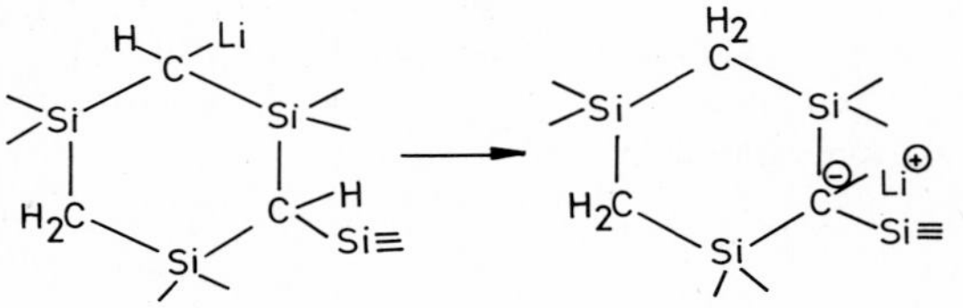

258   256

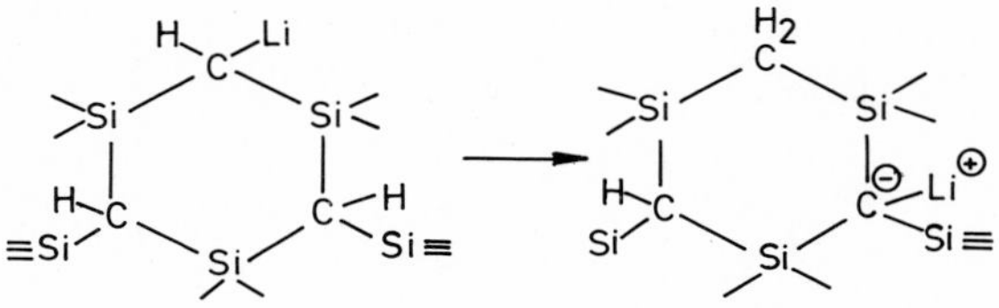

259   257

The compounds 256 and 257 are the thermodynamically-favored carbanions of 258 and 259. The carbanions 256 and 257 are stabilized by the presence of three SiMe$_3$ groups. Rearrangement in TMEDA/hexane is evoked by increase in reaction temperature. That is, 259 is formed at 20 °C but will rearrange at 30–40 °C to yield 257. Subsequent reaction with Me$_3$SiCl yields almost exclusively 255 [102].

1,1,3,3,5,5-Hexamethyl-2-trimethylsilyl-1,3,5-trisilacyclohexane 251 displays primary, secondary and tertiary C-atoms and is therefore a model substance to control the thermodynamic increase in acidity of the compound classes ($\equiv$Si)$_3$CH > > $\equiv$Si—CH$_2$—Si$\equiv$ > $\equiv$Si—CH$_3$. Because MeLi in THF is a weaker base than n-BuLi/TMEDA in hexane, and sterically less pretentious than n-BuLi/TMEDA, reaction with MeLi should be a thermodynamically controlled process. This is confirmed by inspection of the following reaction,

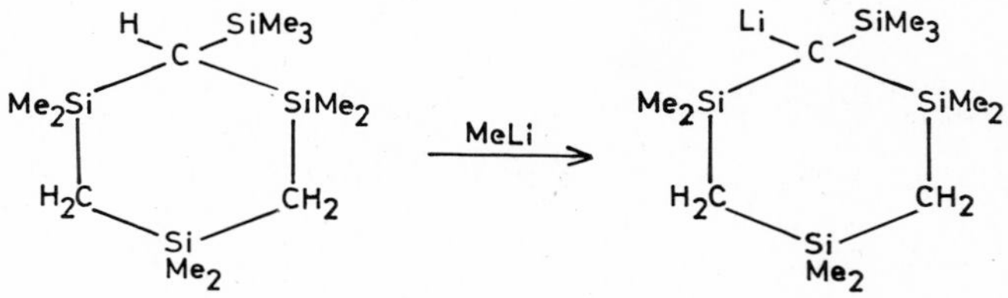

carried out in THF/ether (7:1) for 24 hrs at 20 °C. Further reaction with Me$_3$SiCl proceeded as expected. Metallation of $\equiv$Si—CH$_2$—Si$\equiv$ or $\equiv$Si—Me groups did not occur. The difference in the reactions undertaken using MeLi/THF and BuLi/ TMEDA can be explained by the steric requirements of the carbosilane and the reagents.

4.1.2 Me₄Ad

It also seemed beneficial to investigate the lithiation of 1,3,5,7-tetramethyl-1,3,5,7-tetrasilaadamantane. C-lithiation is favored by using stronger lithiation sources. The most favorable results are obtained when the ratios of reactants are Me_4Ad: n-BuLi:TMEDA:hexane = 1:1,5:1,2:3,5, and the reaction temperature is 40 °C, and reaction time 4 hrs. Gas chromatographic separation of the products obtained by reaction of the lithiation product with Me_3SiCl was carried out, and the results are presented in Table 61. The lithiation of the skeletal C atom occurs in the presence of n-BuLi to the extent of 94%. Dilithiation, as indicated by appearance of the isomeric mixture of $Me_4Ad(SiMe_3)_2$, maintains a subordinate role in the overall reaction [80].

Table 61. Composition of reaction products from the reaction of Me_4Ad with n-BuLi and t-BuLi, and subsequently with Me_3SiCl, after gas chromatographic separation

		Me_4Ad	$Me_3(Me_3SiCH_2)Ad$	$Me_4Ad(SiMe_3)$	$Me_4Ad(SiMe_3)_2$
I.	t-BuLi	1%	28%	58%	13%
II.	n-BuLi	3%	2%	94%	1%

This preferred C-lithiation can be very beneficial in the synthesis of Si-adamantanes with varying substituents on the skeletal C-atoms.

Reaction of $Me_4Ad(SiMe_3)$ with n-BuLi and Me_3SiCl makes synthesis of bissilylated derivatives possible

$$Me_4Ad(SiMe_3) + n\text{-BuLi} \rightarrow Me_4Ad(SiMe_3)Li + n\text{-BuH}$$

$$\downarrow + ClSiMe_3$$

$$Me_4Ad(SiMe_3)_2 + LiCl$$

The isomers of $Me_4Ad(SiMe_3)_2$ form a very complex mixture. To date this mixture has not been separated, but it was established that further lithiation followed by silylation was in fact possible:

$$Me_4Ad(SiMe_3)_2 + n\text{-BuLi} \rightarrow Me_4Ad(SiMe_3)_2Li + n\text{-BuH}$$

$$Me_4Ad(SiMe_3)_2Li + ClSiMe_3 \rightarrow Me_4Ad(SiMe_3)_3 + LiCl$$

By gas chromatography, unreacted $Me_4Ad(SiMe_3)_2$ was separated from the reaction mixture in a yield of 35%. Further reaction of the isomeric mixture of $Me_4Ad(SiMe_3)_3$ with n-BuLi and Me_3SiCl yielded products with a molecular weight of approximately 544 corresponding to $Me_4Ad(SiMe_3)_4$ [80].

The influence of Si-substituents on 1,3,5,7-tetrasilaadamantane when it reacts with organolithium reagents is of interest. While reaction of Me_3IAd with $LiCH_2-$ $-SiMe_2Ph$ enables selective substitution on the bridgehead in a yield of 93%,

$$Me_3IAd + LiCH_2-SiMe_2Ph \xrightarrow[\text{pentane}]{\text{TMEDA}} Me_3(PhMe_2Si-CH_2)Ad + LiI$$

reactions of Me_3ClAd and Me_3BrAd metallate the skeletal C atom:

$$Me_3ClAd + n\text{-}BuLi \xrightarrow[25\,°C]{\text{pentane/TMEDA}} Me_3ClAdLi + n\text{-}BuH$$

$$Me_3ClAdLi + ClSiMe_3 \rightarrow Me_3ClAd(SiMe_3) + LiCl$$

The compounds $Me_3XAd(SiMe_3)$ (X = Cl, Br), arise through lithiation of a skeletal CH_2 group in a position adjacent to the SiX group, followed by silylation. Further substitution of SiX enables the formation of compounds such as $Me_3(Me_3Si-CH_2)Ad(SiMe_3)$ [80].

Skeletal C-substituted derivatives are accessible in a yield of between 50 and 80% by reaction of Me_3ClAd and Me_3BrAd with $LiCH_2-SiMe_3$. These reactions also illustrate why reaction of Cl_4Ad or Br_4Ad with $LiCH_2-SiMe_3$ will produce only to a very limited extent Si-substituted 1,3,5,7-tetrasilaadamantanes. This reaction is determined by initial C-metallation, and byproducts arise from consecutive reactions [80].

4.1.3 Summary

The variation in kinetically and thermodynamically controlled reactions is seen as an advantage in the formulation of further carbosilane systems of two general types:

a) n-BuLi/TMEDA:

$$\equiv Si-CH_2-Si\equiv \rightarrow \equiv Si-CHLi-Si\equiv \xrightarrow{\equiv SiX} (\equiv Si)_3CH$$

$$(\equiv Si)_2CH-\underset{|}{Si}-CH_2-Si\equiv \rightarrow (\equiv Si)_2CH-\underset{|}{Si}-CHLi-Si\equiv$$

$$\xrightarrow{+\,\equiv SiX} (\equiv Si_2CH)_2Si{<}$$

b) MeLi/THF:

$$(\equiv Si)_3CH \rightarrow (\equiv Si)_3CLi \xrightarrow{\equiv SiX} (\equiv Si)_4C$$

$$(\equiv Si)_2CH-\underset{|}{Si}-CH_2-Si\equiv \rightarrow (\equiv Si)_2CLi-\underset{|}{Si}-CH_2-Si\equiv$$

$$\xrightarrow{\equiv SiX} (\equiv Si)_3C-\underset{|}{Si}-CH_2-Si\equiv$$

Further examples come from the metallation of Si-barrelane and of C-silylated 1,3,5,7,9-pentasiladecalin [56].

4.2 Metallation of Bridging C Atoms in CCl- and SiCl-Containing 1,3-Disilapropanes

While no definite products were isolated from the reaction of $(Cl_3Si)_2CH_2$ with t-BuLi at $-78\,°C$, reaction of $Me_3Si—CH_2—SiMe_2Cl$ and $Me_3Si—CH_2—SiPh_2Cl$ with t-BuLi enabled the isolation of compounds explained by metallation of CH_2 groups and butylation of SiCl groups by the pathway

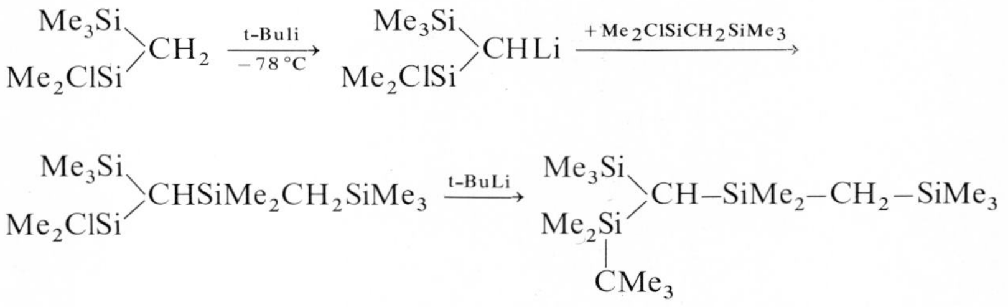

$$\begin{matrix} Me_3Si \\ \diagdown \\ CH_2 \\ \diagup \\ Me_2ClSi \end{matrix} \xrightarrow[-78\,°C]{t\text{-}BuLi} \begin{matrix} Me_3Si \\ \diagdown \\ CHLi \\ \diagup \\ Me_2ClSi \end{matrix} \xrightarrow{+\,Me_2ClSiCH_2SiMe_3}$$

$$\begin{matrix} Me_3Si \\ \diagdown \\ CHSiMe_2CH_2SiMe_3 \\ \diagup \\ Me_2ClSi \end{matrix} \xrightarrow{t\text{-}BuLi} \begin{matrix} Me_3Si \\ \diagdown \\ CH—SiMe_2—CH_2—SiMe_3 \\ \diagup \\ Me_2Si \\ | \\ CMe_3 \end{matrix}$$

4.2.1 Evidence of CCl_2-Lithiation in the Presence of SiCl Groups

Of special significance is the proof for the C-metallation of perchlorinated carbosilanes. Confirmation comes from reactions of $(Cl_3Si)_2CCl_2$ and 249 with MeI or Me_3SiI. Reaction of $(Cl_3Si)_2CCl_2$ with n-BuLi and trapping reagents produced in a yield of 80% the compounds $(Cl_3Si)_2CClMe$ and $(Cl_3Si)_2C(Cl)SiMe_3$. The reaction of 249 occurred as depicted below [126].

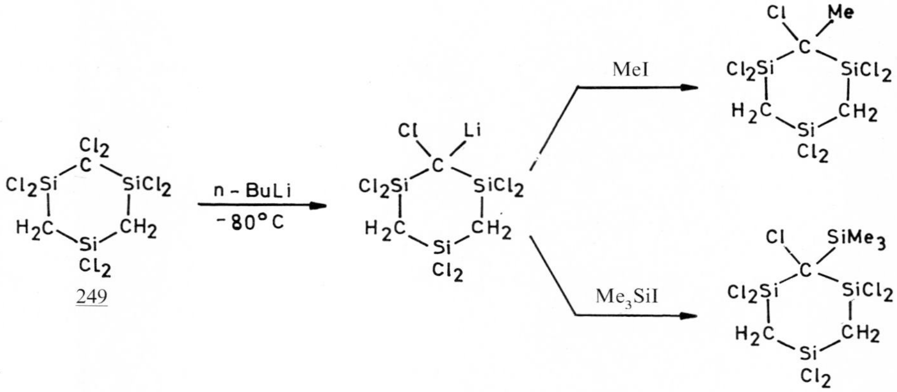

4.2.2 Metallation of $Me_3Si—CCl_2—SiMe_2Cl$

The reactions of lithiated compounds such as $(Cl_3Si)_2CClLi$ at increasing temperature are of considerable interest. As further investigations showed, LiCl precipitated due to cleavage of a Cl atom from the Si—Cl bond. Here $Me_3Si—CCl_2—SiMe_2Cl$ served as a model substance for further investigation because apart from a CCl_2 group it contained only one SiCl group [127].

The synthesis followed the reaction pathway

$$Me_3Si-CCl_2Li + Me_2SiCl_2 \rightarrow Me_3Si-CCl_2-SiMe_2Cl$$

and in the course of doing so produced further reaction with n-BuLi at $-100\ °C$:

$$Me_3Si-CCl_2-SiMeCl + n\text{-BuLi} \rightarrow Me_3Si-CCl(Li)-SiMe_2Cl$$

Reaction with MeI enabled the isolation of a compound containing a C(Cl)Me group in 90% yield:

$$Me_3Si-CCl(Li)-SiMe_2Cl + MeI \rightarrow Me_3Si-CClMe-SiMe_2Cl + LiI$$

On slowly warming the compound produced at $-100\ °C$, $Me_3Si-CCl(Li)-$ $-SiMe_2Cl$, a color change was observed ranging from light yellow through light green, blue green, blue, light green again and eventually to yellow at $-25\ °C$, at which point LiCl precipitated. 2,2,4,4-Tetramethyl-1,3-bis(trimethylsilyl)-2,4-disilabicyclo[1.1.0]-butane was isolated out of the reaction mixture as a white crystalline substance with melting point of 28 °C and structural formula shown below (see also Chapt. IV.1.6.3) [127].

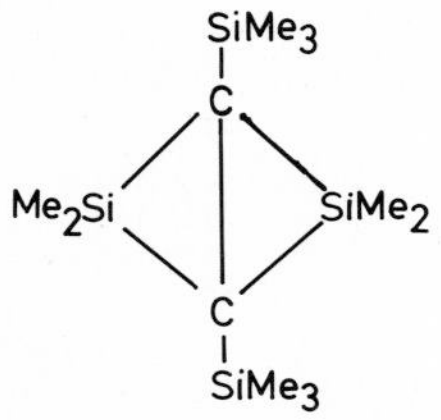

309

In contrast to hexamethylsilirane [128], which decomposes in benzene at 60–70 °C forming tetramethylethylene and dimethylsilylene, 2,4-disilabicyclo[1.1.0]butane shows a relatively higher stability. Solutions of this compound in toluene show no change after 20 hrs at 100 °C. However, its reactivity to oxygen is so large that the powdered compound ignites immediately on exposure to air.

4.2.3 Formation of 2,2,4,4-Tetramethyl-1,3-bis(trimethylsilyl)-2,4-disilabicyclo[1.1.0]-butane

To understand the formation of 2,2,4,4-tetramethyl-1,3-bis(trimethylsilyl)-2,4-disila-bicyclo[1.1.0]butane 309 an investigation was conducted first into the reactivity of 309 towards metallation agents, secondly, into the influence of various functional groups of the starting material $Me_3Si-CCl_2-SiMe_2X$ (where X = F, Cl, Br, OR, Tos) on the course of the reaction, and thirdly into the behavior of $Me_3Si-CCl_2-SiMeCl_2$ under the conditions of the synthesis of 309.

It was proved that the carbenoid $Me_3Si-CClLi-SiMe_2Cl$, formed at $-100\ °C$ in Et_2O by treating $Me_3Si-CCl_2-SiMe_2Cl$ with BuLi, yields the isomeric 1,3-disilacy-clobutanes 310 and 311 in approximately 80% yield [129] upon warming the solution to 20 °C.

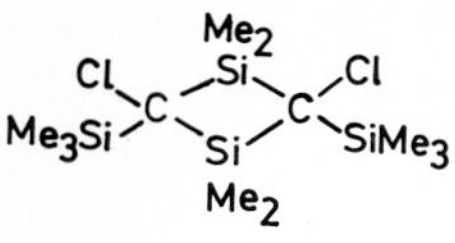

Me₃Si—C(Cl)—Si(Me₂)—C(Cl)(SiMe₃) **310**

Me₃Si—C(Cl)—Si(Me₂)—C(SiMe₃)(Cl) **311**

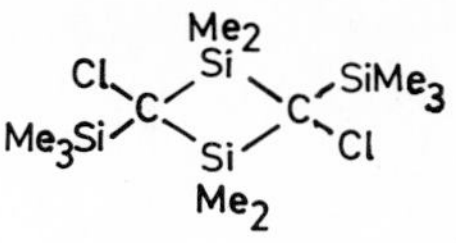

This emphasizes that 310 and 311 have to be considered as possible intermediates in the pathway leading to 309. They are also obtainable by reaction of $Me_3Si—CCl_2—$ $—SiMe_2Br$ with BuLi, provided the all-too-easy halogen exchange to produce $Me_3Si—CClBr—SiMe_2Cl$ is suppressed. Synthesis of the former compound can be achieved through reaction of $Me_3Si—CCl_2—SiMe_2H$ with Br_2 at $-40\,°C$. The direction of the overall reaction pathway is decided by the polarity of the solvent and the nature of the substituent X in $Me_3Si—CCl_2—SiMe_2X$.

Table 62. Compounds from the reaction of the cis- and trans-1,3-disilacyclobutanes 310 and 311 with BuLi in THF

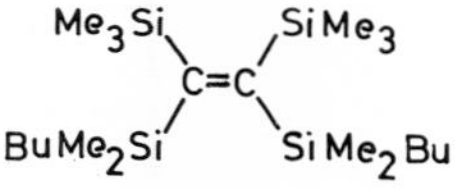

Me₃Si, BuMe₂Si\C=C/SiMe₃, SiMe₂Bu   **312**

BuMe₂Si, Me₃Si\C=C/SiMe₃, SiMe₂Bu   **313**

BuMe₂Si, BuMe₂Si\C=C/SiMe₃, SiMe₃   **314**

[Structures of 1,3-disilacyclobutanes with various H, SiMe₃, Bu substituents]

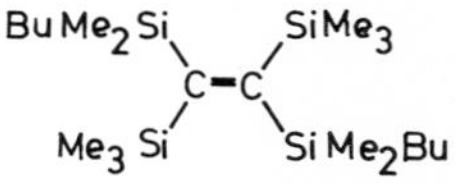

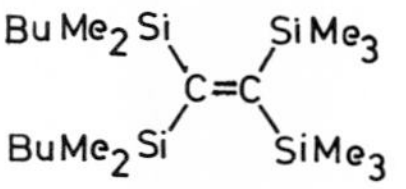

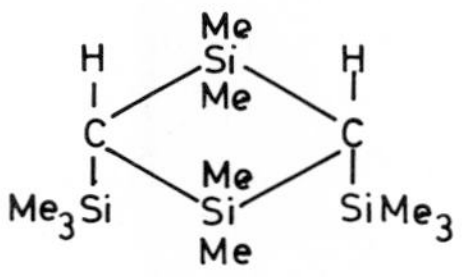

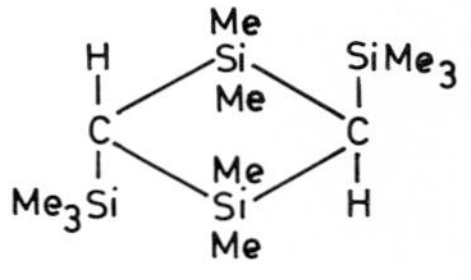

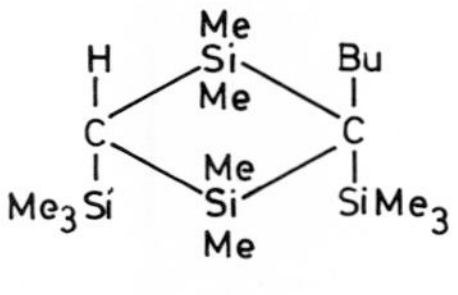

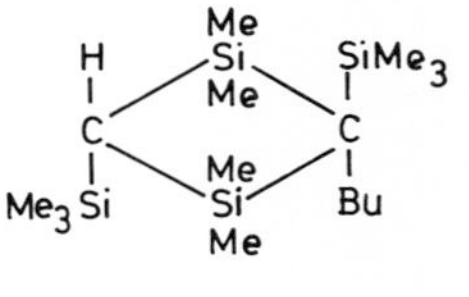

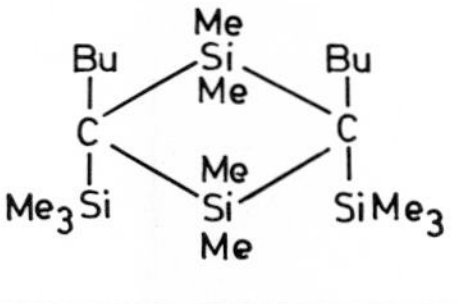

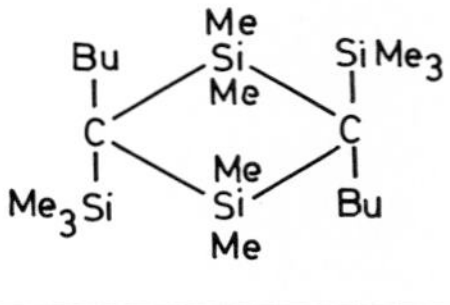

In THF, the reaction of $Me_3Si-CCl_2-SiMe_2Cl$ with BuLi generates preferably the 2,4-disilabicyclo[1.1.0]butane 309. The influence of the substituent X is emphasized in the following examples: $Me_3Si-CCl_2-SiMe_2F$ reacts with BuLi forming $Me_3Si-CCl_2-SiMe_2Bu$, while $Me_3Si-CCl_2-SiMe_2Br$ and $Me_3Si-CCl_2-SiMe_2Tos$ react via lithiation of the CCl_2 group forming the 1,3-disilacyclobutanes 310 and 311.

Attempts to effect the reaction involving the previously isolated isomeric mixture of 310 and 311 and BuLi in THF at $-105\ °C$ resulted in only a 10% yield of the bycyclic compound 309. The products of this reaction are presented in Table 62. The major products obtained were the ethylene derivatives 312, 313 and 314. It was also established that the cis compound 310 was at all times the more reactive isomer.

Thorough investigation showed that the 1,3-disilacyclobutanes 310 and 311 were actually formed as intermediates in the formation of the bicyclic 309, and that the carbenoid $Me_3Si-CClLi-SiMe_2Cl$ acts as a metallation agent in the formation of the bicyclic 309.

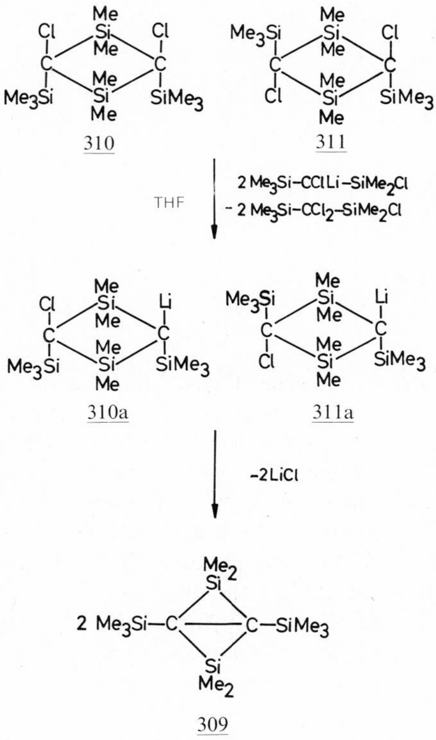

The solvent plays a deciding role. In Et_2O, the reaction of 310 and 311 with the carbenoid $Me_3Si-CCl(Li)-SiMe_2Cl$ does not proceed to the bicyclic 309; rather the carbenoid itself converts to the 1,3-disilacyclobutanes 310 and 311 as final products.

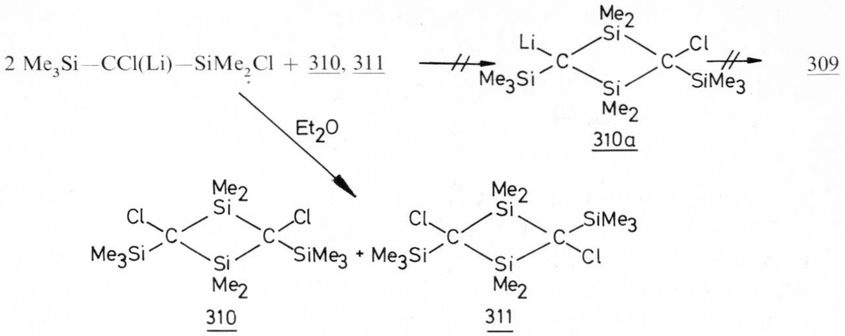

$$2 \ Me_3Si-CCl(Li)-SiMe_2Cl \ + \ \underline{310}, \ \underline{311} \ \longrightarrow\!\!/\!\!/ \quad \underset{\underline{310a}}{\underset{Me_3Si}{\overset{Li}{\diagdown}}\overset{}{C}\underset{}{\overset{Me_2}{\underset{Si}{\diagup}\underset{Me_2}{\diagdown}}}C\overset{Cl}{\underset{SiMe_3}{\diagup}}} \ \longrightarrow\!\!/\!\!/ \quad \underline{309}$$

↓ Et₂O

$$\underset{\underline{310}}{\underset{Me_3Si}{\overset{Cl}{\diagdown}}C\underset{Si}{\overset{Me_2}{\diagup}\diagdown}C\overset{Cl}{\underset{SiMe_3}{\diagup}}} \ + \ \underset{\underline{311}}{Me_3Si\diagdown C\underset{Si}{\overset{Me_2}{\diagup}\diagdown}C\overset{SiMe_3}{\underset{Cl}{\diagup}}}$$

With MeLi, the 1,3-disilacyclobutanes <u>310</u> and <u>311</u> react to produce the silylated ethylene derivative $(Me_3Si)_2C=C(SiMe_3)_2$ <u>315</u>.

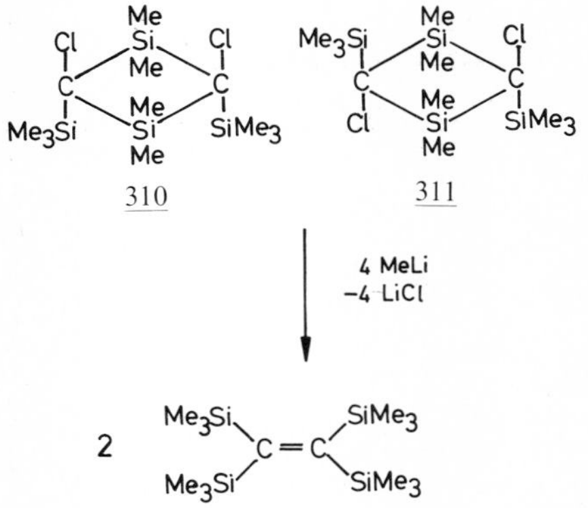

<u>310</u>   <u>311</u>

↓ 4 MeLi
  −4 LiCl

$$2 \quad \underset{Me_3Si}{\overset{Me_3Si}{\diagdown}}C=C\overset{SiMe_3}{\underset{SiMe_3}{\diagup}}$$

<u>315</u>

It is critical for clarification of the mechanism of formation that the bicyclic compound <u>309</u> on reaction with BuLi in THF should produce the same compounds which result from the reaction of the 1,3-disilacyclobutanes <u>310</u> and <u>311</u> and which are presented in Tab. 62. Further, the bicyclic <u>309</u> reacts with MeLi yielding $(Me_3Si)C=C(SiMe_3)_2$. This shows that <u>310</u> and <u>311</u> are proven intermediates for the formation of <u>309</u>, and that they are lithiated by $Me_3Si-CCl(Li)-SiMe_2Cl$ generating the bicyclic system <u>309</u> through LiCl elimination. Performing the reaction with n-BuLi in THF or MeLi produces the bicyclic <u>309</u> according to the reaction pathway just described. As soon as the bicyclic <u>309</u> forms, though, it will degrade, principally to a number of silylated ethylene derivatives [130].

4.2.4 The Reactive Behavior of 2,2,4,4-Tetramethyl-1,3-bis(trimethylsilyl)-2,4-disilabicyclo[1.1.0]butane

Reaction with 2 mole equivalents of HBr in pentane proceeds at −90 °C according to the equation

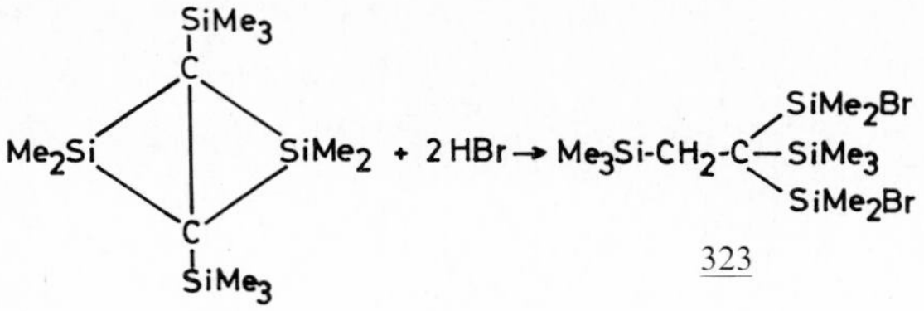

From the almost exclusive formation of 323 it can be assumed that cleavage of the three-membered ring proceeds in stages, and that the silacyclopropane 324 will appear as an intermediate. Compound 324 is indeed formed from 309 and 0.7 mole equivalents of HBr [127].

In this three-membered ring, cleavage of the Si—C bond is a sterically easier process than in the starting material. In addition to this, the SiBr group will influence the polarization of the neighboring Si—C bonds in such a way that cleavage of the second silacyclopropane ring only yields the asymmetric Me_3Si—CH_2— —$C(SiMe_3)(SiMe_2Br)_2$, 323.

In a totally corresponding way, reaction with methanol at −20 °C in THF follows the pathway

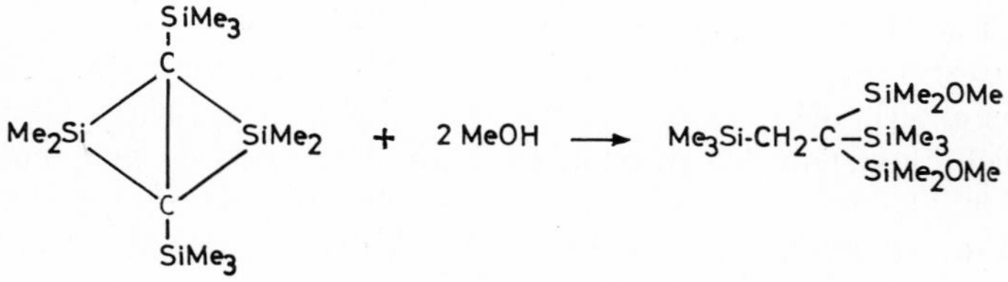

The bicyclic compound reacts violently with Br_2. However, this reaction is adequately controlled in pentane at −78 °C. The reaction products detected were Me_3SiBr, Me_2SiBr_2 and $BrMe_2Si$—$C\equiv C$—$SiMe_2Br$ in 52%, 15% and 17% yield, respectively.

The reaction pathway presented below shows that Si—C bond cleavage occurs by addition of Br_2 followed by β-elimination [127].

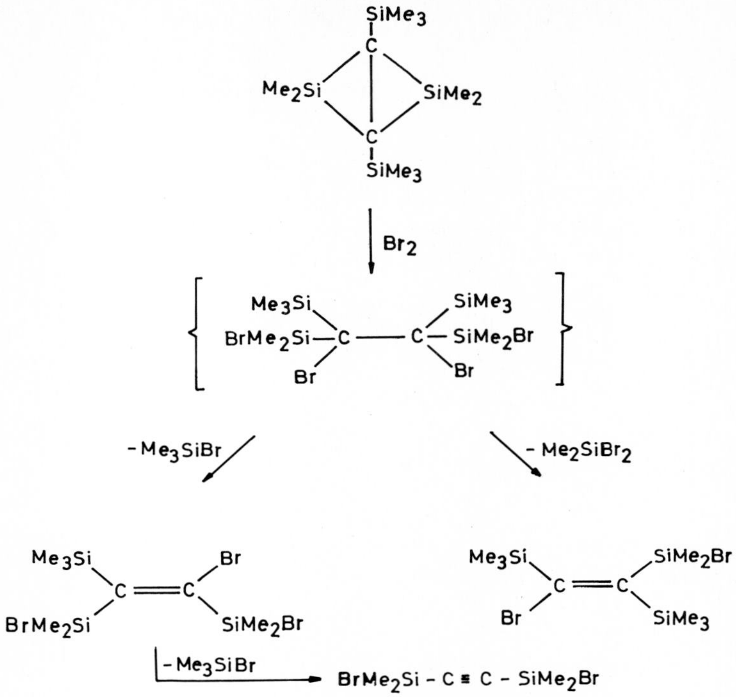

4.2.5 Lithiation of Me₃Si—CCl₂—SiMeCl₂

4.2.5 Lithiation of $Me_3Si-CCl_2-SiMeCl_2$

Reaction of $Me_3Si-CCl_2-SiMeCl_2$ with lithium in a mole ratio of 1:2 in THF at
−50 °C proceeds to the ethylene derivatives 316 and 317 in 74% yield. These compounds are converted to 318 and 319 on reaction with $LiAlH_4$ [130].

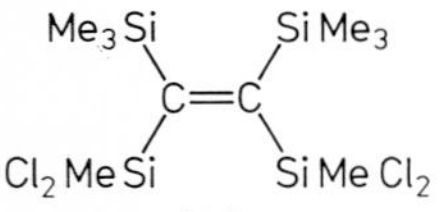

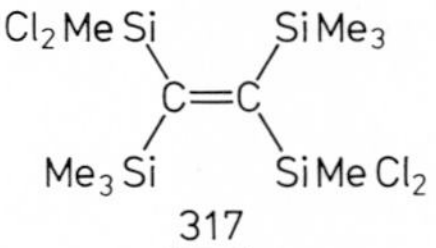

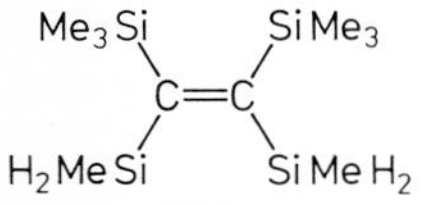

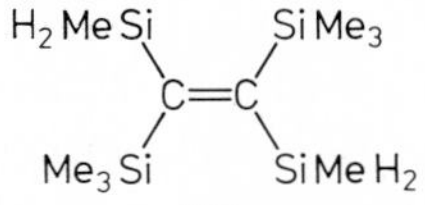

In fact, this shows the different behavior of the carbenoids $Me_3Si-CCl(Li)-$
$-SiMe_2Cl$ and $Me_3Si-CCl(Li)-SiMeCl_2$. The formation of these ethylene derivatives does not proceed via a carbene reaction. Rather, it is based on an intermolecular
elimination of LiCl, in which one carbenoid molecule acts as a nucleophile, the other
one as an electrophile, forming initially the intermediate 320:

$$
\begin{array}{c}
\text{Me}_3\text{Si} \overset{\text{Cl}}{\underset{\text{SiMeCl}_2}{\diagdown\!\!\!\!\text{C}\!\!\!\!\diagup}} \text{Li}
\\[4pt]
\text{Cl}_2\text{MeSi}\overset{\text{Li}}{\underset{\text{SiMe}_3}{\diagdown\!\!\!\!\text{C}\!\!\!\!\diagup}}\text{Cl}
\end{array}
$$

$-\text{LiCl}$

$$
\begin{array}{c}
\text{Cl}_2\text{Me Si}\diagdown \qquad \diagup\text{SiMe Cl}_2 \\
\text{Me}_3\text{Si}-\text{C}-\text{C}-\text{SiMe}_3 \\
\diagup \qquad \diagdown \\
\text{Cl} \qquad \text{Li}
\end{array}
$$
320

β–elimination
$-\text{LiCl}$

$$
\underset{\textbf{316}}{
\begin{array}{c}
\text{Me}_3\text{Si}\diagdown \qquad \diagup \text{Si Me}_3 \\
\text{C}=\text{C} \\
\text{Cl}_2\text{Me Si}\diagup \qquad \diagdown \text{Si MeCl}_2
\end{array}}
\qquad
\underset{\textbf{317}}{
\begin{array}{c}
\text{Cl}_2\text{Me Si}\diagdown \qquad \diagup \text{Si Me}_3 \\
\text{C}=\text{C} \\
\text{Me}_3\text{Si}\diagup \qquad \diagdown \text{Si MeCl}_2
\end{array}}
$$

The presence of the intermediate 320 could be demonstrated by means of a trapping reaction with MeI and by the isolation of 321 and 322.

$$
\underset{\textbf{321}}{
\begin{array}{c}
\text{Me}_3\text{Si}\diagdown \qquad \diagup \text{Si Me}_3 \\
\text{Cl}_2\text{Me Si}-\text{C}-\text{C}-\text{Si Me Cl}_2 \\
\diagup \qquad \diagdown \\
\text{Me} \qquad \text{Cl}
\end{array}}
\qquad
\underset{\textbf{322}}{
\begin{array}{c}
\text{Me}_3\text{Si}\diagdown \qquad \diagup \text{Si Me}_3 \\
\text{Cl}_2\text{Me Si}-\text{C}-\text{C}-\text{Si Me Cl}_2 \\
\diagup \qquad \diagdown \\
\text{Me} \qquad \text{H}
\end{array}}
$$

The formation of compound 322 begins with a C-metallation of 321, followed by an ether cleavage reaction.

4.3 The Si-Metallation of 1,3,5-Trisilacyclohexanes with Transition Metal Complexes

The investigations on the Si—C skeleton of 1,3,5-trisilacyclohexanes and 1,3,5-trisilapentanes presented previously showed that Si-substituents (H, halogens, alkyls and aryls) and C-substituents (Cl, Br, Li) influence not only the reactivity of remaining substituents but also the stability of the Si—C skeleton. The formation of transition metal complexes of 1,3,5-trisilacyclohexanes therefore was undertaken. The formation of bonds between Si atoms and the transition metals in such complexes could be generated via known methods, such as elimination of NaCl from the SiCl group and sodium carbonylates, or the elimination of hydrogen from the SiH group of a silane and a carbonyl hydride [161, 162].

4.3.1 Si-Metallation via Salt Elimination

4.3.1.1 Singly Metallated Trisilacyclohexanes

These reactions are understood to operate through nucleophilic substitution of the brominated Si atom, whereby the carbonylate anion attacks the electropositive Si atom. Here 1-bromo-1,3,5-trisilacyclohexane was chosen for this reaction, because it has a SiBr group suitable for salt elimination and because the remaining Si—H groups in the reaction product are still capable of undergoing further reactions. The various transition metal carbonyl complexes formed are shown in Table 63 [163]. Just as in the preparation of the simple Si transition metal compounds of group VI, the reactions were run in methylcyclohexane [164].

The reaction of 1-bromo-1,3,5-trisilacyclohexane with $KFe(CO)_2cp$ in THF at $-50\,°C$ generated byproducts which impeded the separation of $\underline{375}$. Similarly, the reaction of $KCo(CO)_4$ in Et_2O at $-20\,°C$ produced a large number of byproducts, favored because of the cleavage of Si—Co bonds in polar solvents [165] and leading to $HCo(CO)_4$ and $Co_2(CO)_8$. These can attack the SiH groups on the ring system, thereby excluding the possibility of the formation of a $Co(CO)_4$-substituted 1,3,5-trisilacyclohexane in homogeneous solution. On the other hand, the heterogeneous formation of compound $\underline{376}$ is possible.

If the stabilities of all the compounds shown in Tab. 63 are compared, $\underline{375}$ is found to be the most stable to oxidation. Considerably more reactive are $\underline{372}$, $\underline{373}$ and especially $\underline{376}$, while their $Mn(CO)_5$- and $cp(CO)_3W$-substituted derivatives have an intermediate reactivity. Compounds $\underline{371}$, $\underline{372}$ and $\underline{373}$ are inherently more stable against more polar solvents such as Et_2O and THF, and also against irradiation, than R_3Si—$M(CO)_3cp$ (M = W, Mo, Cr) [164]. NMR and IR investigations indicate that because of metal substitution, a weakening of the Si-H bond of the metal-substituted Si atom will occur.

4.3.1.2 Doubly Metallated Trisilacyclohexanes

The formation of disubstituted 1,3,5-trisilacyclohexanes from the reaction between metal carbonylates and 1,3-dibromo-1,3,5-trisilacyclohexanes was investigated to determine to what extent the already described electronic effects in compounds $\underline{371}$–$\underline{376}$ are affected by an increase in the degree of substitution. Because the influence exerted by carbonylate substituents on the ring system of 1,3,5-trisilacyclohexane does not depend on the particular transition metal, as can be seen from the previous section, reactions were undertaken using only transition metal carbonylates containing different numbers of carbonyl groups, such as $NaW(CO)_3cp$, $KFe(CO)_2cp$, $KMn(CO)_5$ and $KCo(CO)_4$, in order to determine exactly what steric factors are involved in these reactions. The 1,3-dibromo-1,3,5-trisilacyclohexane was used as a cis, trans isomeric mixture [64]. The reactions carried out in methylcyclohexane generated Si-metallated compounds $\underline{377}$, $\underline{378}$, and $\underline{379}$, as shown in Table 64.

Depending on how nucleophilic the anion is, the reaction takes between 2 to 4 days to complete. The chemical work-up of $\underline{377}$ was without problems because almost no byproducts were formed, and because of the fact that a large difference in solubility exists between the substituted carbosilane and $Mn_2(CO)_{10}$ in methylcyclohexane. Isolation of $\underline{378}$, however, is not so easy. The toilsome working up of $\underline{379}$ is explained first by its thermal instability and secondly by the tendency of this compound to react with byproducts.

It is surprising that the yield of the crystalline disubstituted compounds $\underline{377}$, $\underline{378}$ and $\underline{379}$ does not exceed the percentage contribution of the cis 1,3-dibromo-1,3,5-trisilacyclohexane in the isomeric mixture of approximately 33%. This suggests that

Table 63. Compounds from the reaction of 1-bromo-1,3,5-trisilacyclohexane with metal carbonylates

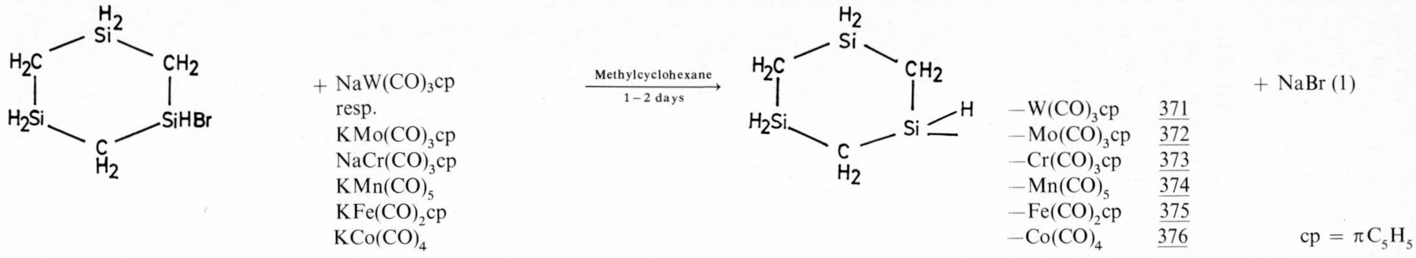
H2 Si
H2C CH2
H2Si SiHBr
C H2
+ NaW(CO)3cp
resp.
KMo(CO)3cp
NaCr(CO)3cp
KMn(CO)5
KFe(CO)2cp
KCo(CO)4
Methylcyclohexane
1–2 days
H2 Si
H2C CH2
H2Si Si H
C H2
—W(CO)3cp 371
—Mo(CO)3cp 372
—Cr(CO)3cp 373
—Mn(CO)5 374
—Fe(CO)2cp 375
—Co(CO)4 376
+ NaBr (1)
cp = πC5H5

Compound	mp. °C	Yield with respect to the carbosilane
371 white needles	86–87	43%
372 pink needles	80–83	39%
373 impure green liquid		
374 yellow needles	36	41%
375 bright yellow blocks	34	45%, in THF 20%
376 brown viscous liquid	−5 to −10	32%

Table 64. Compounds generated from the reaction of 1,3-dibromo-1,3,5-trisilacyclohexane with metal carbonylates [163]

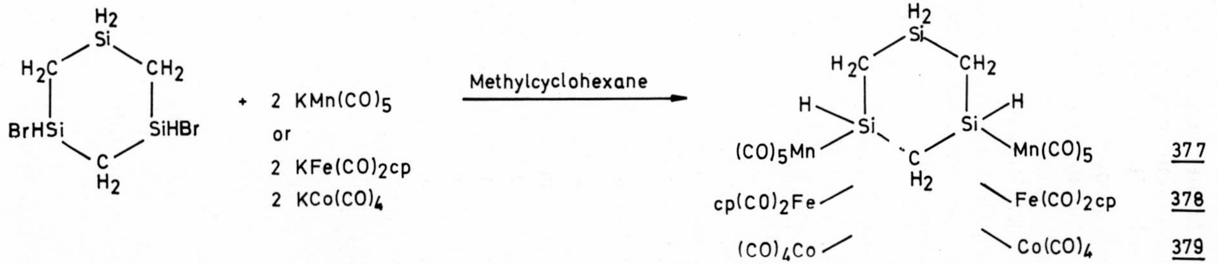
H2 Si
H2C CH2
BrHSi SiHBr
C H2
+ 2 KMn(CO)5
or
2 KFe(CO)2cp
2 KCo(CO)4
Methylcyclohexane
H2 Si
H2C CH2
H Si Si H
(CO)5Mn Mn(CO)5 377
cp(CO)2Fe Fe(CO)2cp 378
(CO)4Co Co(CO)4 379
C H2

Compound	mp. °C	Yield with respect to the carbosilane
377 white needles	76–78	38%
378 yellow needles	157	28%; in THF 17%
379 metallic golden lamina	69 (decomposition)	33%

the metal-substituted compounds 377, 378 and 379 exist in the cis arrangement. The somewhat higher yields of the cobalt and manganese compounds may possibly indicate that in comparison to the $Fe(CO)_2cp$ substituent the smaller $Mn(CO)_5$ and $Co(CO)_4$ groups satisfy the more compact space requirement of the axial positions on the ring. These substituted trans isomers increase the overall yield to more than 33%.

In comparison to the previous section, the ^1H-NMR spectra of the dimetal substituted 1,3,5-trisilacyclohexanes show no exceptional characteristics. The compounds maintain a fixed molecular conformation in solution. The IR spectra of 378 and 379 show that the SiH-metal stretching band is shifted to a lower wave number, indicating a further weakening of the Si—H bond.

While in compounds 371 to 376 a medium to strong SiH_2 band appeared, in compounds 377, 378 and 379 this peak was much weaker and is obviously split into a doublet. However, a distinctly clear absorption pattern was observed for the remaining peaks in the spectra of 371 to 376, as well as 377 and 379. Differences nevertheless are observed in the frequencies of similar infrared vibrations for comparable compounds [163].

4.3.1.3 Triply Metallated Trisilacyclohexanes

Reactions of $(HBrSi—CH_2)_3$ with $KFe(CO)_2cp$ or $KMn(CO)_5$ generate dimetal substituted derivatives, and only in the case of $KCo(CO)_4$ is the formation of the trimetallo species 383 achieved, along with a tetrasubstituted 1,3,5-trisilacyclohexane species. The compounds are presented in Table 65. The thermal stability of these $Co(CO)_4$-substituted 1,3,5-trisilacyclohexanes decreases as the number of $Co(CO)_4$ substituents coordinated to the ring increases.

4.3.1.4 Reaction of $(Cl_2Si—CH_2)_3$ with $KFe(CO)_2cp$

Reaction with $(Cl_2Si—CH_2)_3$ is of particular interest because the products obtained still maintain active functional groups. The Si atoms bear two functional groups, so that as a result of nucleophilic attack sterically favored positions on this ring can be selected. The reaction of $(Cl_2Si—CH_2)_3$ with 1 mole equivalent of $KFe(CO)_2cp$ in unpolar solvents generated compound 384, which was isolated in the form of light yellow crystals of mp. 113 °C. The side reactions observed frequently in etheral solvents did not occur.

The reaction performed with a mole ratio of 1:2 in methylcyclohexane yielded similarly 384. In the more polar solvent THF, the tendency for the metal carbonylate to cause nucleophilic attack increases to such an extent, that a two-fold substitution generated 385, which was isolated in the form of yellow crystals with mp. 134–136 °C.

384   385

Table 65. Compounds produced by the reaction of $(BrHSi-CH_2)_3$ with metal carbonylates

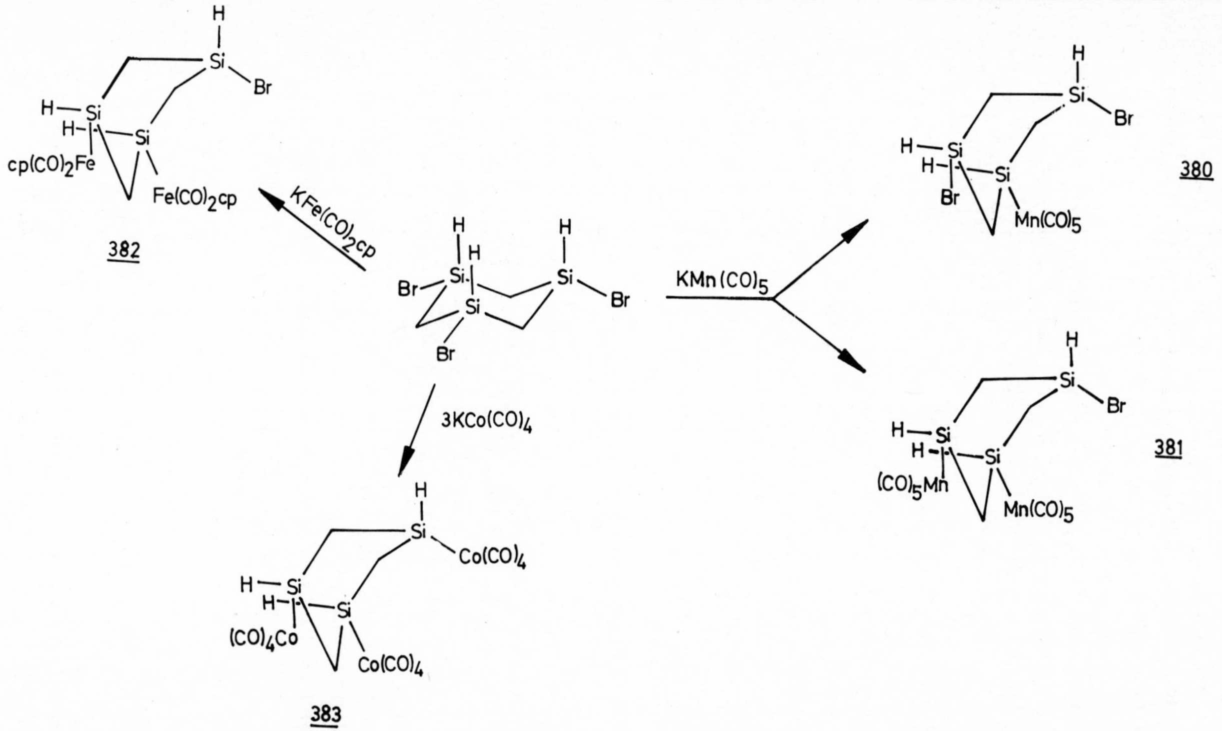

Compound	mp. °C	Yield with respect to the carbosilane
380, 381 white amorphous substances		
382 yellow squared crystals	164–166	19 %; in THF 13 %
383 brown crystals	106–109	22 %

4.3.2 Cobalt-Substituted 1,3,5-Trisilacyclohexane Complexes

The investigations described so far indicate that reactions of metal carbonylates in general lead only to the coordination of two metal carbonyl groups to the 1,3,5-trisilacyclohexanes. From the chemistry of silylcobalt compounds it is known that silanes of the type $R_{4-n}SiH_n$ (n = 1–3) cleave dinuclear $Co_2(CO)_8$, producing silyl-cobalt-carbonyl complexes [166, 167, 171].

Utilizing the method just described, higher metal-carbonyl-substituted 1,3,5-trisilacyclohexanes can be achieved, as the reactions of compounds 383, 379 and 376 with $Co_2(CO)_8$ show [163].

4.3.2.1 Multiply Cobalt-Carbonylate-Substituted Trisilacyclohexanes

Reaction of 383 with $Co_2(CO)_8$ in a mole ratio of 2:1 at 0 °C in pentane generates compound 386 in the form of a yellow amorphous powder, mp. 95 °C:

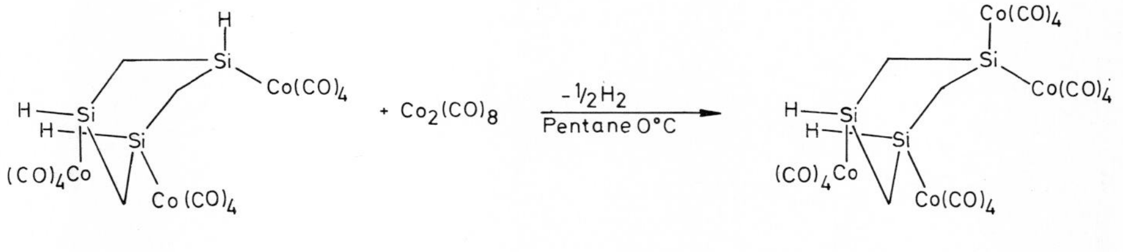

383 386

A higher stage of substitution cannot be achieved. On the contrary, because two SiH_2 groups exist in compound 379, the possibility of generating two different molecules exist.

$$>SiH_2 \xrightarrow{Co_2(CO)_8} \begin{cases} >Si[Co(CO)_4]_2 + H_2 \\ >SiCo_2(CO)_7 + CO + H_2 \end{cases}$$

Reaction of 379 with $Co_2(CO)_8$ in a mole ratio of 1:1 proceeded almost exclusively to 386:

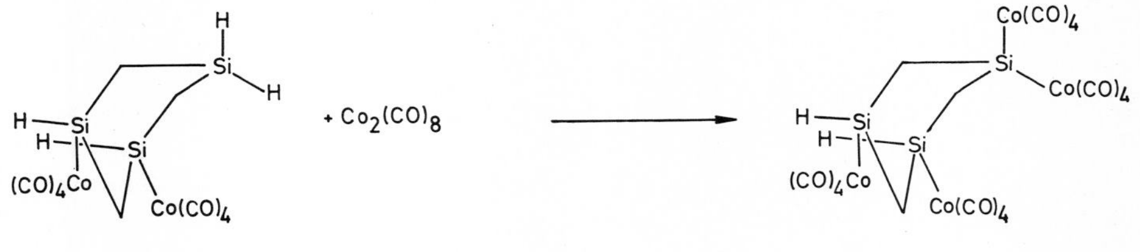

379 386

The reaction of 376 with $Co_2(CO)_8$ in a mole ratio of 2:3 proceeded to 386 and 387:

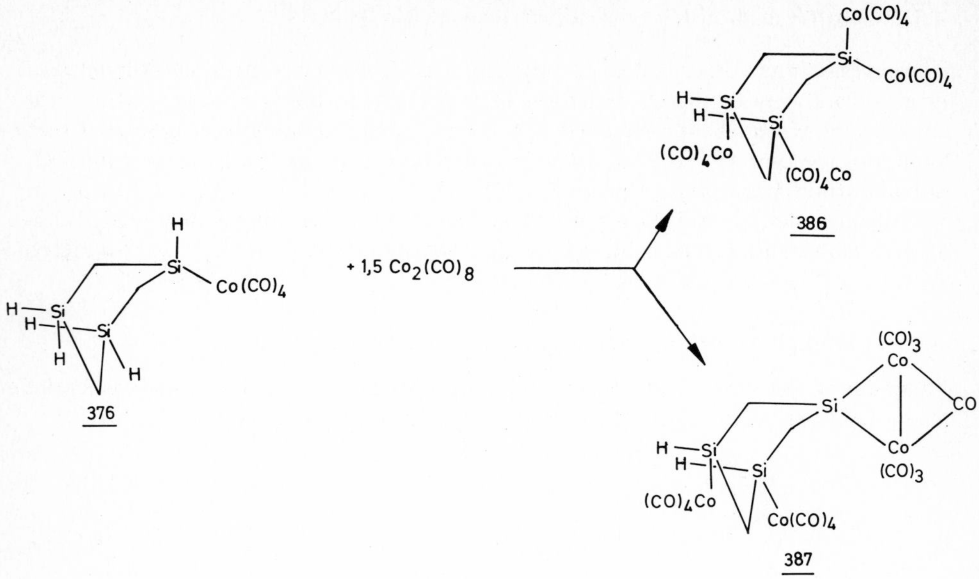

386

387

+ 1,5 Co$_2$(CO)$_8$

376

4.3.2.2 Reactions of (H$_2$Si—CH$_2$)$_3$ with Co$_2$(CO)$_8$

It was desired to make compound 376 accessible via this reaction pathway and at the same time to determine to what extent the Co$_2$(CO)$_7$-substituted derivative 388 is generated.

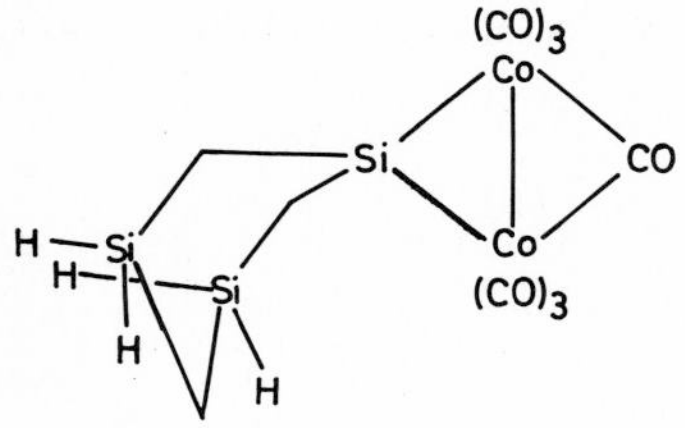

388

Compounds 379, 376 and 388 were produced in the approximate mole ratio of 2:3:5, showing that the formation of the heptacarbonyl compound is preferred.

In all of our experiments, the hydrogen-elimination method produced various substituted 1,3,5-trisilacyclohexanes if two or three SiH$_2$ groups existed originally in the starting material. The compounds formed can be classified simply as Co(CO)$_4$- and Co$_2$(CO)$_7$-substituted 1,3,5-trisilacyclohexanes. Figure 18 presents the compounds which to date have been identified or isolated.

Compound 386 is the product of reactions proceeding exclusively via compounds containing Co(CO)$_4$ groups. The formation of Co$_2$(CO)$_7$ units via elimination of CO did not occur. The formation of heptacarbonyl derivatives was favored only if two SiH$_2$ groups existed in the starting compound. It seems that (H$_2$Si-CH$_2$)$_3$ reacts more rapidly to produce 388 than to generate 376. Compound 376 itself formed the preferred 387 and only small amounts of 379. Beginning with 379 a chain of reactions occurred, producing eventually the tetra-substituted compound 386.

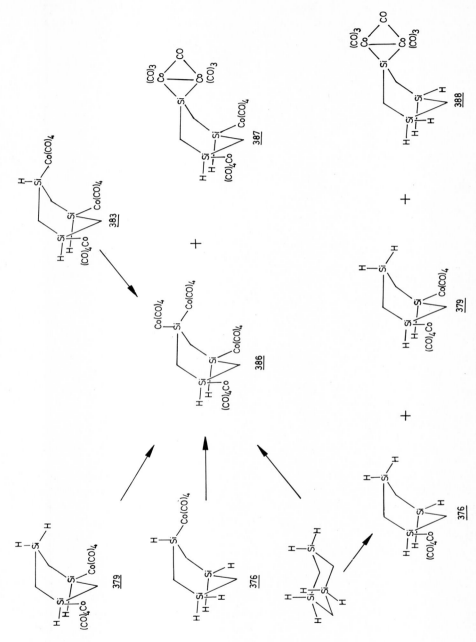

Fig. 18. Reactions with $Co_2(CO)_8$ starting from $(H_2Si—CH_2)_3$ and from the compounds $\underline{376}$, $\underline{379}$ and $\underline{383}$

4.3.3 Influence of cp(CO)₂Fe Groups on the Reactive Behavior of Trisilacyclohexanes

The reactions undertaken with n-BuLi were those involving $(H_2Si-CH_2)_3$, $(Cl_2Si-CH_2)_3$ and $(F_2Si-CH_2)_3$, along with the cp(CO)₂Fe-substituted derivatives 375, 389 and 384.

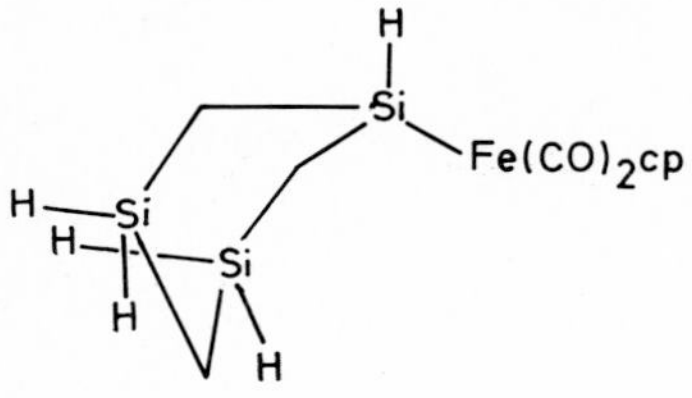

375

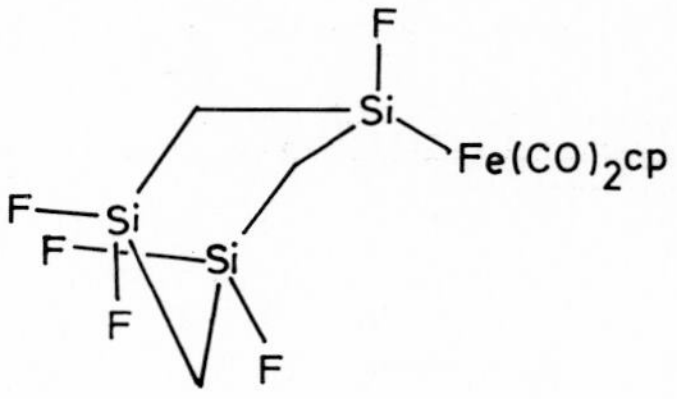

389

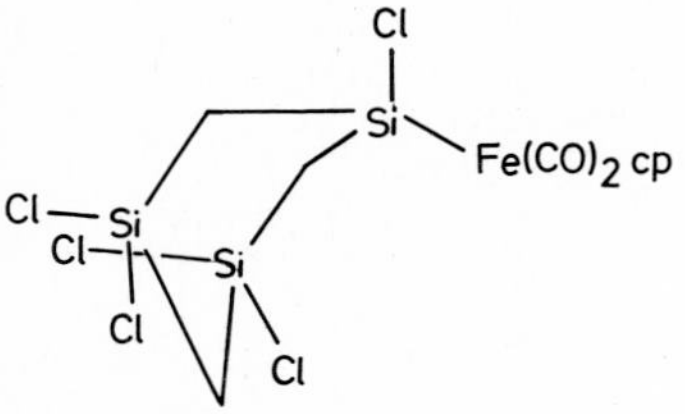

384

The reactions were undertaken in order to determine the effects of substituents on the reactivity of the above-mentioned compounds. The compounds $(H_2Si-CH_2)_3$ and $(F_2Si-CH_2)_3$ always underwent Si-substitution, because the acidity of the C—H bond near the SiH₂ and SiF₂ groups was too low. C-metallation could not be proved to occur even at —100 °C.

In the Si-functional (SiH, SiCl, SiF) 1,3,5-trisilacyclohexanes 249, 289 and 302 containing a CCl₂ group, KFe(CO)₂cp always attacks the CCl₂ group, followed by a stepwise ether cleavage yielding first the CHCl and then the CH₂ group.

The Si-chlorinated cp(CO)₂Fe-substituted compound 384 reacts with LiAlH₄ by means of complete Cl/H exchange without Fe—Si bond cleavage to produce the Si-hydrogenated compound 375. Similarly, 384 reacts with ZnF₂ to yield the Si-fluorinated 389, which is also accessible from $(F_2Si-CH_2)_3$ interaction with KFe(CO)₂cp [168].

Compared to $(H_2Si-CH_2)_3$, incorporation of a cp(CO)₂Fe group into 375 leads to a higher Si—H bond reactivity. Butylation of the Si atom is observed at —105 °C, which occurs, however, at the expense of Si—Fe bond cleavage. Compound 389 readily undergoes butylation at —80 °C without any significant Si—Fe bond cleavage. Such cleavage occurs only with an excess of n-BuLi. By this means it was possible to convert one or two SiF₂ groups to Si(F)n-Bu groups without causing disruption to the Si—Fe bond. Subsequent F/H exchange occurs on reaction with LiAlH₄, producing the Si-hydrogenated compounds 390 or 390a. Compound 390a is accessible also via the butylation of 389, but in only very small yield.

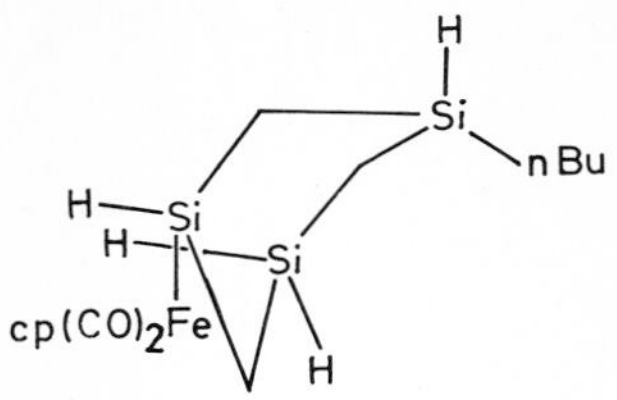

390

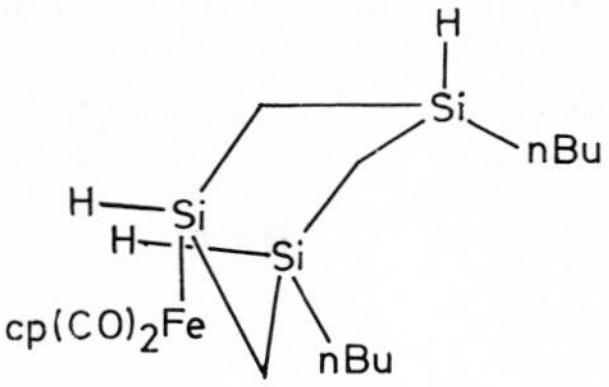

390a

The Si-chlorinated $cp(CO)_2Fe$-substituted compound 384 reacts with n-BuLi at $-78\ °C$ in THF via C-metallation to yield 391:

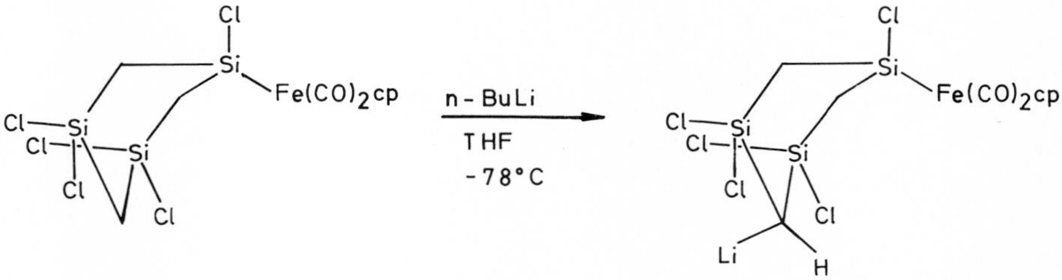

384

391

This reaction direction could not be proved using $(Cl_2Si-CH_2)_3$, but the lithiation of compound 384 was proven by reaction of 391 with Me_3SiCl or MeI:

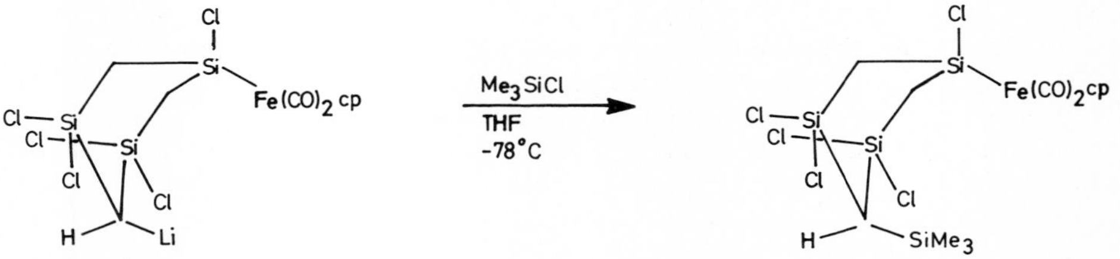

391

392

The $cp(CO)_2Fe$ substituent in 384 obviously deactivates the SiCl groups to such an extent that protection of these functional groups via methylation or phenylation before metallation of the CH_2 group is not necessary. Hence, this reaction provides an alternative pathway to the photochlorination of a CH_2 group in Si-chlorinated 1,3,5-trisilacyclohexanes. It appears unusual that the CH_2 group in position 4 of 384 is lithiated, while those in positions 2 and 6 remain unaffected. If 392 reacts with $LiAlH_4$ or ZnF_2, the $cp(CO)_2Fe$-substituted C-silylated compounds 393 and 394 are generated, which have so far not been accessible by other means [168].

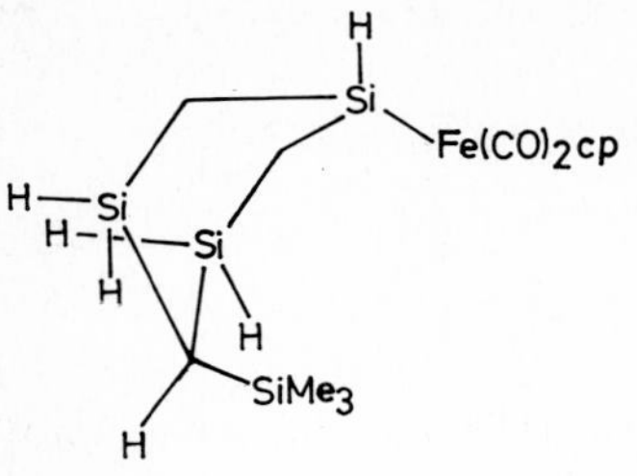

393

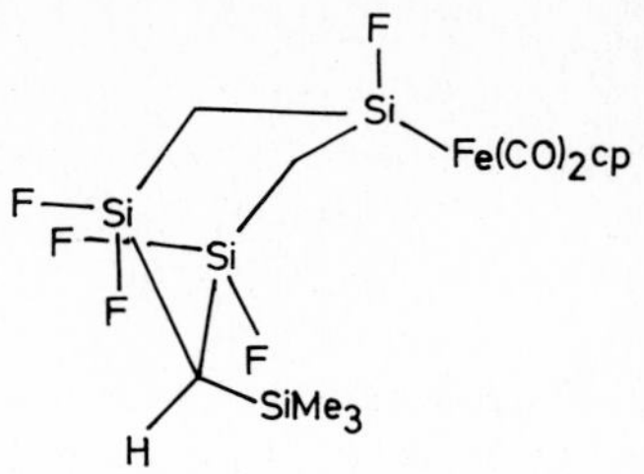

394

5. Reactions of C-Chlorinated Carbosilanes with Silylphosphanes

5.1 Formation and Reactions of the Ylide $(Cl_3Si)_2CPMe_2Cl$ and the Effect of Substituents on Ylide Formation

In perchlorinated carbosilanes, the reactivity of neighboring SiCl groups is so reduced due to chlorination of bridging C atoms that characteristic reactions of the SiCl group disappear. This is illustrated by reactions of $(Cl_2Si-CCl_2)_3$ and $(Cl_3Si-CCl_2)_2SiCl_2$ with MeMgCl and alkyllithium reagents. What now must be described are the reactions of perchlorinated carbosilanes such as $(Cl_3Si)_2CCl_2$, $(Cl_3Si-CCl_2)_2SiCl_2$ and $(Cl_2Si-CCl_2)_3$ with $LiPMe_2$ and $Me_3Si-PMe_2$, which begin by attack on the CCl_2 group followed by spontaneous rearrangement, shown in the case of an Si-chlorinated compound, yielding the corresponding ylides:

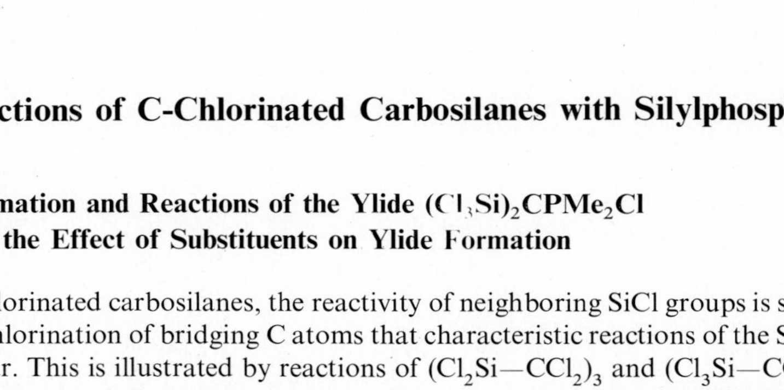

Crystal structure analysis confirms the ylidic form of 325 and shows that both C—P and C—Si bonds are notably shortened [132]. Compound 325 also is formed from $(Cl_3Si)_2CCl_2$ on cleavage of P_2Me_4:

$$(Cl_3Si)_2CCl_2 + Me_2P-PMe_2 \rightarrow (Cl_3Si)_2CPMe_2Cl + Me_2PCl$$

The formation of the ylide 325 is favored through Si-chlorination. In the reaction of $(Me_2ClSi)_2CCl_2$ with $Me_3Si-PMe_2$ or $LiPMe_2$ in a lesser mole ratio (0.05 to

0.2 mole), the yield of $(Me_2ClSi)_2CPMe_2Cl$ could be determined by NMR spectroscopy to be only 1%, while C- and Si-substitution represented the major part of the reaction. The $(Me_3Si)_2CCl_2$ reacts with $Me_3Si-PMe_2$ via C-phosphorylation

$$(Me_3Si)_2CCl_2 + Me_3Si-PMe_2 \rightarrow Me_3Si-CCl(PMe_2)-SiMe_3 + Me_3SiCl$$

Ylide formation did not occur in this case.

In contrast to this, $MeCl_2Si-CCl_2-SiCl_3$ readily reacts with Me_2PLi to form ylides:

$$MeCl_2Si-CCl_2-SiCl_3 + LiPMe \rightarrow (MeCl_2Si)(Cl_3Si)CPMe_2Cl + LiCl$$

The same influence exerted by Si- and C-chlorination is shown even in simple silanes. Thus Me_3SiCH_2Cl reacts with $LiPMe_2$ to produce $Me_3Si-CH_2-PMe_2$, and correspondingly Cl_3Si-CH_2Cl reacts with $Me_3Si-PMe_2$ to produce $Me_2PSiCl_2-CH_2Cl$. These reactions yield no ylides, while $Cl_3Si-CCl_3$ with $LiPMe_2$ reacts immediately forming an ylide:

$$Cl_3Si-CCl_3 + LiPMe_2 \rightarrow (Cl_3Si)(Cl)CPMe_2Cl + LiCl$$

Ylide formation is possible for 1,3,5-trisilapentanes in a way corresponding to the previously-described 1,3-disilapropanes. For instance, $Cl_3Si-CH_2-SiCl_2-CCl_2-SiCl_3$ reacts with $LiPMe_2$ forming an ylide according to

$$Cl_3Si-CH_2-SiCl_2-CCl_2-SiCl_3 +LiPMe_2 \rightarrow (Cl_3Si-CH_2-SiCl_2)(Cl_3Si)CPMe_2Cl$$

and $(Cl_3Si-CCl_2)_2SiCl_2$ reacts with 0.1 mole equivalents of $LiPMe_2$

$$(Cl_3Si-CCl_2)_2SiCl_2 + LiPMe_2 \rightarrow (Cl_3Si-CCl_2-SiCl_2)(Cl_3Si)CPMe_2Cl + LiCl$$

while $(Cl_3Si-CCl_2)_2SiCl_2$ forms a diylide on reaction with 2 mole equivalents of $Me_3Si-PMe_2$ [132]:

$$(Cl_3Si-CCl_2)_2SiCl + 2 Me_3Si-PMe_2 \rightarrow$$
$$\rightarrow Cl_3Si-C(PMe_2Cl)-SiCl_2-C(PMe_2Cl)-SiCl_3 + 2 Me_3SiCl$$

In these Si- and P-chlorinated ylides, the P—Cl bond is the most reactive one. It reacts readily with MeLi:

$$(Cl_3Si)_2CPMe_2Cl + MeLi \rightarrow (Cl_3Si)_2CPMe_3 + LiCl$$

Substitution of the P—Cl group also occurs in the reactions with $LiPMe_2$, with $Me_3Si-PMe_2$ and with Me_2P-PMe_2, all yielding the same product [132]:

$$(Cl_3Si)_2CPMe_2Cl + \begin{Bmatrix} LiPMe_2 \\ Me_3Si-PMe_2 \\ Me_2P-PMe_2 \end{Bmatrix} \rightarrow (Cl_3Si)_2CPMe_2PMe_2$$

5.2 The Triylide $(Cl_2Si—CPMe_2Cl)_3$

Reactions of $LiPMe_2$ and $Me_3Si—PMe_2$ with linear perchlorinated carbosilanes, producing ylides [212] apply also to reactions involving cyclic carbosilanes:

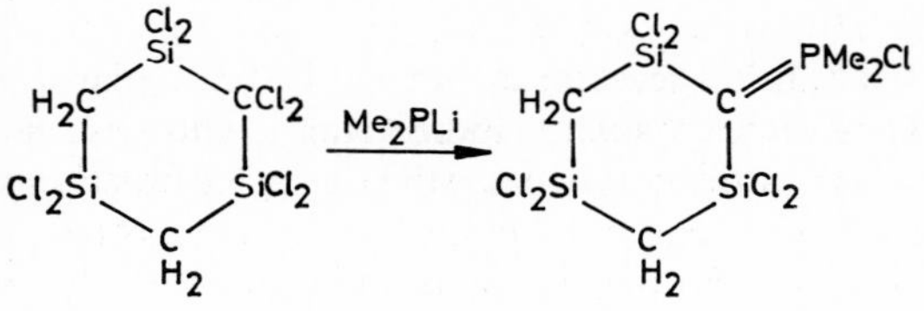

326

Of special interest are the reactions of $(Cl_2Si—CCl_2)_3$ with 3 mole equivalents of $Me_3Si—PMe_2$ or 3 mole equivalents of $LiPMe_2$ [132]:

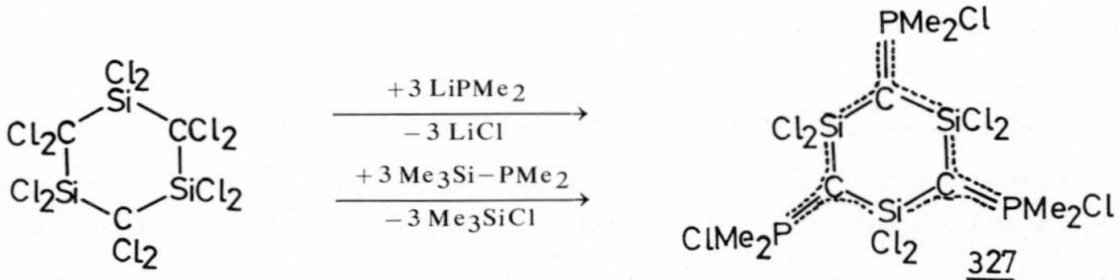

Taking both directions, the triylide 327 still ends up being the main product. The compound is planar, as shown by X-ray analysis of crystal structure [132]. The P—C bond to the carbosilane ring is distinctly shorter (170 pm) than a normal P—C bond, and in the same manner the Si—C bond length in the ring is reduced from 188 to 180 pm.

5.3 Influence of Si-Substitution on the Formation, Structure and Rearrangement of Ylides

The influence of Si-substituents on the formation and structure of the ylides of 1,3,5-trisilacyclohexanes were investigated in the products from reactions of $Me_3Si—PMe_2$ with C-chlorinated or C-brominated 1,3,5-trisilacyclohexanes containing only one CX_2 group [133]:

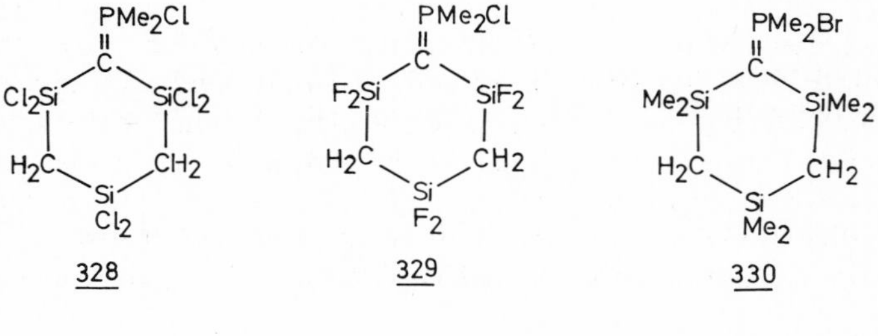

328   329   330

The X-ray analysis of crystal structure (Chapt. IV.1.1.4) shows that the atoms of the ylide moiety of 328 lie in a plane, with reduced C—P and C—Si bond lengths. The remaining endocyclic C—Si bond lengths in 328 are left unchanged on ylide formation. Only the C—Si—C endocyclic angles adjacent to the ylide group are extended. In 330, the C—P bond length is still shorter, while the P—Br distance is unusually long, and has a bond order of only 0.63.

There is also a distortion of the ylide C-atom out of the ylide plane. The bond lengths and angles of the Si-methylated six-membered ring in this case are normal. The ring of 330 exists in the boat conformation, while the ring in 328 exists in a flattened chair form [133]. In the ylide 330, double bond contributions are clearly localized in the P—C bond. A distance of 164.6 pm means one of the shortest P—C single bond lengths ever observed. In comparison with normal P—C single and P≡C triple bond lengths, the experimental value is remarkably similar to the P=C double length of 165 pm [134].

While a shortening of the Si—C bond length is observed in SiCl-containing ylides, the SiMe-containing ylides show a normal Si—C bond length but an additional shortening of the C—P bond distance indicating an increased basicity of the ylidic C-atom. Structural differences in the monoylides of the 1,3,5-trisilacyclohexanes were also recognizable in their chemical behavior. While $(Cl_3Si)_2CCl_2$ and Me_3Si——PMe_2 react readily at less than 0 °C to form the ylide $(Cl_3Si)_2CPMe_2Cl$ [132], the methylated analog $(Me_3Si)_2CCl_2$ does not lead to ylide formation. Rather, on extended reaction time or on heating, $(Me_3Si)_2C(H)PMe_2$ is formed.

A detailed investigation into this irregular behavior produced the following results: all 1,3-disilapropanes containing CCl_2 groups react with Me_3Si—PMe_2 in a molar ratio of 1:1 according to the reaction course shown below. All individual stages of this mechanism were experimentally determined through individual trapping reactions [135].

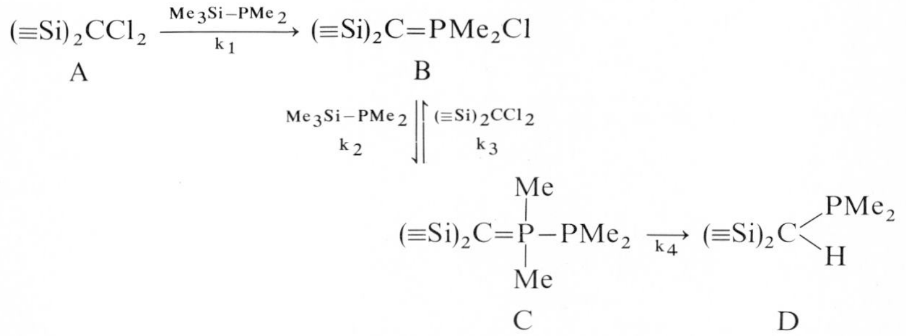

$$(\equiv Si)_2CCl_2 \xrightarrow[k_1]{Me_3Si-PMe_2} (\equiv Si)_2C=PMe_2Cl$$

$$A \qquad\qquad\qquad\qquad B$$

$$Me_3Si-PMe_2 \Bigg\updownarrow \begin{matrix} k_2 \\ \end{matrix} \begin{matrix} (\equiv Si)_2CCl_2 \\ k_3 \end{matrix}$$

$$\underset{\displaystyle \underset{Me}{|}}{\overset{\displaystyle \overset{Me}{|}}{(\equiv Si)_2C=P}}-PMe_2 \xrightarrow{k_4} (\equiv Si)_2C\!\!\underset{H}{\overset{PMe_2}{<}}$$

$$C \qquad\qquad\qquad\qquad D$$

The first identifiable stage of this reaction is always the formation of the phosphorus ylide B.

The next stage could involve the formation of the phosphino-substituted phosphorus ylide C. Whether or not this reaction actually occurs depends on the difference in reactivity of the CCl_2 group in the original carbosilane A, which in turn depends on the extent of Si-chlorination, compared to the PMe_2Cl group in the phosphorus ylide B. As far as the phosphino-substituted ylide C is concerned, it can react with excess starting material A to yield the phosphorus ylide B, or rearrange via a competitive reaction to produce the phosphane D. The extent to which these competitive reactions occur, and hence the ratio of the products formed, is determined by the substituents bound to the Si atom of the starting material. Three different reaction courses have to be considered when starting with the reactants in a 1:1 molar ratio.

a) $k \simeq k_3 \gg k_2 \gg k_4$
— with Si-perchlorinated C-brominated carbosilanes.
— with Si-permethylated C-brominated carbosilanes.

In both cases, the phosphorus ylide (B), $(\equiv Si)_2C=PMe_2X$, where X=Cl, Br, is the kinetically and thermodynamically controlled reaction product. The formation of the phosphino-substituted ylide C and the phosphane D was not observed.

b) $k_2 \gg k_1 > k_3 \gg k_4$
— with partly Si-chlorinated C-chlorinated carbosilanes.

The phosphino-substituted ylide $(\equiv Si)_2C=PMe_2-PMe_2$ (C) is the kinetically controlled reaction product, while the formation of the phosphorus ylide $(\equiv Si)_2C= =PMe_2Cl$ (B) represents the thermodynamically controlled reaction.

c) $k_2 \gg k_1 > k_3 \simeq k_4$
— with Si-permethylated C-chlorinated carbosilanes.

The formation of the phosphino-substituted ylide $(\geq Si)_2C=PMe_2-PMe_2$ (C) is a kinetically controlled reaction in this case. Under thermodynamic control a competing reaction occurs in which the phosphorus ylide $(\geq Si)_2C=PMe_2Cl$ (B) and, by thermal rearrangement, the phosphane $(\geq Si)_2C(H)PMe_2$ (D) are produced.

Si-chlorinated and Si-methylated ylides such as $(Cl_3Si)_2C=PMe_2-PMe_2$ and $(Me_3Si)_2C=PMe_2-PMe_2$ or $(Me_3Si)_2C=PMe_2-P(Me)SiMe_3$ all have characteristic differences in their chemical behavior. While $(Cl_3Si)_2C=PMe_2-PMe_2$ is stable on heating for 15 hrs up to 180 °C, $(Me_3Si)_2C=PMe_2-PMe_2$, $(Me_3Si)_2C=PMe_2- -P(Me)SiMe_3$ and the corresponding phosphino-substituted ylides rearrange readily at room temperature to form phosphanes such as $(Me_3Si)_2C(H)PMe_2$. The H atom in the $>C(H)PMe_2$ group originates from the CH_3 group on the trivalent P atom of the ylide. This rearrangement proceeds according to the following mechanism.

R= CH₃ , SiMe₃

This is also substantiated in ylides in which a CD_3 group is bound to the trivalent phosphorus atom and which rearrange producing the $>C(D)PMe_2$ group. The reason for the differing behavior of these Si-chlorinated and Si-methylated derivatives is derived without doubt from the chemical qualities of the ylidic C atom. The results of the X-ray analysis of crystal structure of the corresponding derivatives provide the final explanation for this behavior. This investigation showed that Si-chlorinated ylides such as $(Cl_3Si)_2C=PMe_2Cl$, $(Cl_2Si-C=PMe_2Cl)_3$ and <u>328</u> all exhibited a shortened Si—C bond length of 180.1 pm on the ylidic C atom, while in the Si-methy-lated ylide <u>330</u> no shortening of the Si—C bond occurs (Si—C: 186.2 pm) [133]. However, the P—C bond length in compound <u>330</u> is shortened to 164.6 pm in comparison to the P—C bond length in the previously-mentioned ylides. This means that an electron delocalization occurs in the Si—C bonds of the Si-chlorinated ylides which causes a weakening of the carbanionic nature of the ylidic C atom. In Si-methylated ylides, electron density is localized in the P—C bond, without exerting any effect on the Si—C bond, causing the ylidic carbon atom to maintain a stronger basic character. The observed chemical behavior of the ylide confirms this reasoning. The following mesomeric forms can be formulated for the Si-methylated ylide, which suggest a possible means for its rearrangement.

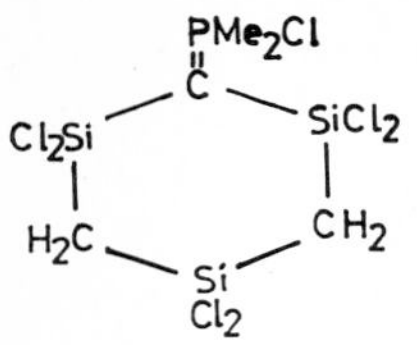

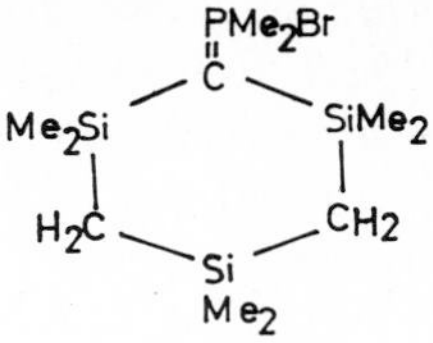

The situation for the Si-chlorinated ylides is also made distinctly clear by inspection of its mesomeric forms. These structures show that suitable rearrangement is not favored.

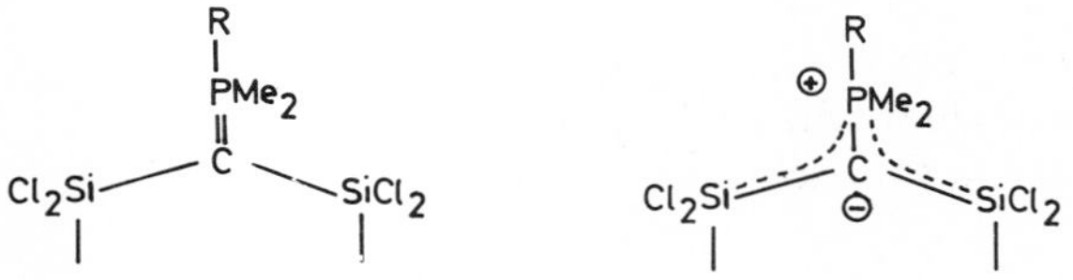

Experiments show that rearrangement to produce a $C(H)PMe_2$ group originates at a much lower temperature from a P—H group on the trivalent P-atom of a P—P ylide than from the CH_3 group adjacent to this P-atom. This preference may be explained by the differing bond energies of P—H and C—H bonds.

6. The Reactive Behavior of Further Cyclic Carbosilanes

6.1 1,3-Disilacyclobutanes

By comparing the thermal and chemical stabilities of the Si-methylated cyclic carbosilanes $(Me_2Si—CH_2)_3$ and $(Me_2Si—CH_2)_2$ 22, it is quickly recognized that the 1,3-

disilacyclobutane is much less stable. For example, 22 very readily undergoes ring opening on reaction with HBr or Br$_2$ producing the corresponding linear compounds, while the six membered ring remains intact [136, 137]. For instance, reaction with HBr produces Me$_3$Si—CH$_2$—SiMe$_2$Br. This differing chemical behavior is at first explained by the ring strain occurring in 22. In addition to this, however, the Si-substituents have a considerable influence on the stability of this ring. This is illustrated by investigations carried out on (Cl$_2$Si—CH$_2$)$_2$ [138]. In general, the conversion of SiCl groups to SiH groups in carbosilanes by reaction with LiAlH$_4$ leaves the molecular skeleton unchanged. However, this rule cannot be extended to 1,3-disilacyclobutanes [139]. Also the Si-chlorinated asterane Si$_6$Cl$_8$C$_6$H$_8$ shows an unusual behavior [114]. Furthermore, although (Cl$_2$Si—CH$_2$)$_2$ 331 is converted to (H$_2$Si—CH$_2$)$_2$ on reaction with LiAlH$_4$, from the hydrogenation carried out with an excess of LiAlH$_4$ a mixture of (H$_2$Si—CH$_2$)$_2$ and MeH$_2$Si—CH$_2$—SiH$_3$ arises. Reactions carried out with $\geqslant$SiCl and 1/4 LiAlH$_4$ in a 1:1 ratio in n-dibutyl ether produce almost pure (H$_2$Si—CH$_2$)$_2$ 332 after fractional condensation [139].

The chemical qualities of (H$_2$Si—CH$_2$)$_2$ are exemplified by the typical reactions of SiH groups, as well as by the preference of disilacyclobutanes to undergo ring cleavage reactions. Thus (H$_2$Si—CH$_2$)$_2$ reacts explosively at —30 °C by passing a 1:1 gas mixture of Cl$_2$/N$_2$ over the undiluted liquid compound. On the other hand, if a solution of Cl$_2$ saturated at room temperature in CCl$_4$ is added dropwise at —10 °C to —15 °C to a solution of (H$_2$Si—CH$_2$)$_2$ in CCl$_4$, all stages of chlorination of 332 can be seen in the ^{1}H-NMR spectra, without allowing halogen cleavage to exceed a value of 1.5%:

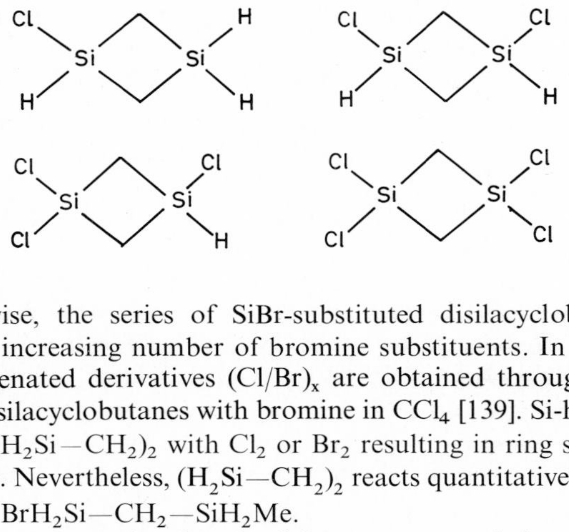

Likewise, the series of SiBr-substituted disilacyclobutanes can be synthesized with an increasing number of bromine substituents. In addition, all possible mixed Si-halogenated derivatives (Cl/Br)$_x$ are obtained through reaction of partly chlorinated disilacyclobutanes with bromine in CCl$_4$ [139]. Si-halogenation occurs via reaction of (H$_2$Si—CH$_2$)$_2$ with Cl$_2$ or Br$_2$ resulting in ring stabilization, faster than ring cleavage. Nevertheless, (H$_2$Si—CH$_2$)$_2$ reacts quantitatively with HBr via ring cleavage to yield BrH$_2$Si—CH$_2$—SiH$_2$Me.

Further information is gained from a systematic investigation of the influence of Si-substituents on 1,3-disilacyclobutanes. The simplest path to the synthesis of appropriate derivatives seemed to be through (Cl$_2$Si—CH$_2$)$_2$, which was available through pyrolysis. This (Cl$_2$Si—CH$_2$)$_2$ reacts with PhMgBr as expected to yield (Ph$_2$Si—CH$_2$)$_2$, 333. The formation of (Cl$_2$Si—CH$_2$)$_2$ via Cl$_3$Si—CH$_2$Cl and Mg has been described [140], but could not be confirmed by us. It was also not possible to achieve the synthesis of $(MeClSi-CH_2)_2$ through reaction of $MeCl_2Si-CH_2Br$ with Mg. Because the synthesis of $(Me_2Si-CH_2)_2$ proceeds without difficulty via $Me_2ClSi-CH_2Cl$ and Mg (in the same synthesis also $(Me_2Si-CH_2)_3$ is formed) [141], it was intended to use the SiPh group as a shielding group in the synthesis of Si-phenylated 1,3-disilacyclobutanes [139]. The cleavage of $(Ph_2Si-CH_2)_2$ had to be investigated in the next stage of development, in order to establish whether Si-Ph bond cleavage or ring opening occurs as the initial step in the replacement reactions of SiBr-containing 1,3-disilacyclobutanes. All further derivatives would then be accessible if these reaction products were obtained. The synthesis of $(Ph_2Si-CH_2)_2$ from $Ph_2ClSi-CH_2Cl$ and Mg has previously been reported [140], and some recent results [142] are identical with ours [139]. Reaction of $(Ph_2Si-CH_2)_2$ with HBr results in an immediate ring cleaving yielding $BrPh_2Si-CH_2-SiPh_2Me$. If bromine is used in equimolar amounts, a mixture arises comprising $PhBrSi \diamondsuit SiPh_2 + (PhBrSi-CH_2)_2$ besides negligible amounts of starting material. The same reaction with double the amount of Br_2 leads to the synthesis of $(PhBrSi-CH_2)_2$. In both cases, and on a small scale, ring cleavage appears at the beginning of the reaction, which is not the case with the Si-brominated derivatives. The results obtained on the formation of $(PhBrSi-CH_2)_2$ correspond to results obtained from earlier investigations on Si-Ph bond cleavage [68] in which it was established that Si-Ph cleavage was hindered by increasing the number of negative substituents on silicon. No further Si-Ph cleavage could be achieved.

Accordingly, $(PhMeSi-CH_2)_2$ was synthesized [140]. In this case, ring cleavage was favored in preference to Si-Ph cleavage, so that its reaction with Br_2 produces $PhMeBrSi-CH_2-SiPhMe-CH_2Br$. The limit to syntheses of this type and to the formation of Si-functional 1,3-disilacyclobutanes by means of Si-Ph cleavage is clearly laid out by consideration of the following examples: BrH_2Si-CH_2Cl is not converted to $(H_2Si-CH_2)_2$, but $(HMeSi-CH_2)_2$ is accessible through reaction of $BrMeHSi \cdot CH_2Br$ with Mg and can be brominated to produce $(BrMeSi-CH_2)_2$. On the other hand, $Me_2Si \diamondsuit SiPh_2$ is obtained from the reaction of $BrMe_2Si-CH_2-SiPh_2-CH_2Cl$ and Mg, but on reaction with Br_2 ring cleavage appears to be the only process undertaken [140].

1,3-disilacyclobutanes can be divided into a number of groups based on their behavior to HBr and Br_2 [139] (Table 66).

Group 1 contains compounds having strong electronegative Si-substituents, and as a result no ring cleavage occurs. Compounds in group 2 undergo ring cleavage with HBr, while substitution on silicon occurs when they react with Br_2. In group 3 ring cleavage is the primary reaction towards both reagents.

Of course, ring strain is a very important ground for the tendency of 1,3-disilacyclo-butanes to be cleaved. This is supported by the fact that compounds such as $(Me_2Si-CH_2)_3$ or linear carbosilanes do not succumb to this reaction. The bond polarization $\overset{\delta+}{Si}-\overset{\delta-}{C}$ is a further cause for the differing chemical behavior of compounds of similar molecular geometry such as $(Cl_2Si-CH_2)_2$, which is stable against Br_2, and $(Me_2Si-CH_2)_2$, which is easily cleaved by Br_2. Strongly electron-withdrawing substituents on the Si atom reduce the polarization of the Si-C bond, which in turn impedes nucleophilic attack on silicon and electrophilic attack on carbon. Nucleophilic attack on the Si atoms of $(Cl_2Si-CH_2)_2$, which occurs through a pentacoordinate

Table 66. Behavior of 1,3-disilacyclobutanes to Br_2 and HBr

Group	Compound	Reaction with HBr	Reaction with Br_2
1	$(Cl_2Si-CH_2)_2$ $(Br_2Si-CH_2)_2$ $[(C_6H_5)BrSi-CH_2]_2$	none	none
2	$[(C_6H_5)_2Si-CH_2]_2$ $(H_2Si-CH_2)_2$ $[(C_6H_5)HSi-CH_2]_2$ $(MeHSi-CH_2)_2$	Ring cleavage	no ring cleavage Si-substitution
3	$[Me(C_6H_5)Si-CH_2]_2$ $(Me_2Si-CH\ )_2$	Ring cleavage	Ring cleavage

transition state, does not lead to ring cleavage; rather, by longer exposure to Br_2, it leads to a halogen exchange producing $(BrClSi-CH_2)_2$ in a limited yield. More electronegative Si-substituents consolidate this four-membered ring system.

The compounds $(F_2Si-CH_2)_2$ and $(MeFSi-CH_2)_2$ are synthesized by reaction of their corresponding Cl-containing analogues with AgF_2 or AgF, respectively [143]. If SbF_3 or ZnF_2 are used instead, ring cleavage is favored.

6.2 1,3-Disilacyclobutane Rings in Hexasilaasteranes

Substituent effects observed in 1,3-disilacyclobutanes are also reflected in larger carbosilane skeletons incorporating such rings.

The Si-asteranes 335, 336, 337 and 338 maintain the same four-membered ring systems, and differ only in the substituents on silicon [144].

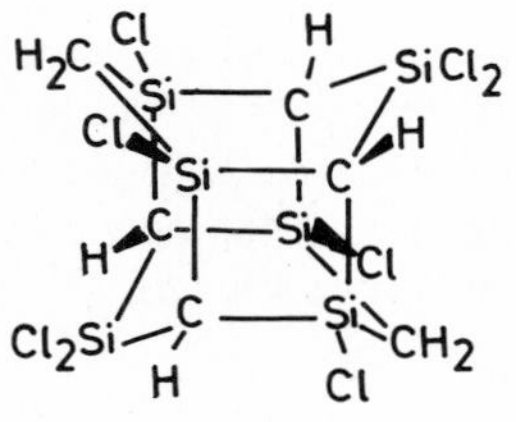

335

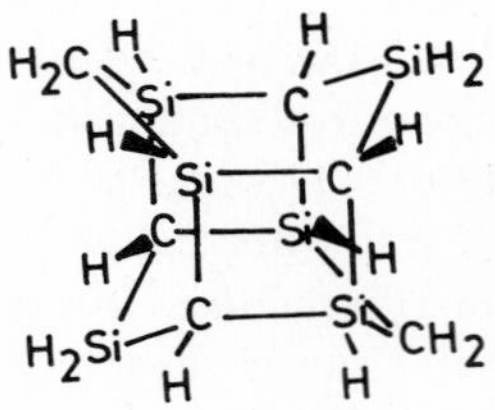

336

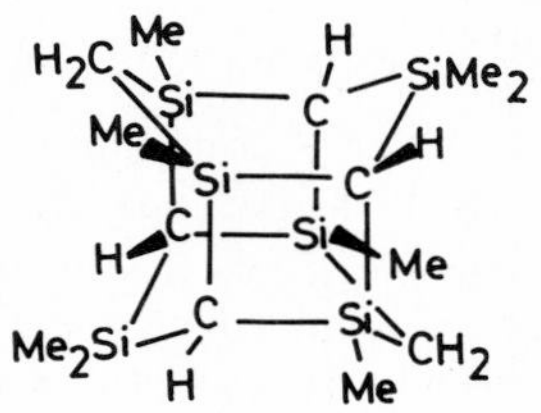

337

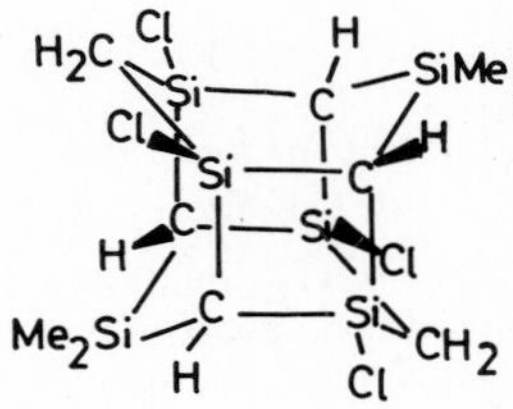

338

The influence of Si substituents on the stability of the skeleton is evident from the behavior of 335, 336, 337 and 338 towards halogens and the corresponding hydrogen halides. The Si-chlorinated derivative 335 reacts neither with chlorine nor bromine, nor with HBr in CCl_4 at 80 °C. Similarly, attack by hydrogen fluoride leaves the asterane skeleton unaffected. Only halogen exchange is observed which enables eventually the formation of $Si_6Cl_4F_4C_6H_8$ via mono-, di- and trifluorinated asteranes. This reaction corresponds to the reaction of 335 with ZnF_2 and SbF_3 [89].

The Si-hydrogenated asterane 336 and its partly-hydrogenated SiCl-containing derivatives react with chlorine via chlorination of Si—H bonds to produce 335. The ring system is cleaved neither by Cl_2 nor by the HCl produced in the SiH substitution reaction. The Si-methylated compound 337 reacts with 2 moles of HBr via clevage of both four-membered rings to generate 339, which contains only 1,3,5-trisilacyclohexane rings [144].

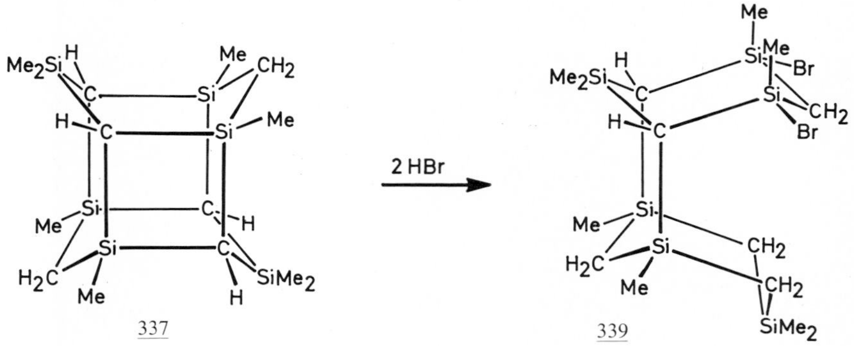

In contrast to the fully methylated compound 337, compound 338 containing 4 tertiary SiCl groups can be cleaved neither by Br_2 nor by HCl.

6.3 1,3-Disilacyclopentenes

1,3-disilacyclopentenes are formed from the pyrolysis of $SiMe_4$, and are involved in further reaction processes of the gas phase pyrolysis. Hence, the chemical qualities pertaining to these compounds are of special interest.

6.3.1 Tetramethyl-1,3-disilacyclopentenes

The reaction of 412 with HBr between -80 °C and -30 °C depends on the HBr concentration. By employing a mole ratio of 1:1, the following reaction products are generated:

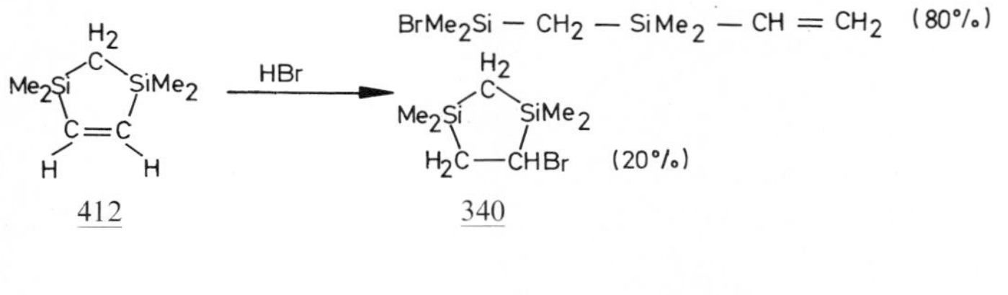

Here $BrMe_2Si$—CH_2—$SiMe_2$—CH_2—CH_2Br is not formed [145]. If the reaction is carried out with 2 mole equivalents of HBr, then under otherwise similar conditions $BrMe_2Si$—CH_2—$SiMe_2$—CH_2—CH_2Br in an 80% yield and $BrMe_2Si$—CH_2——$SiMe_2Br + CH_2=CH_2$ in a 20% yield are obtained. Under other reaction conditions, namely by passing a mixture of HBr/N_2 through a solution of 412 it was confirmed that originally HBr is added to 340, followed by cleavage of the ring skeleton.

Cleavage of a Si—CH_2—Si group does not occur under any circumstances. Ring opening and formation of a vinyl group are explained by a β-elimination. In the same way C_2H_4 and $(BrMe_2Si)_2CH_2$ result from 341. Although HCl was found to be considerably less active, the reaction proceeded along the same lines [145].

Reaction of 412 with bromine in a 1:1 molar ratio in CCl_4 has been studied meticulously. First, bromine is added to the double bond:

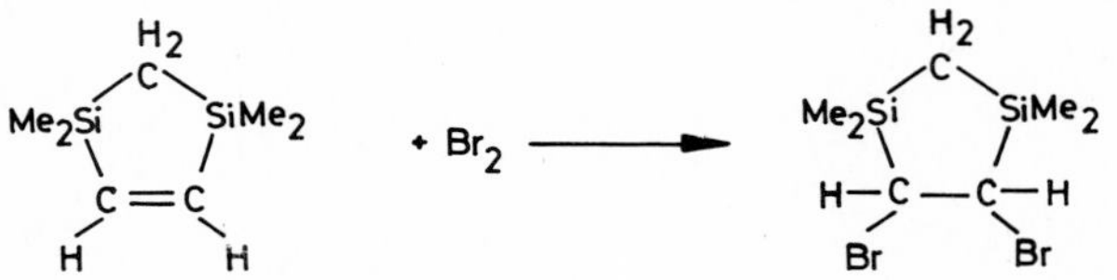

The compound 342 is isolated in the form of colorless crystals. In the presence of a suitable catalyst, 342 rearranges producing $BrMe_2Si$—CH_2—$SiMe_2$—$CH=CHBr$

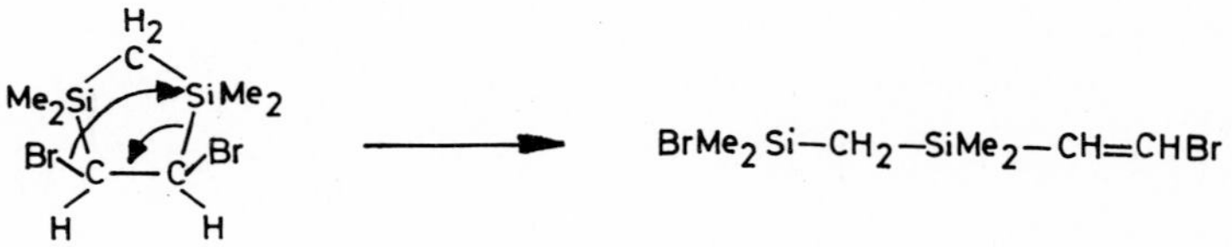

After addition of Br_2 is complete, an intramolecular β-elimination occurs. This is also the case in the reaction with bromine in the absence of solvent, which proceeds likewise to $BrMe_2Si$—CH_2—$SiMe_2Br$ and trans-$BrHC=CHBr$ as additional products [145].

Reaction with chlorine at −80 °C in the absence of solvent causes a violent explosion. If chlorine is passed through a solution of 412 in CCl_4 at 20 °C, the addition product 343 is formed.

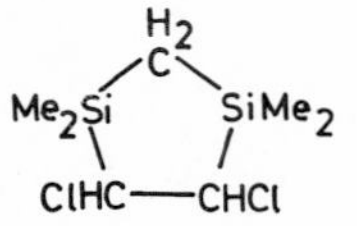

Ring cleavage does not occur. Compound 343 can be isolated in the form of colorless crystals.

6.3.2 4-Trimethylsilyl-tetramethyl-1,3-disilacyclopentene

Reaction with HBr in the absence of solvent was initially carried out at −78 °C. After the reaction eased off slightly, it was completed by increasing the temperature.

The products are shown in Table 67. Only linear carbosilanes were detected; simple HBr addition to the double bond did not appear. However, such an addition cannot be ruled out as a probable initial reaction step [146].

It is conceivable that compounds 344 and 345 are formed via cleavage by HBr between atoms 3 and 4, or 5 and 1, respectively. However, it seems more probable that the HBr addition process occurs before that of ring cleavage, and that this ring cleavage proceeds via a β-elimination. Also, all further reactions utilizing higher HBr concentrations can be understood to proceed through addition of HBr followed by β-elimination.

Table 67. Reaction of 1,1,3,3-tetramethyl-4-trimethylsilyl-1,3-disilacyclopentene with HBr at −78 °C, followed by heating of the reaction solution to 90 °C

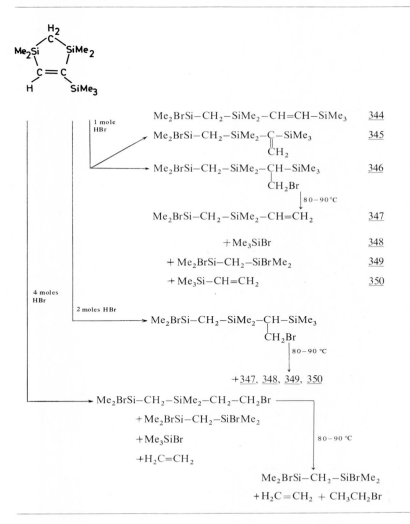

6.4 Tetrachloro-1,3-disilacyclopentane

6.4.1 Stepwise Photochlorination

All investigations into 1,1,3,3-tetrachloro-1,3-disilacyclopentane 351 are related to the work done on $(Cl_2Si-CH_2)_3$. The 351 is accessible by reaction of Cl_3Si-CH_2- $-CH_2-SiCl_2-CH_2Cl$ with Mg [147]. Photochlorination of 351 proceeds through a number of stages:

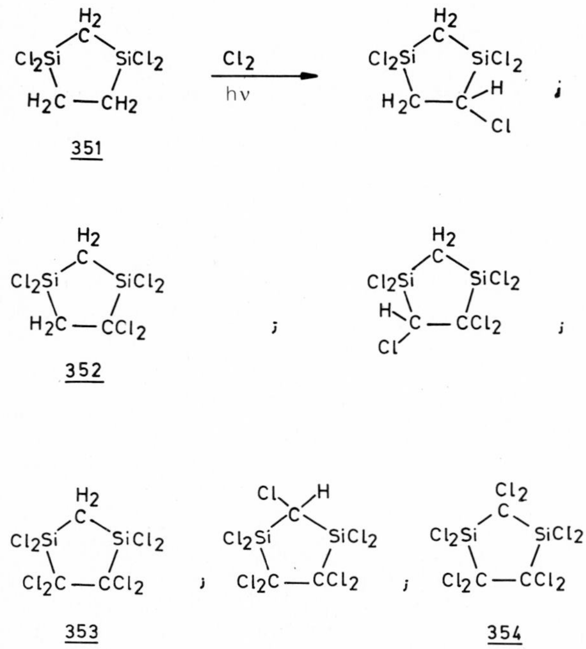

The perchlorinated 1,3-disilacyclopentane 354 was obtained in approximately 70% yield. Progressive chlorination of 351 was carefully monitored using NMR spectroscopy, and in the course of doing so the appearance and disappearance of various compounds in the reaction mixture was determined. By intercepting the chlorination reaction at points along the way, compounds 352 and 353 could be isolated.

Radical chlorination of 351 always begins on C atoms 4 or 5, while chlorination of C atom 2 will only proceed when all available positions on C atoms 4 and 5 are substituted. Therefore, if was not possible to synthesize 1,1,2,2,3,3-hexachloro-1,3-disilacyclopentane in this way [148].

6.4.2 Reaction of 1,3-Disilacyclopentanes with MeMgCl

Reaction of 352 with 2 mole equivalents of MeMgCl in Et_2O at 20 °C generates 355:

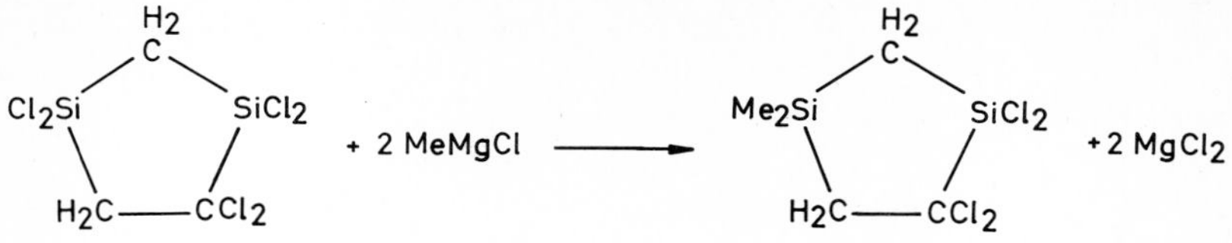

On subsequent warming, compound 356 is produced via HCl-elimination:

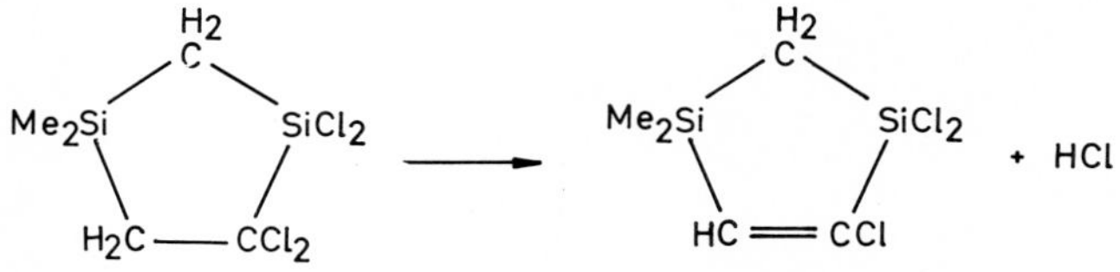

At temperatures exceeding 140 °C, a ring cleavage occurs via β-elimination producing the linear vinyl compound 357.

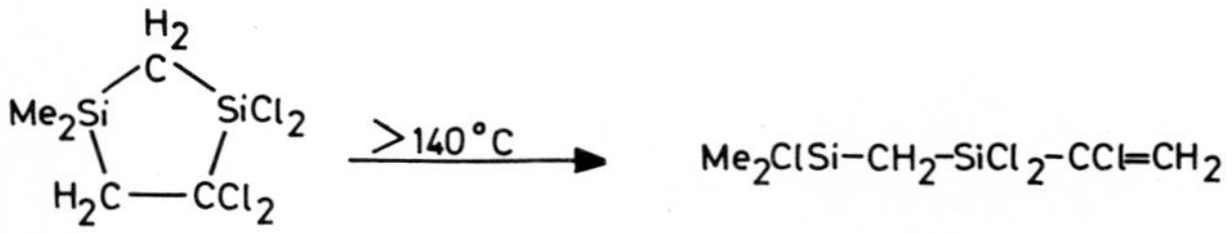

Obviously Si-methylation begins on the $SiCl_2$ group not adjacent to the CCl_2 group.

The reaction of 352 with 6 mole equivalents of MeMgCl also produces methylation of Si atom 3, and yet the main products of the reaction arise from ring cleavage occurring by means of β-elimination. An example of this is the formation of $H_2C=CCl-SiMeCl-CH_2-SiMe_2Cl$ [148].

Compound 352 reacts at first with MeMgCl in Et_2O under preservation of the ring skeleton via methylation of the Si atom not adjacent to the CCl group. The CCl_2 group is not incorporated into the reaction. Subsequently, Si-methylation at higher temperatures is followed by ring opening, again via β-elimination. The distinctly

Table 68. Products from the reaction of 354 with 10 mole equivalents MeMgCl in Et_2O

Compounds		Percentage[a]
$Cl_2C=CCl-SiMe_2-C(=CH_2)-SiMe_3$	359	4%
$Cl_2C=CCl-SiMe_2-CHCl-SiMe_3$	360	12%
$Cl_2C=CCl-SiMe_2-CCl_2-SiMe_3$	361	66%
$Cl_2C=CCl-SiCl_2-CMeCl-SiMe_2Cl$	362	11%

[a] The remaining 7% of compounds were not identified.

different reactivity of these $SiCl_2$ groups is also reflected in their ^{29}Si-NMR chemical shifts, for the $SiCl_2$ group adjacent to the CCl_2 group displaying a lower reactivity shows a high field shift.

Compound 358 is produced exclusively on reaction of 353 with 6 mole equivalents of MeMgCl in Et_2O. This reaction could occur via MeMgCl attack on the CCl_2 group according to the following chemical pathway.

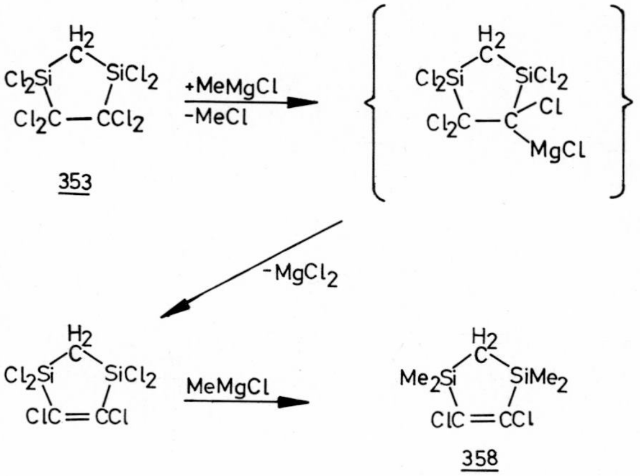

However, it is not certain whether Si-methylation yielding 358 occurs before or after C-metallation of 353.

The reaction of the perchlorinated ring 354 with 10 mole equivalents of MeMgCl in Et_2O at -20 °C generates the compounds shown in Table 68.

All compounds arise through ring cleavage, and with the exception of 362, are permethylated on silicon. This means, of course, that apart from ring cleavage, Si-methylation is the major process in the reaction pathway. The isolated CCl_2 group present may also be incorporated into the reaction, as shown by the synthesis of compounds 359, 360 and 362. An understanding of the behavior of this CCl_2 group has been obtained from a consideration of the behavior of $(Cl_3Si)_2CCl_2$ [113]. Reaction of 354 with MeMgCl generates at first Si-methylation, as proved by reactions with 4 mole equivalents of MeMgCl. The products of this latter reaction comprise a mixture of partly Si-methylated 1,3-disilacyclopentanes along with a small amount of linear C-methylated derivatives such as 362. Compounds containing CHCl or $C=CH_2$ groups do not appear under these conditions. It can therefore be concluded that Si-methylation will occur first, followed by ring cleavage. Further reactions on the Si and C atoms follow this ring cleavage reaction.

The reaction of compound 354 with MeMgCl in pentane proceeds exclusively via methylation of all SiCl groups to the C-chlorinated 1,3-disilacyclopentane [148].

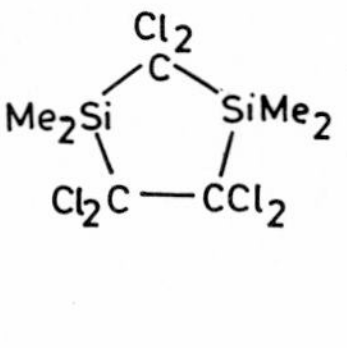

6.5 Tetrasilabicyclo[3.3.0]oct-1(5)ene

Because of the crystallizing capability of 363, this compound was successfully isolated from the pyrolysis products of $MeSiCl_3$ and converted to the Si-methylated compound 364 by reaction with MeMgI. This compound is then in position to convert to 365 on addition of bromine. Through the influence of $LiAlH_4$ the C=C bond can be reformed, producing 364. Thermolysis of the Br-containing compound 365 results in a cleavage of the five-membered rings, generating the linear $BrMe_2Si-CH_2-$ $-SiMe_2-C\equiv C-SiMe_2-CH_2-SiMe_2Br$ 366 [149].

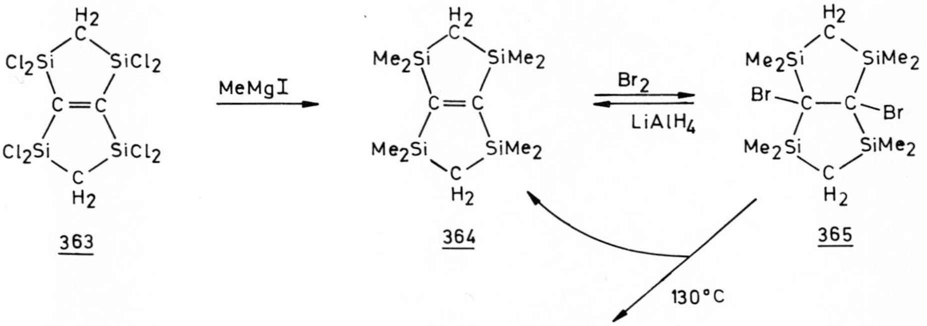

$BrMe_2Si - CH_2 - SiMe_2 - C \equiv C - SiMe_2 - CH_2 - SiMe_2Br$ 366

This ring cleavage is understood to occur via β-elimination, and is illustrated by the arrow mechanism below. This reaction course is confirmed by formation of SiBr groups from CBr groups.

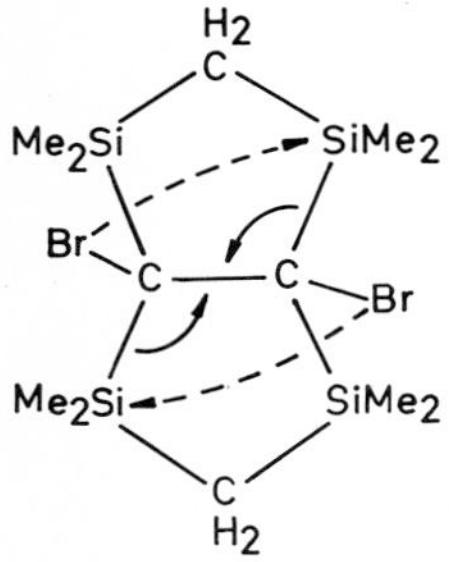

Formation of the SiH-containing compound 367 was possible by means of the following reaction.

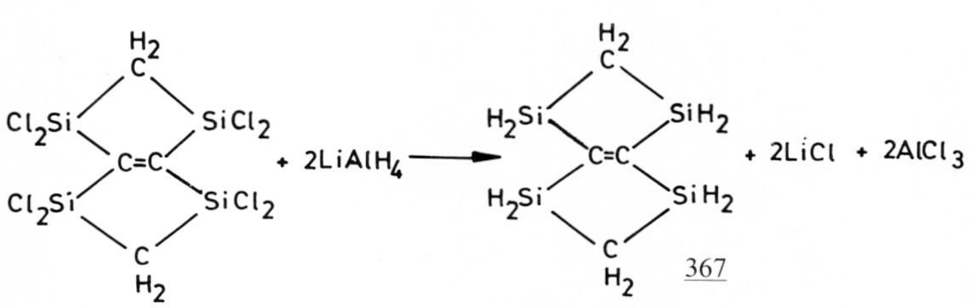

The skeleton of tetrasilabicyclo[3.3.0]oct-1(5)enes is planar, as shown by the X-ray crystal structure determination of 364 [150]. Formation of radical cation and anion derivatives was investigated using photoelectron as well as ESR spectroscopy [151].

7. Investigations into the Cleavage of Si—Me Bonds in Carbosilanes

An understanding of the chemical stability of the Si—C bond and the possibilities for the cleavage of this bond yielding functional groups on silicon is important in a number of aspects. Substituent effects on the Si—C bond can be understood through changes in polarity. It is known for example that the tendency of the Si—C bond to undergo alkaline cleavage increases with the degree of chlorination of the C atom [152]. This reaction,

$$\equiv Si-CHCl_2 \xrightarrow[H_2O]{OH^-} \equiv SiOH + CH_2Cl_2$$

proceeds to completion, and as a result a quantitative determination of the groups $\equiv Si-CH_2Cl$, $\equiv Si-CHCl_2$, $\equiv Si-CCl_3$ and $\equiv Si-CCl_2-Si\equiv$ was formulated [99]. The resulting chloromethanes were quantitatively analysed by gas chromatography. Cleavage of the Si—Ph group by an aqueous/alcoholic HCl solution has also been described [153]. All these methods have the associated disadvantage that an Si—OH group forms on cleavage of a Si—C bond, and subsequent condensation causes a complete loss of the active moiety of the molecule. Therefore, the cleavage of the Si—Ph bond with anhydrous hydrogen halides became an important factor in the organometallic synthesis of carbosilanes. Cleavage of the Si—Me bond by Me_3SiCl or acetyl chloride using $AlCl_3$ as catalyst has already been described by Kumada [154, 155]:

$$Me_3Si-(CH_2)_n-SiMe_3 + 2\ Me_3SiCl$$

$$\xrightarrow{AlCl_3} ClMe_2Si-(CH_2)_{2n}-SiMe_2Cl + 2\ SiMe_4 \qquad (n = 1-4)$$

$$SiMe_4 + CH_3COCl \xrightarrow{AlCl_3} Me_3SiCl + CH_3COCH_3$$

Cleavage by H_2PtCl_6 has been demonstrated by Benkeser [156] and extended by Hengge to the Si-chlorination of methylated polysilanes [157].

$$R_3Si-CH_3 + HSiCl_3 \xrightarrow{H_2PtCl_6} R_3SiCl + HSiMeCl_2$$

Calas reported on the formation of Me_3SiCl by reaction of $SiMe_4$ with ICl: [158]

$$SiMe_4 + 2\ ICl \rightarrow Me_3SiCl + MeCl + I_2$$

Common to all these reactions is the fact that an Si—Me group can be converted to a Si—Cl group.

7.1 Si-Chlorination of $(Me_2Si-CH_2)_3$ with $HSiCl_3/H_2PtCl_6$

Extending such known cleavage reactions to carbosilanes, it was intended to establish to what extent Si—C bonds in the ring are attacked leading to degradation of the carbosilane. Out of this investigation arose the fact that only Si—Me groups were cleaved from the ring skeleton, and depending upon reaction conditions the following derivatives were obtained from $(Me_2Si-CH_2)_3$:

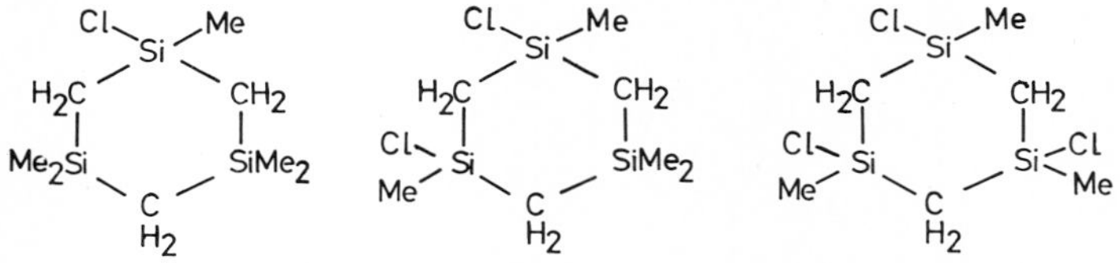

Apart from the formation of $HMeSiCl_2$ and HMe_2SiCl only such Si-chlorinated derivatives arise, containing one chlorine substituent on every Si atom. These 1,3,5-trisilacyclohexane derivatives react with $LiAlH_4$ producing their SiH-containing analogues for a more convenient work-up and identification [159].

7.2 Cleavage with ICl

7.2.1 Hexamethyl-1,3,5-trisilacyclohexane

The reaction of $(Me_2Si-CH_2)_3$ with ICl in CCl_4 at room temperature generated the compounds 368, 369 and 370, their percentage yields being dependant on the mole ratio of $(Me_2Si-CH_2)_3$: ICl used. The reaction proceeded, for example, in the way shown below.

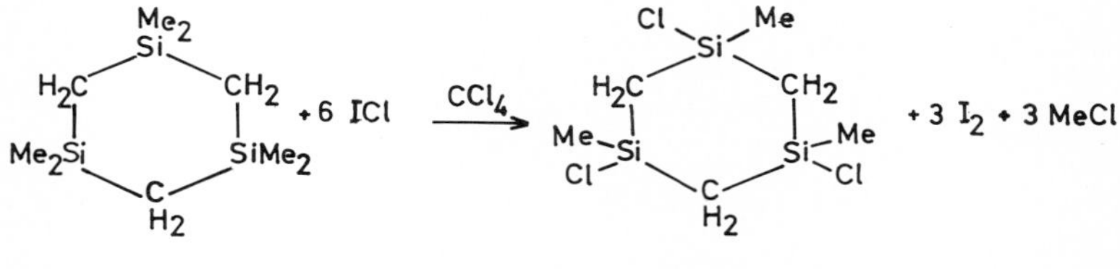

In no case was a higher degree of chlorination of 1,3,5-trisilacyclohexane observed, and likewise no cleavage of ring Si—C bonds. The reaction was also possible in CCl_4 containing some I_2 present as catalyst by passing chlorine through the solution [160].

7.2.2 Linear Carbosilanes

Similarly, linear carbosilanes react to produce their respective SiCl-containing analogues

$$Me_3Si-CH_2-SiMe_2-CH_2-SiMe_3 \xrightarrow{2\ ICl} ClMe_2Si-CH_2-SiMe_2-CH_2-SiMe_3$$
$$+ Me_3Si-CH_2-SiMeCl-CH_2-SiMe_3$$

The compound present with the highest degree of chlorination is $(ClMe_2Si-$ $-CH_2)_2SiMeCl$ [154, 160].

$$(Me_3Si-CH_2)_2SiMe_2 \xrightarrow[CCl_4]{Cl_2/I_2} (ClMe_2Si-CH_2)_2SiMeCl$$

7.2.3 1,3,5-Trisilacyclohexanes with Side Chains

Also, in Si-methylated 1,3,5-trisilacyclohexanes containing a carbosilane side chain on a skeletal Si atom or on a skeletal C atom, reaction with ICl forming Si—Cl bonds occurred:

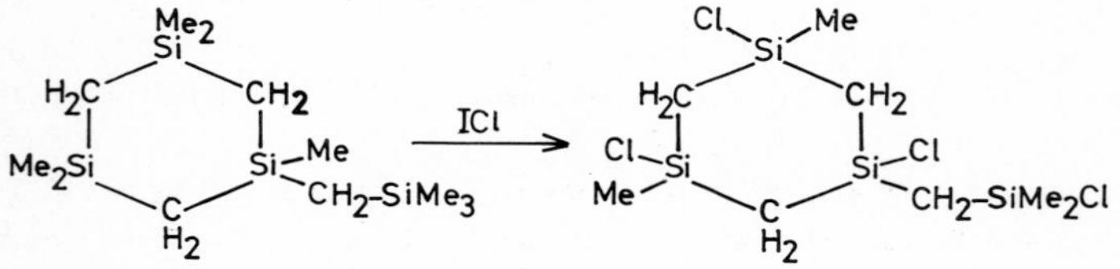

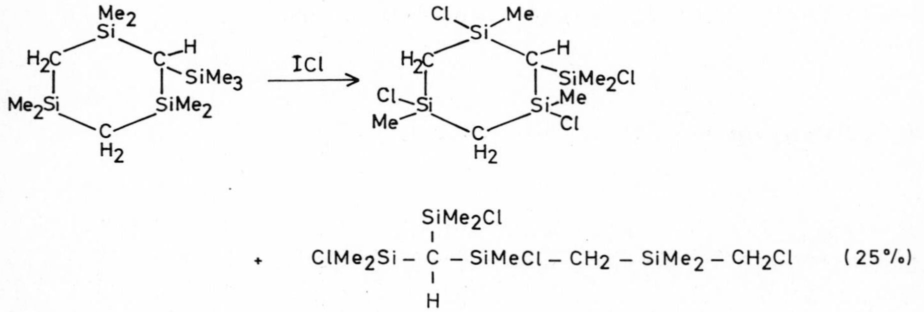

No cleavage of the side chain occurred, and on every Si atom not more than one Si—Cl bond resulted [160].

7.2.4 Adamantanes

The reaction of Si-methylated 1,3,5,7-tetrasilaadamantane with ICl generated the following products:

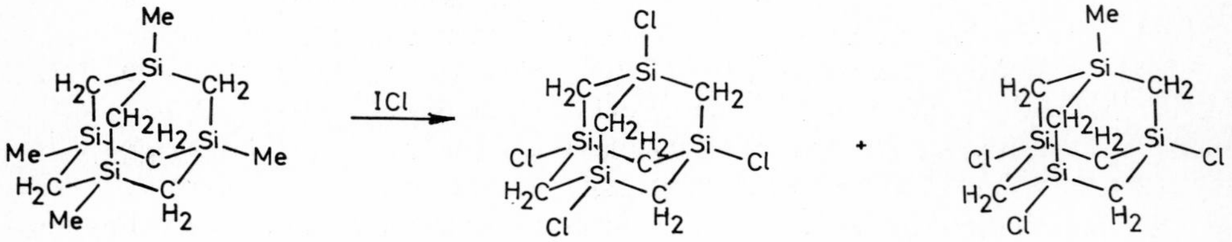

and therefore showed that even the Si—Me bond in a bridgehead position can be cleaved, in the course of which no Si—C cleavage in the molecular skeleton appeared [160].

7.2.5 Adamantanes with Side Chains

Similarly, on an Si-adamantane containing a side chain, such as that shown, no Si—C bond cleavage appears in the molecular skeleton or in the side chain [160].

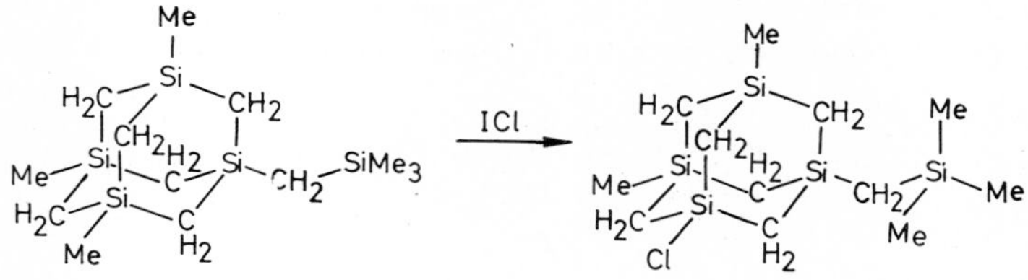

No cleavage of the molecular skeleton was observed; rather the incorporation of one Si—Cl bond onto each Si atom resulted.

Further Si-chlorination of the Si-methylated carbosilanes such as $(ClMeSi-CH_2)_3$ 370 was possible by using Cl_2/I_2 in the presence of $AlCl_3$. Chlorination proceeded then from compound 370 through all possible intermediate stages to yield eventually $(Cl_2Si-CH_2)_3$. Hence in this reaction path the cleavage of all Si—Me groups was achieved without change in the carbosilane skeleton [160].

8. Substituent Effects in Carbosilanes

Attempts to understand the immense amount of experimental data obtained, in terms of general phenomenological ground rules concerning the Si—C bond polarization in carbosilanes, have led to a reliable concept. The substituents on silicon and on carbon have a distinct influence on the reaction possibilities of the remaining groups on the molecular skeleton, as well as on the overall stability. The following points should be stressed:

1. The reaction of linear and cyclic Si-methylated carbosilanes with $HSiCl_3/H_2PtCl_6$, or with ICl alone results in only single Me group elimination from any one Si atom: $(ClMeSi-CH_2)_3$ is formed from $(Me_2Si-CH_2)_3$. However, $(Cl_2Si-CH_2)_3$ is produced from the reaction of $(Me_2Si-CH_2)_3$ with ICl in the presence of $AlCl_3$. Hence, a second Me group is removed from the Si atom under these conditions, but the Si—C—Si skeleton of the molecule still is not attacked.
2. The four-membered ring $(Me_2Si-CH_2)_2$ reacts with HBr via cleavage of an C—Si bond of the ring, forming $BrMe_2Si-CH_2-SiMe_3$, while $(Cl_2Si-CH_2)_2$ is stable under the same conditions. The six-membered ring $(Me_2Si-CH_2)_3$ shows no reaction under comparable conditions.
3. The Si—Ph bond in $PhSiMe_3$ is easily cleaved on reaction with HBr, yielding Me_3SiBr and benzene. Negative substituents on the Si atom retard this bond cleavage; $PhSiCl_3$ is stable under comparable conditions.
4. The strongly polar SiCl group is easily converted to an SiMe group on reaction with organometallic reagents such as MeMgCl. In comparison, the CCl group is normally not so reactive.

5. In carbosilanes such as $(Cl_3Si—CCl_2)_2SiCl_2$ or $(Cl_2Si—CCl_2)_3$ the reaction with MeMgCl begins by metallation of the CCl_2 group, followed by further characteristic reactions. The C atom of $(Cl_3Si)_2CCl_2$ is lithiated on reaction with n-BuLi at about $-100\,°C$, without the SiCl groups participating in the reaction. Si-alkylation occurs only at higher temperatures.

6. In carbosilanes containing CCl_2 groups, one observes an easy cleavage of a skeletal C—Si bond, as in the attempt to reduce $(Cl_3Si)_2CCl_2$ with $LiAlH_4$ to form $(H_3Si)_2CCl_2$, or in the corresponding reaction of $(Cl_2Si—CCl_2)_3$ to yield $(H_2Si——CCl_2)_3$. Such cleavages of the Si—C bond in the molecular skeleton are observed also in the reactions of SiH- and CCl-containing carbosilanes with organometallic reagents, as in the reaction of $(H_2Si—CCl_2)_3$ with MeMgCl. Similar cleavage reactions of related CH_2-containing derivatives such as $(Cl_3Si)_2CH_2$ or $(Cl_2Si——CH_2)_3$ do not appear.

7. The SiH groups in carbosilanes such as $(H_2Si—CH_2)_3$ react with MeMgCl in Et_2O only slowly by means of Si-alkylation, such reactions occurring relatively more easily if the reaction is carried out with Li alkyls in better-solvating ethers. If an SiH group is positioned adjacent to a CCl_2 group, it will react forming an SiMe group without attack on the CCl_2 group.

8. Alkylation of the SiF group always occurs in reactions of C-chlorinated Si-fluorinated carbosilanes, without change in the CCl_2 group.

9. The SiCl group is capable of cleaving the Si—P bond in silylphosphanes:

$$HSiCl_3 + Me_3Si—PMe_2 \rightarrow Me_3SiCl + Cl_2HSi—PMe_2$$

On the other hand, reaction with $(Cl_3Si)_2CCl_2$ occurs through the CCl_2 group producing $(Cl_3Si)_2CCl(PMe_2)$ which immediately rearranges to form the ylide $(Cl_3Si)_2CPMe_2Cl$.

10. Linear Si-methylated carbosilanes, or Si-methylated cyclic carbosilanes bearing an Si-methylated side chain, rearrange under the influence of $AlBr_3$ to generate six-membered rings along with the elimination of $SiMe_4$ or CH_4. This cyclization reaction does not occur with the corresponding partly Si-chlorinated compounds. These findings can be explained almost entirely by an enhanced or reduced Si—C bond polarization through the substituents.

In the reactions described in 1. it is remarkable that only the bond of the SiMe group is cleaved and not the bonds of the Si—CH_2—Si group. This is appropriately explained by a comparably smaller bond polarization in the carbosilane skeleton. This is supported by the fact that only one methyl group is cleaved from any one silicon atom. Introducing an electron-withdrawing Cl atom onto a Si atom reduces the bond polarization in the remaining SiMe group, which then becomes more resistant to subsequent attack. The conversion of the second SiMe group to an SiCl group with $ICl/AlCl_3$ results in a further reduction of bond polarization in the carbosilane skeleton, and hence a corresponding increase in molecular stability.

Apart from existing ring strain, polarization of the Si—C bond is obviously responsible for the cleavage occurring in the reaction of $(Me_2Si—CH_2)_2$ with HBr. The SiCl substitution in $(Cl_2Si—CH_2)_2$ might not have a significant influence on the ring strain but it certainly reduces the Si—C bond polarization in the ring which then impedes the cleavage induced by a polar reagent.

In a corresponding way, an explanation is derived for the impeding influence of electron-withdrawing substituents bound to the Si atom on Si—Ph cleavage.

The increased reactivity of the CCl_2 group in Si-chlorinated carbosilanes is understood on a similar basis. Introducing two Cl atoms onto a carbosilane skeletal C atom must cause the adjacent Si atoms to become more electropositive. However, this is well balanced by the electron transfer occurring in the Si—Cl bond, which reduces the reactivity of Si against nucleophilic attack. This occurs in relation to the fact that SiH groups having an adjacent CCl_2 group more readily undergo substitution reactions, and that the reactivity of SiF groups is not decreased by placing CCl_2 groups adjacent to them.

The increased polarization of the Si—C—Si bonds in the skeleton through C-chlorination readily explains also the ring cleavage and chain shortening occurring in reactions of perchlorinated carbosilanes with $LiAlH_4$.

The ring-closing reactions occurring in rearrangements of Si-methylated carbosilanes are induced by the Lewis acid $AlBr_3$. They are suitably explained by the Lewis activity of $AlBr_3$ and the polarization of the Si—Me bond. Partly Si-chlorinated carbosilanes do not follow this reaction path, because Si-chlorination reduces the Si—Me bond polarization required for a successful coordination of $AlBr_3$, hence ruling out a ring-closing reaction.

Regarding the polarity of the Si—C bond, discussion of substitution reactions in carbosilanes has generally to consider either nucleophilic attack on silicon or electrophilic attack on carbon. Only a limited number of reagents can be assigned to a single reaction type; most depend on the extent of bond polarization in the carbosilane in determining which reaction path to adopt. Hence, reactions such as the cleavage of $(Me_2Si—CH_2)_2$ by Br_2 to produce $BrMe_2Si—CH_2—SiMe_2—CH_2Br$ occurs through electrophilic attack on the methylene C atom of the ring. On the other hand, the halogen exchange SiCl$_2$/SiClBr resulting from a futile attempt to cleave $(Cl_2Si—CH_2)_2$ by Br_2 is classified as occurring through nucleophilic attack of Br_2 on the Si atom of the ring.

Organolithium reagents and Grignard compounds, and also $LiAlH_4$, can react in either direction depending on the bond polarization of the substrate. Therefore two simple overall applicable rules can be introduced:

1. Electrophilic attack on the C atom of a carbosilane causes Si—C bond cleavage, because the silyl group adjacent to the reaction center is an excellent leaving group.

The qualities of the particular carbosilanes determine whether the electrophile attacks the C atom of a substituent or a C atom of the ring.

2. Nucleophilic attack on the Si atom of a carbosilane causes no change to the molecular skeleton, because the adjacent C group to the reaction center is a very poor leaving group.

In such cases one observes exclusively substituent exchange. A halogen atom on silicon for example, is easily replaced by an alkyl or aryl group or even by another halogen atom.

Of course, these rules do not account for all of the effects and influences appearing throughout the experiments, but on the other hand they provide an uncontradictory and useful guide which clarifies the apparently confusing finding that the same reagent under the same conditions causes substitution in one carbosilane but causes complete rearrangement or even degradation in another one, containing just a slightly different group. For some carbosilanes the choice between both reaction possibilities is so balanced, that simply lowering the reaction temperature from $+20\ °C$ to $-100\ °C$ is a sufficient means to switch from a nucleophilic to an electrophilic attack on the same carbosilane molecule. These examples have been dealt with extensively.

9. Hydrosilylation in Carbosilane Chemistry

9.1 Formation of Cyclic Carbosilanes Through Hydrosilylation

The synthesis of cyclic carbosilanes and related compounds through Si—H addition has already been reviewed [60]. Si—H addition catalysed by H_2PtCl_6 can be used in cyclization, as shown by the following example [71]:

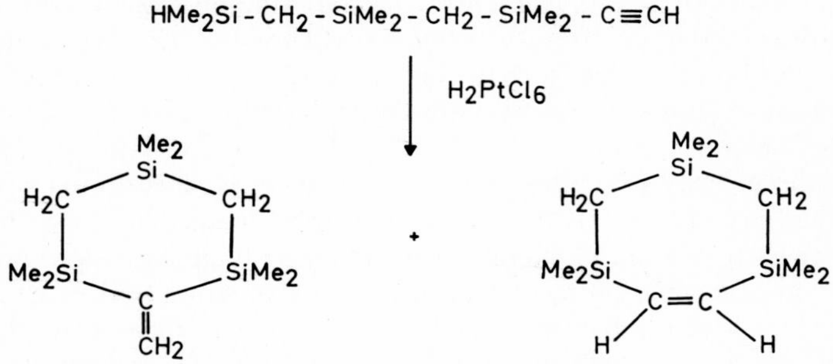

$$HMe_2Si - CH_2 - SiMe_2 - CH_2 - SiMe_2 - C{\equiv}CH$$

The catalytic Si—H addition of 1,3,5-trisilacyclohexanes to monosilylethynes discussed in this section occurs in polycyclic carbosilane synthesis. The scheme below illustrates the principles of synthesis, whereby only the 1,1-disilylolefine is shown as product, since the 1,2-disilylolefine does not lead to ring formation.

While only 1,2-disilaalkanes emerge from the catalytic addition of Si—H groups to Si-substituted $C{=}C$ double bonds [72], 1,1- as well as 1,2-disubstituted products arise from the analogous addition to alkynes [73]. The corresponding reaction of 1,1,3,3,5-pentamethyl-1,3,5-trisilacyclohexane with $HC{\equiv}C—SiMe_2—CH_2Cl$ proceeds without difficulty to

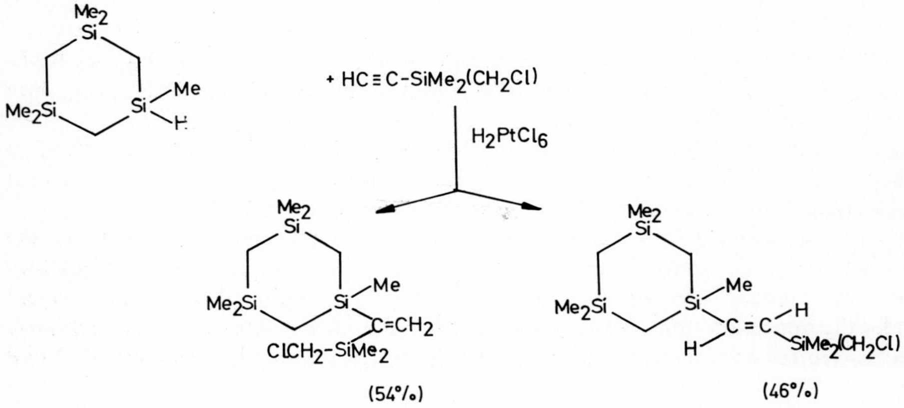

The emerging addition product is not applicable to further ring closure, because no Si-functional groups exist in the compound. Reactions of derivatives containing several Si—H groups, for example (MeHSi—CH$_2$)$_3$ or (H$_2$Si—CH$_2$)$_3$, with HC$\equiv$C—SiMe$_2$—CH$_2$Br yielded mostly polymeric compounds. It was not possible to stop the reaction at any intermediary stage as long as the catalyst existed in the reaction mixture. Polymer formation was favored by increasing the number of Si—H groups in the starting material.

If the reaction mixture is run over an Al$_2$O$_3$ chromatographic column, H$_2$PtCl$_6$ is sufficiently removed [74], and formation of highly viscous polymeric reaction products is prevented. Yet, even under these conditions and with a mole ratio of monosilylethyne to 1,3,5-trimethyl-1,3,5-trisilacyclohexane of 1:1, it was not possible to prevent completely the formation of bis- and trissubstituted 1,3,5-trisilacyclohexanes [76].

It was therefore obvious to prepare and use as starting material a trisilacyclohexane which contained five protective substituents and one reactive SiH-substituent. After successful hydrosilylation these groups could be cleaved from the molecule, enabling ring closure. From the previous investigations a number of suitable compounds were available. The reaction of 1,1,3,3,5-pentaphenyl-1,3,5-trisilacyclohexane with HC$\equiv$C—SiMe$_2$—CH$_2$Br was carried out. The highly viscous reaction products consisted of 1,1- and 1,2-disilylolefins, as shown:

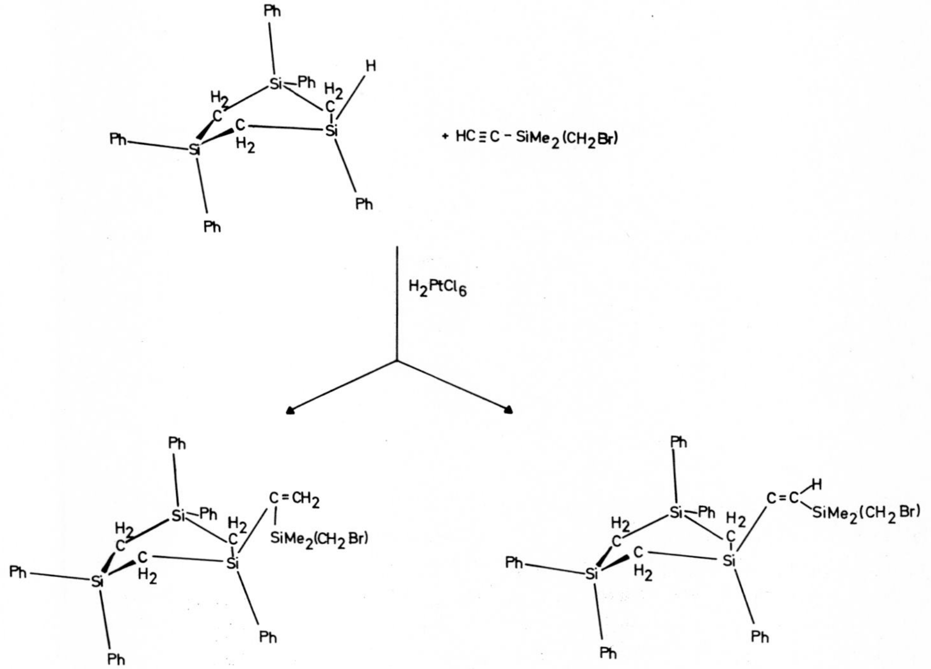

Only after separation of both isomers by means of HPLC could they be satisfactorily characterized by ^{1}H-NMR spectroscopy. Considering the synthesis of bicyclic carbosilanes, a larger yield of 1,1-disilylolefine 407 was desired. Investigations carried out on the influence of solvent on the isomer ratio established that the use of a CCl_4/cyclohexane mixture (1 : 1) produced the best yield, increasing the yield of 1,1-disilylolefine 407 from 38 % (reaction done in hexane) to 72 % (CCl_4/cyclohexane) [76]. In order to achieve a ring closing reaction with a 1,1-disilylolefine, it is first necessary to exchange Si—Ph for Si-halogen substituents. This reaction must be performed under as mild conditions as possible because exchange of the phenyl substituents by means of halogen or hydrogen chloride can cause simultaneous addition to the vinyl group, resulting in a β-elimination of the organosilyl moiety $SiMe_2$—CH_2Br. Therefore, bromine had to be ruled out as a cleavage agent because it requires more severe reaction conditions. Also, treatment of the isomeric disilylolefine mixture with $HCl/AlCl_3$, even under mild conditions, proceeded to degradation of the potential ring closing group.

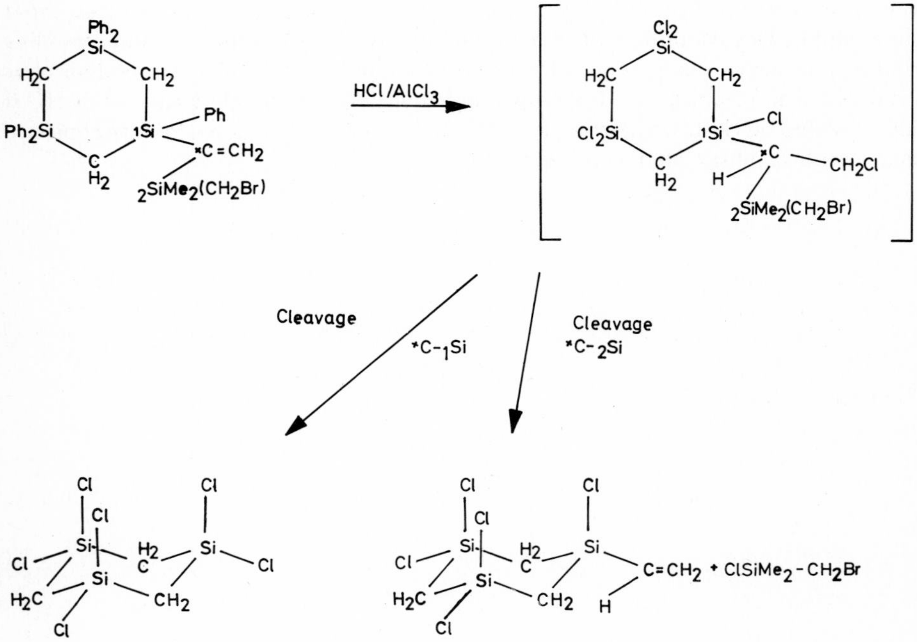

The intended phenyl cleavage in the 1,1-disilylolefine 407 was finally achieved by reaction with HBr (52 hrs at −78 °C) with preservation of the $SiMe_2$—CH_2Br group. The reaction product is free of $BrSiMe_2$—CH_2Br. Because of the absence of peaks in the vinyl group region in the ^{1}H-NMR spectrum, it was recognized that only hydrobrominated derivatives of the 1,1-disilylolefine remained. Reaction of the products with $LiAlH_4$ showed just one phenyl group per silicon atom to be lost. In order to obtain a successful ring closure, the side chains must be axially situated and can couple with a likewise axially situated Si—H group on the ring. Possibilities for a successful ring closure were therefore greatly limited. Hydrogenation after the attempted coupling reaction showed that apart from the desired bicyclic system, the following compound was also formed which was clearly derived from the 1,1-disilyl-olefine 407:

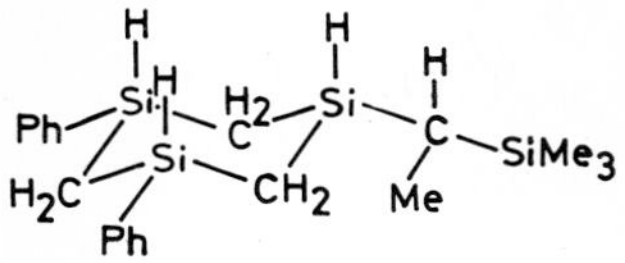

Hence the results show that hydrosilylation is still not a suitable means for the synthesis of polycyclic carbosilanes, and that further elaboration of systems which cleave Si—C bonds more specifically is necessary.

9.2 Linking of Linear SiH-Containing Carbosilane Units over $HC \equiv CH$

The reactions of CH_2Cl_2 with Cu catalysed silicon, described in Chapt. II.2.1.1, lead to linear carbosilanes which, upon treatment with $LiAlH_4$, yield SiH-containing derivatives such as $(H_3Si-CH_2-SiH_2)_2CH_2$ 121. Hydrosilylation of these enables linking of terminal SiH groups in such molecules through a H_2C-CH_2 group by reaction with $HC \equiv CH/H_2PtCl_6$. The addition occurs preferably at the 1,2-position, so that bridging spans such as

$$... H_2Si-CH_2-SiH_2-CH_2-CH_2-SiH_2-CH_2-SiH_2 ...$$

form. Hence reactions of $(H_3Si-CH_2-SiH_2)_2CH_2$ 121 with $HC \equiv CH$ produced molecules containing 8, 12, or 16 Si atoms. These chain molecules consist of two, three or four structural units of 121 which are bound together through one, two or three CH_2-CH_2 bridges. In these, SiH-addition results preferentially on terminal Si atoms.

$$(H_3Si-CH_2-SiH_2)_2CH_2$$

$$\downarrow \quad \begin{array}{l} HC \equiv CH/H_2PtCl_6\text{-} \\ \text{activated carbon} \end{array}$$

$H_3Si-(CH_2-SiH_2)_3-CH_2-CH_2-(SiH_2-CH_2)_3-SiH_3$ Si_8

$H_3Si-(CH_2-SiH_2)_3-CH_2-(CH_2-SiH_2)_4-CH_2-CH_2-(SiH_2-CH_2)_3-SiH_3$

Si_{12}

$H_3Si-(CH_2-SiH_2)_3-CH_2-(CH_2-SiH_2)_4-CH_2-(CH_2-SiH_2)_4-CH_2-$

$-CH_2-(SiH_2-CH_2)_3-SiH_3$ Si_{16}

IV. Results of Structural Investigations of Carbosilanes

Associated with the determination of the exact molecular structures of the carbosilanes was the question of the chemical behavior of the compounds, which in turn is a question of the influence of the substituents on conformation and bond behavior. The most important results of these studies are now presented.

1. X-Ray Investigations of Crystal Structures

1.1 1,3,5-Trisilacyclohexanes with Different Substituents

1.1.1 $(Cl_2Si—CH_2)_3$ 83 and $(Cl_2Si—CCl_2)_3$, 402 [169]

The structural determinations of $(Cl_2Si—CH_2)_3$ and of $(Cl_2Si—CCl_2)_3$ show that the chair form exists in both cases. Chlorine substituents are bonded both axially and equatorially to the ring plane, as shown in Fig. 19 [169].

The influence of the chloro substituents on the Si—C bond lengths in compound 83 is of special interest. The average bond length d(Si—Cl) = 204.1 pm and the Si—C bond length d(Si—C) = 184.5 pm are in agreement with the Si-substituted derivative of 83, namely with 1-cp(CO)$_2$Fe-1,3,3,5,5-pentachloro-1,3,5-trisilacyclohexane 384. The structural determination of $(Cl_2Si—CCl_2)_3$ shows a large variation in bond lengths due to disorder of the trisilacyclohexane rings in the crystal. Nevertheless, the average Si—C bond length obtained was 189 pm, which was also observed in other related compounds. The Si—Cl bond lengths measured were too inaccurate to be useful.

The question of whether the C—Cl bond lengths and subsequently in a similar fashion the Si—CCl$_2$ bond lengths could be changed through substituent effects cannot be answered.

1.1.2 $(Ph_2Si—CH_2)_3$, 403 [65, 170]

In 1,1,3,3,5,5-hexaphenyl-1,3,5-trisilacyclohexane, 403, the central trisilacyclohexane system exists in a skewed boat conformation which resembles a twisted boat conformation more than a symmetrical boat form. The bond angles of Si ($\bar{\alpha} = 110°$) and C ($\bar{\beta} = 117.9°$) shown in Fig. 20 [170] produce an average value of α, β = 114.0° which, in comparison to the tetrahedral angle is considerably increased towards the value of 120° displayed by a planar six-membered ring. The cyclohexane ring is

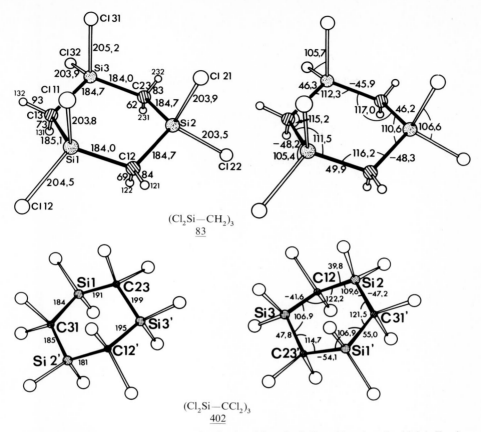

$(Cl_2Si—CH_2)_3$
83

$(Cl_2Si—CCl_2)_3$
402

Fig. 19. The trisilacyclohexane skeleton showing bond lengths (left) and bond angles (right). Torsion angles of the skeleton are also given. The different labelling of atoms (below left and right) indicates the two disordered positions of the trisilacyclohexane rings of **402**

distinctly flattened. The dihedral angle sequence (18°; 32°; −57°; 22°; 38°; −62°) shows the considerable distortion from the ideal twisted boat conformation (33.2°; 33.2°; −70.7° etc.) as well as the flattening of the six-membered ring. This occurs mostly at the cost of the configuration on the C atoms ($\bar{\beta} = 117.9°$), while the Si atoms (as in other carbosilanes [172, 127]) again tend to maintain the tetrahedral configuration. By making $\alpha < \beta$, the Si atoms are ultimately shifted further away from each other (320 pm) in comparison to the C atoms of the six-membered ring (306 pm). This is an effect that may be attributed to the bulky substituents. The steric influence of the 6 phenyl groups is enough to favor a twisted boat conformation, because in this case repulsion of axial substituents can be drastically reduced (Fig. 21). Mutual orientation of the phenyl ring planes shows a further steric optimization.

The Si—C bond lengths in the six-membered ring measured $\bar{a} = 187.1(5)$ pm, which is somewhat smaller than the bond lengths to the substituent atoms ($\bar{d} = 187.9(3)$ pm). However, they lie within the limits of previous measurements made on cyclic carbosilanes [172, 173, 174].

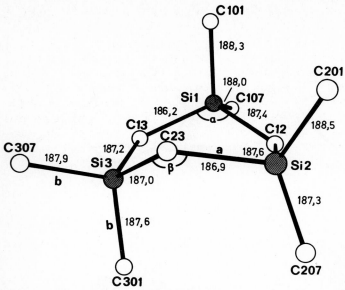

Fig. 20. Trisilacyclohexane skeleton with the directly bound C atoms of the phenyl rings. Si atoms are shown in black. Average bond lengths and angles are: $\bar{a} = 187.1(5)$ pm; $\bar{b} = 187.9(3)$ pm; $\bar{\alpha} = 110.0(3)°$; $\bar{\beta} = 117.9(6)°$

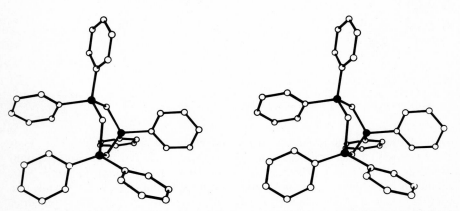

Fig. 21. Stereochemical pair of molecules shown without indicating H atoms. The molecule has been rotated approx. 90° in comparison to the molecule shown in Fig. 20. Si atoms are drawn in black, Si 1 right, Si 2 below

1.1.3 1-Cyclopentadienyl-dicarbonyl-iron-1,3,3,5,5-pentachloro-1,3,5-trisilacyclohexane $\underline{384}$ and 1,3-Bis(cyclopentadienyl-dicarbonyl-iron)-1,3,5,5-tetrachloro-1,3,5-trisilacyclohexane, $\underline{385}$ [163, 174]

The influence of bulky substituents on the 1,3,5-trisilacyclohexane skeleton and their effect on bond behavior were established. As can be seen in Fig. 22, the six-membered rings of both complexes exist in the twisted conformation wherein the attachment of the bulky cp(CO)$_2$Fe ligand to the 1,3,5-trisilacyclohexane ring transforms the skeletal configuration from a chair form to a twisted boat form. On the other hand, less bulky substituents, such as Me and Cl, for example, favor the chair configuration. A comparison of the Si—Cl and Si—C bond lengths in (Cl$_2$Si—CCl$_2$)$_3$ shows no remarkable differences. A discussion of these values is presented in connection with Table 69.

The bond lengths in (Me$_2$Si—CH$_2$)$_3$ <u>35</u> and (Cl$_2$Si—CH$_2$)$_3$ <u>83</u> can be referred to in the discussion of bonds in other six-membered carbosilane rings which are part of larger systems and which in turn succumb to the extra effects of steric influences. The Si—C bond lengths for cyclic carbosilanes are presented in Table 69. The average value for all comparable Si—C bond lengths in these compounds is $\bar{d}$(Si—C) =

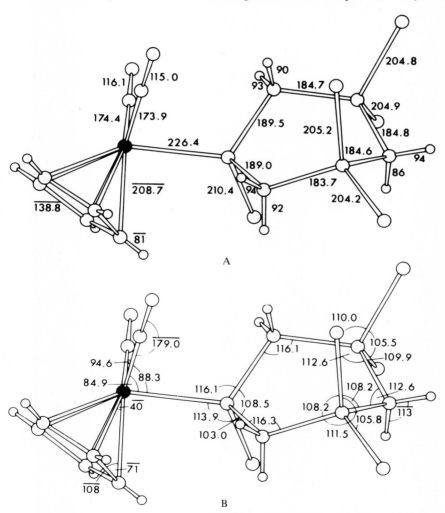

Fig. 22. **A, B:** <u>384</u>; **C, D:** <u>385</u>. The bond lengths (top) and relevant bond angles (bottom) are shown for both compounds

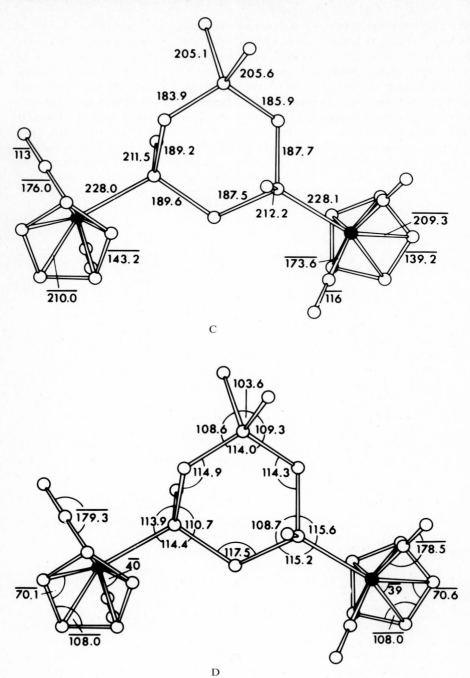

C

D

Fig. 22. C, D

188.4 pm. Individual values vary in the range of 186 pm $< d <$ 192 pm in the cases where H, Me and Ph are bound to silicon atoms as substituents. However, Cl substituents bound to silicon appear to reduce the Si—C bond length considerably: $d(C—SiCl) = 182–185$ pm. This trend also applies to Si—C bonds in acyclic compounds [200]. Especially striking is the fact that the Si—Cl lengths are even smaller in such systems than expected from the Schomaker-Stevenson [176] corrected radii summation $\Sigma\, r_{corr} = (217 - 7) = 210$ pm. Obviously, inductive effects of substituents can also accumulate. These effects can best be accounted for by assuming that the effective covalent radius of the central atom is decreased by introducing more electronegative substituents onto silicon. This is equivalent to an assumed decrease in electronegativity of the central atom. The single-bond distances d_1^*, obtained from consideration of known structural data collected to date [200], are of interest here in analysis of the silyl groups (d_1^* = sum of covalent radii):

a) for —SiC$_3$ $d_1^* = (d_1 - 6)$ pm

b) for —SiC$_2$Cl $d_1^* = (d_1 - 8)$ pm

c) for —SiCCl$_2$ $d_1^* = (d_1 - 11)$ pm

d) for —SiCl$_3$ $d_1^* = (d_1 - 14)$ pm

Furthermore, it is evident from Table 69 that in trisilacyclohexane systems of similar substitution the bond lengths depend upon the conformation of the six-membered ring. They are slightly smaller in the chair form than in the twisted boat conformation: $\bar{d}$(Si—C) (chair form) $\simeq$ 186.6 pm; $\bar{d}$(Si—C) (twist form) $\simeq$ 188.5 pm. A difference of 2 pm between the average values lies of course within experimental error. However, they correspond well with the different bond lengths d(Si—C) in cubic and hexagonal (2 H) silicon carbide, whose structures have been precisely determined [201]. In cubic silicon carbide, the bond length is d(Si—C) = 188.7 pm; the condensed six-membered rings exist exclusively in the chair conformation.

Hexagonal SiC yields bond lengths of d_s = 187.5 pm for the six-membered rings in chair conformation (parallel to (001)), while the bond length of the "ideal" boat-oriented six-membered rings comes to d_w = 193.0 pm. In connection with this, the fact is noteworthy that the sum of the bond orders is exactly $\Sigma\, n = 4$ for the atoms in hexagonal SiC ($3 \times 1.035 + 1 \times 0.895$ with d_1 = 188.8 $\approx$ 189 pm).

The six-membered rings of compounds 83, 402, 35, 39, 40 and 209 (Table 69) all exist in the chair form, while compounds 430, 384, 385, 198, 200, 58 and 405 all contain skeletons in a distorted boat form. The largest deviation from the usual mode exists in compound 58 which maintains an especially rigid skeleton. The Si—CH$_3$ bond distance in this case (187.4 pm) is smaller than the corresponding Si—CH$_2$ distance (188.9 pm).

The Fe—Si bond distances in compounds 384 and 385, 226.4 pm and 228.1 pm respectively, are essentially smaller than the value for the sum of the covalent radii ($\Sigma\, r$ = 235 pm) and are among the shortest Fe—Si bond lengths observed to date. They do correspond, however, to the general trend observed for relevant bond distances in other cyclopentadienyl complexes of iron [202–206]. The Fe—C and C—O bond distances lie within the range observed for other such measured bond lengths.

The bond lengths in the complexes 384 and 385 allow for interesting estimations to be made on bond order. The short bond length of chlorine-substituted Si atoms,

Table 69. Comparison of Si—C bond lengths in cyclic carbosilanes and in cubic and hexagonal silicon carbide (SiC) (S = chair form; W = boat form; TW = twisted boat form)[a]

Compounds		Conformation and configuration	d(Si—C) (pm)	Refs.
83	$(Cl_2Si—CH_2)_3$	6-Ring, S	184.5	[169]
402	$(Cl_2Si—CCl_2)_3$	6-Ring, S	189.1	[169][b]
35	$(Me_2Si—CH_2)_3$	6-Ring, S	(197.4)	[207][c]
430	$(Ph_2Si—CH_2)_3$	6-Ring, TW	187.1(6)	[170]
384	$Si_3C_3H_6Cl_5Fe(CO)_2 \pi cp$	6-Ring, TW	186.1(15)	[174]
385	$Si_3C_3H_6Cl_4[Fe(CO)_2 \pi cp]_2$	6-Ring, TW	187.3	[174]
327	$(Cl_2Si—CPMe_2Cl)_3$	6-Ring (Ylide)	179.5	[132]
39	$Si_4C_{10}H_{24}$	6-Ring, S (Adamantane)	186.6(7)	[182]
40	$Si_7C_{16}H_{36}$	6-Ring, S (Diadamantane)		
		Si—CH	188.6(3)	[183]
		Si—CH$_2$	187.0(9)	
69	$Si_8C_{18}H_{40}$	6-Ring, S (Adamantane part)	187.8	[185]
		6-Ring, W (Wurtzitane part)	190.0	
191	$Si_4C_{11}H_{28}$	6-Ring, TW (Scaphane)	188.1(14)	[172]
200		6-Ring, TW (Scaphane)	188.4	[56]
58	$Si_8C_{27}H_{36}$	6-Ring, TW (Scaphane)	188.9	[190]
405	$Si_7C_{19}H_{48}$	6-Ring, TW (Propellane)	188.7	[173]
209	$Si_6C_{16}H_{40}$	6-Ring, S (Hexasila-perhydrophenalene)		[199]
240	$Si_8C_{22}H_{56}$	Dispirotetradecane 6-Ring, S	191.8	[191]
		4-Ring	192.2	
310	$Si_4C_{12}H_{30}Cl_2$	1,3-Disilacyclobutane 4-Ring	191.9	[129]
309	$Si_4C_{12}H_3$	2,4-Disilabicyclobutane		[127]
		3-Ring Si—C	183.6	
		3-Ring C—C	178.1	
335	$Si_6C_6H_8Cl_8$	Asterane 4-Ring	188.7	[67]
		6-Ring, W (=CH$_2$)	189.7	
		6-Ring, W (=SiCl$_2$)	181.5	
364	$Si_4C_{12}H_{28}$	5-Ring, Bicyclic with C=C	189.7	[150]
404	$Si_4C_{12}H_{32}$	8-Ring	189.7	[184]
	SiC (cubic)	6-Ring, S	187.5	[201]
	SiC (2 H)	6-Ring, W	193.0	

[a] S from German: Sessel; W: Wanne; TW: Twist-Wanne
[b] poor structure determination
[c] disorder of C_3Si_3 skeleton in the solid state

as well as the short Fe—Si bond distances, have already been mentioned. For single bonds (n = 1) one expects from the summed covalent radii and after consideration of the polar contribution by the method of Schomaker and Stevenson [176] $d_1^*(Si—C)$ = (195 — 6) = 189 pm, $d_1^*(Si—Cl)$ = (217 — 7) = 210 pm and $d_1(Fe—Si)$ = (235 — 0) = 235 pm. In this case the value for the Fe—Si bond is taken from the so-called metallic single bond radius for iron (117 pm), which was calculated by Pauling from the Fe—Fe distance in the metal with some assumptions being made (CN = 12). Other appraisals will be considered along the following lines:

a) The summation of the bond orders (Σ n) for Si atoms of trisilacyclohexane ligands is assumed to be Σ n = 4.

b) The effect discussed above relating to the effective Si-contraction through chlorine substitution is considered (reduced single-bond distances d* are proportional to the number of Cl substituents on silicon). Then the bond orders n are sufficiently and accurately described by the Donnay-Allmann relationship $n = (d_1^*/d_n)^5$ [175]. This latter projection determines at first the limits for the Si atoms in the $SiCl_2$ groups of both complexes $3.98 \leq \Sigma\, n(Si) \leq 4.05$, the basic values for these limits being $r(C) = 77$ pm, $r(Si) = 118$ pm, $r(Cl) = 99$ pm, $d_1(Si—C) = 195$ pm, $d_1(Si—Cl) = 217$ pm, $d_1^*(Si—C, —SiCl_2—) = 184$ pm, and $d_1^*(Si—Cl, —SiCl_2—) = 206$ pm. Similarly, it is to assume that for the SiCl groups bound to Fe atoms $d_1^*(Si—C, —SiCl—) = 187$ pm and $d_1^*(Si—Cl, —SiCl—) = 209$ pm. The bond orders can be calculated from these bond lengths as the difference to $\Sigma\, n(Si) = 4$, including an average degree of bonding for the Fe—Si bond of $\bar{n}(Fe—Si) = 1.14$. With the average value of $\bar{d}(Fe—Si) = 227.5$ pm for the observed bond lengths, one obtains then $d_1(Fe—Si) = 234$ pm as a result for the single bond, and $r(Fe) = 116$ pm for the covalent radius of iron. If Fe—Si bonds were in a similar fashion influenced by Cl substituents as previously considered, the values should increase to $d_1(Fe—Si) = 242$ pm and $r(Fe) = 124$ pm. In any case, the values coincide with the range existing from Pauling's estimations.

1.1.4 The Ylides $(Cl_3Si)_2CPMe_2Cl$ 325, $(Cl_3Si)_2CPMe_3$ 328 and $(Cl_2Si—CPMe_2Cl)_3$ 327 [132] as well as $Cl_2Si(CH_2—SiCl_2)_2CPMe_2Cl$ 326 and $Me_2Si(CH_2—SiMe_2)_2CPMe_2Br$ 330 [133]

The results of structural determinations of the ylides of disilaphosphorane type $(Cl_3Si)_2CPMe_2Cl$ 325, $(Cl_3Si)_2CPMe_3$ 328 and $(Cl_2Si—CPMe_2Cl)_3$ 327 are shown in Fig. 23 [132]. The ylides 325 and 328 exhibit a "propeller type" C_{3h}-symmetry if the difference between Me and Cl substituents is left out of consideration. Prominence is given to the fact that in each case one substituent lies in the ylide plane, and that this position on the P atom will be occupied at all times by a methyl group.

The structure of the triylide 327 deviates only slightly from C_{3v}-symmetry. In this molecule none of the substituents on Si or P is situated in the ylide plane.

On the other hand, the polar structure resulting from the unilateral axial orientation of all Cl(P) atoms is remarkable [132]. The atoms of the central six-membered ring deviate from the balanced plane at the most by about 5 pm. However, all 3 P substituents lie about 25–35 pm in the same direction away from the plane of the six-membered ring and lie on the same side as the axially-oriented Cl substituents. This effect might be caused by the influence of solvent molecules. The configuration of the ylide C atoms is in principle trigonal planar; however, in all four investigated ylide molecules they lie significantly out of the (Si, Si, P) plane (325 : 7 pm; 328 : 9 pm; 327 : 3 pm (2 ×), 8 pm). The bond lengths are surprisingly short and indicate distinctly higher bond orders at the central P—C bonds, as well as at the Si—C bonds. Peripheral Si—Cl bonds likewise show bond length shortening, and exhibit known changes on increasing Cl-substitution $d_{SiCl}(SiCl_3) < d_{SiCl}(SiCl_2)$ [174]. Corresponding to the covalent radii, for these ylides applies $d(P—C) < d(Si—C)$. However, changes occurring on substituting $—PMe_3$ for $—PMe_2Cl$ cannot be overlooked.

The bond angles on the ylidic C atom deviate only slightly from 120°. Deviations of 1° on average depend mainly on the aberration of the C atom out of the (Si, Si, P) plane. The average bond angles on Si and P atoms are close to the tetrahedral bond

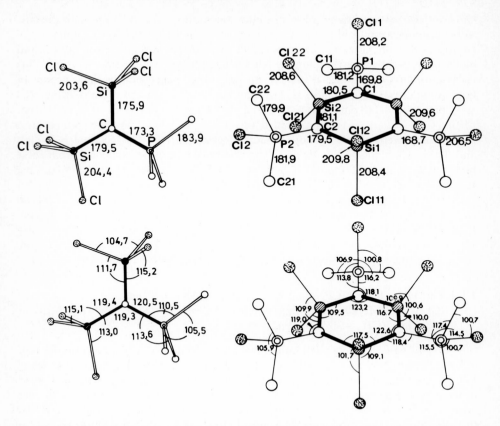

Fig. 23. Left: General structures of ylides <u>325</u> and <u>328</u>; with bond distances (above) and bond angles (below) for <u>328</u>. Right: Ylide <u>327</u>

angle of 109.5°. The particular values deviate from the average by up to 6°. Hence it is remarkable that the C—Si—X and C—P—X angles, where X represents the coplanar substituent, in <u>325</u> and <u>328</u> actually increase to 111°–114°, while the angles Y—Si—Y' and Y—P—Y', where Y and Y' represent the remaining peripheral substituents, actually reduce. Obviously the coplanar position of the substituents X leads to a distinct repulsion between neighboring substituents. It is assumed that a connection exists between the noticeable configuration of ylide substituents, with the higher bond orders n(C—Si) and n(C—P) favored by the coplanar adjustment of the substituents X. The situation is different in the triylide <u>327</u> because the Si atoms are members of the six-membered ring and hence determine the configuration of the Cl(Si) substituents. The orientation of the substituents on the P atoms adapts to this particular situation.

An insight into bond strengths was gained from a calculation of bond orders n_i in the way already described [174]. The observed bond lengths d_i were compared with the reduced bond lengths for single bonds d_i^* ($n = 1$). The calculation of n_i is obtained from the Donnay-Allmann [175] relationship $n_i = (d_i^*/d_i)^5$. The values for

d_1^* result from the summation of the covalent radii after consideration is given to polar contributions according to the method of Schomaker and Stevenson [176]. The influence of the Cl substituents on shortening peripheral bond lengths is also taken into consideration [174]: $d_1^*(Si—C) = 189$ pm; $d_1^*(P—C) = 183$ pm; $d_1^*(Si—Cl) = 204$ pm or 206 pm; $d_1^*(P—Cl) = 201$ pm. The results of the calculations are presented in Table 70.

Table 70. Average bond orders $\bar{n}_i$ in the ylides 335, 328 and 327

Type	Ylide 335	Ylide 328	Ylide 327
$\bar{n}(Si—C^-)$	1.28	1.36	1.26
$\bar{n}(P—C^-)$	1.55	1.31	1.49
$\bar{n}(Si—Cl)$	1.02	1.00	0.93
$\bar{n}(P—C)$	1.00	0.97	1.06
$\bar{n}(P—Cl)$	1.00	—	0.86
$\Sigma\,n(C^-)$	4.11	4.03	4.01
$\Sigma\,n(Si)$	4.34	4.36	4.38
$\Sigma\,n(P)$	4.55	4.22	4.47

The average bond orders of individual bonds $\bar{n}_i$, as well as their summed values $\Sigma\,\bar{n}_i$, show interesting features. In all cases $\Sigma\,n(C) \simeq 4$ for the ylidic carbon atom and $\Sigma\,n(Si, P) \simeq 4.33$ for the silicon or phosphorus atoms. The higher bond orders of Si—C and P—C bonds are on average $\Sigma\,n(C—Si, P) \simeq 1.33$ and are relatively evenly distributed. The charge of the formal carbanion is obviously delocalized over the Si—C bonds. This accounts for the observation that characteristic ylide reactions such as adduct formation with BMe_3 fail to appear. Furthermore, the values $\Sigma\,n(Si, P) \simeq 4.33$ show that the Si and P atoms do not equilibrate the stronger bond to the ylidic C atom ($n \simeq 1.33$) through weakening of bonds of the peripheral substituents; rather they display extra bond contributions from $(d_\pi — p_\pi)$ interactions [177].

On inspection of the ylides 325, 328 and 327, a distinct strengthening of the P—C bond at the expense of Si—C bonds is seen if a PMe_3 group is replaced by a PMe_2Cl group.

If the equation above relating n_i with d_1^*, d_i using $d_1^*(P—C) = 183$ pm is applied to the (tris(trimethylphosphonio)methanide)$^{2+}$ cation [178], then surprisingly $n(P—C) = 1.25$, $\Sigma\,n(C) = 3.75$ and $\Sigma\,n(P) = 4.60$ is obtained. The smaller bond order on the ylidic C atom can be explained by an intensified electron transfer onto the cationic P atoms. Further enlargement of $\Sigma\,n(P)$ accounts for increased $(d_\pi — p_\pi)$ interactions in comparison to ylide 327 ($\delta P = +19.6$ pm in the di-cation [178]; $\delta P = +13.7$ pm in 327).

Regarding the structural investigations made on the ylides 330 and 326, the effect of one ylidic ring carbon atom of 1,3,5-trisilacyclohexane on the molecular structure and the effect of Si substituents (Me or Cl) on bonding conditions were sought. Figure 24 presents some most relevant data in connection with these molecules [133]. The structure of the ylide 330 shows striking features in comparison to those compounds described beforehand.

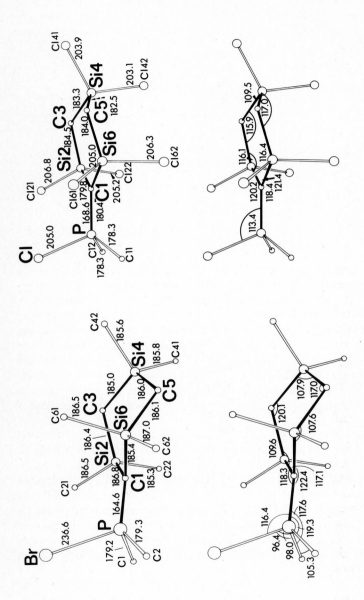

Fig. 24. Ylide 330 (left) and ylide 326 (right). For both molecules only the most relevant bond lengths (above) and bond angles (below) are given

a) The double bond contributions in this molecule are distinctly localized in the P–C bond. The distance $d(P—C) = 164.6$ pm is one of the shortest observed bond lengths to date, and in comparison to $d(P—C) = 185$ pm and $d(P≡C) = 154$ pm [134] this value corresponds well to the value for the $P=C$ double-bond length of 165 pm.

b) The bond distance $d(P—Br) = 236.6$ pm is unusually large and, in comparison to the single bond length of 216 pm, maintains a bond order of only $n = 0.63$.

c) The formal ylidic carbon atom C1 lies out of the P/Si2/Si6 ylide plane by about 17 pm just opposite the position occupied by the Br substituent on the P atom.

d) The C—Si bond distances in the ylidic system correspond well to C—Si single bond lengths. In the vicinity of C1, no atom exists with a shortened van der Waals contact, from which inter- or intramolecular interactions could occur. The existing system corresponds almost entirely to an alkylidene phosphonium salt $Br^- R_2 P^+ = C <$. This holds in connection with the stronger planarity on the P atom and the unchanged bond lengths and angles in the Si_3C_3 ring. The conformation of the six-membered ring corresponds to a slightly distorted boat form (torsion angles $\gamma = -6°$, $50°$, $-35°$, $-23°$, $61°$, $-44°$ respectively for the bonds C1—Si2/Si2—C3/ etc.). The distortion mainly applies to the C3, Si4, C5 and Si6 part of the molecule.

The ylide systems of 326 in all three crystallographically independant molecules are planar within experimental limits. The bond lengths in these three independant molecules are also completely equivalent. The average values obtained are $\bar{d}(P—C1) = 168.6$ pm, $\bar{d}(C1—Si) = 180.1$ pm and $\bar{d}(P—Cl) = 205.0$ pm (Fig. 24). These distances also correspond to the triylide 327. The Si_3C_3 six-membered ring in 326 shows an expected difference: it exists in a flattened chair form (average endocyclic bond angle $\bar{\beta} = 116°$, with torsion angles of $-14°$, $35°$, $-52°$, $50°$, $-31°$ and $12°$ corresponding to the bonds C1—Si2/Si2—C3/etc.). The individual endocyclic bond angles show that the leveling-out of the six-membered ring subsides more and more with increasing distance from the ylidic centre.

The bond orders were calculated, as in [132], from the following values: $d_1(Si—C) = 188$ pm, $d_1(C—SiCl_2) = 183$ pm, $d_1(Si—Cl) = 204$ pm, $d_1(P—Cl) = 201$ pm, $d_1(P—Br) = 216$ pm. The value selected for $d_1(P—C) = 185$ pm is taken from gaseous PMe_3 [180]. The bond distances $d_1(P—C) = 185$ pm, $d_2(P=C) = 165$ pm and $d_3(P≡C) = 154$ pm can be generally produced from the Donnay-Allmann relationship [175]. By making $n = 6$ for P—C bonds and $n = 5$ for all other bond distances, the following values arise for the molecules presently investigated: $\Sigma \bar{n}(P) = 5.1$, $\Sigma \bar{n}(C) = 4.0$, $\Sigma \bar{n}(Si) = 4.1$. Particularly in 326 and 330, the bond orders $n(P — C^-)$ and $n'(P—Hal)$ add to $n + n' = 2.66$; the partial bond weakening of the Br substituent in 330 is counterbalanced by a strengthening of the $P—C^-$ bond.

1.2 Structures of Si-Adamantanes

1.2.1 1,3,5,7-Tetrasilaadamantane, $Si_4C_{10}H_{24}$, 39 [182]

The results of the crystal structure analysis of compound 39 are presented in Fig. 25, including bond lengths and a projection of the unit cell to (010).

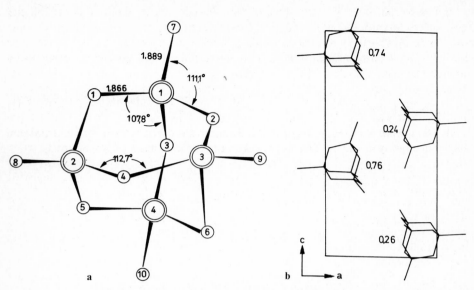

Fig. 25. a Average bond lengths and angles in the 1,3,5,7-tetramethyl-tetrasilaadamantane. **b** Projection of the structure to (010). The numbers correspond to the y-coordinate of the molecular centers

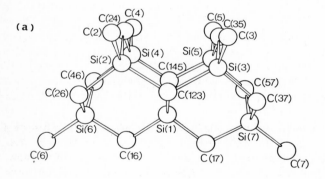

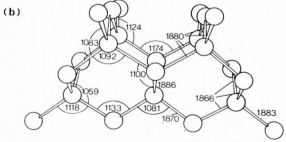

Fig. 26. Hexamethyl-heptasila-hexacyclo-heptadecane. **a** gives the designation of the atoms in the molecule as used in the discussion, **b** the average bond distances (pm) and bond angles (°)

In compound <u>39</u>, if we regard the Si_4 tetrahedron and the C_6 octahedron as crystallographically separated polyhedra, a common center exists. The variance of the angles on the Si or C atoms of these crystallographically-independent polyhedra is very small: $0.25°$. Compound <u>39</u> culminates in perfect T_d symmetry within experimental limits. The crystal packing does not cause distortions of this molecule. The Si—C bond lengths correspond well with related values in other molecules (Tab. 69). The Si—CH_3 bond distances (185.9–190.6 pm) differ from the values obtained for Si—CH_2 (183.2–190.0 pm). The difference between the Si—CH_3 mean values of 189.9(10) pm and 186.6(7) pm for Si—CH_2 is statistically significant. The tetrahedral coordination on Si is slightly distorted in comparison with the exocyclic CH_3—Si—CH_2 angle (mean = 111.1°) and the endocyclic CH_2—Si—CH_2 angle (mean = 197.8°).

According to available results, the H atoms of the CH_3 group exist in a staggered conformation with respect to the neighboring methylene groups. In only one CH_3 group the H atoms appear to be disordered. Both these appearances, namely, staggered conformation to the methylene groups and a rotation of CH_3 groups resulting in crystallographic disorder of the H atoms, have recently been observed in other cases [181].

1.2.2 Hexamethyl-heptasilahexacyclo-heptadecane, $Si_7C_{16}H_{36}$, <u>40</u> [183]

This molecule consists of two condensed 1,3,5,7-tetrasilaadamantanes connected through atoms C(123), Si(1) and C(145) (all designations of atoms in this section refer to Fig. 26a, and not to IUPAC nomenclature).

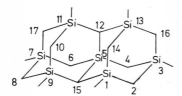

1,3,7,9,11,13-Hexamethyl-
1,3,5,7,9,11,13-heptasilahexacyclo-
$[7.5.1.1^{3,13}.1^{7,11}.0^{5,12}.0^{5,15}]$heptadecane

Regarding these as fused non-distorted adamantane skeletons with tetrahedral angles and bond distances of Si—C = 188 pm, the distances between the methylene C atoms C(24) and C(35), as well as those between the methyl groups on Si atoms Si(2) and Si(3) or Si(4) and Si(5) would be considerably too small. This would cause proton contacts of about 130 pm. It was therefore important and informative to measure the distortion in the region of the Si bridged eight-membered ring Si(2), C(123), Si(3), C(35), Si(5), C(145), Si(4), C(24) and its effect on the total adamantane skeleton.

The results of the X-ray structure analysis are presented in Fig. 26. It shows the molecular structure of compound <u>40</u> in the designation of atoms (a) and mean bond lengths and bond angles (b). All six-membered rings exist in the chair form. The torsion angles in these rings depend on the bond lengths and bond angles and lie between 54.0 and 66.0°. In the above described eight-membered Si-bridged ring in the saddle conformation, the mean Si—C bond distance is 188.0(6) pm. In an undistorted carborundane, the atoms Si(2), Si(3), Si(5) and Si(4) in this ring should form a square with a side length of 310 pm corresponding to the mean distance of carbon-bridged Si atoms. The distances of C(24), C(35) and C(123), C(145) should also amount to

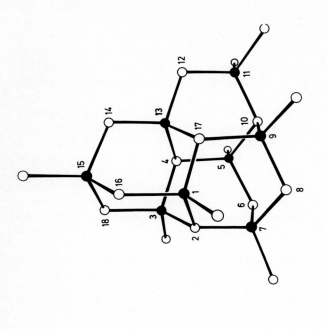

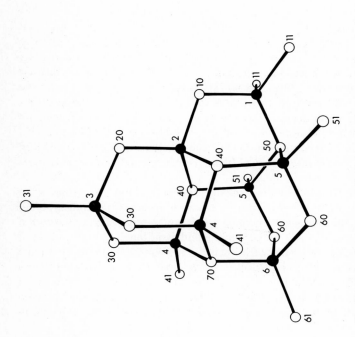

Fig. 27. Molecular structure of $Si_8C_{18}H_{40}$ $\underline{69}$. Si atoms are drawn in black, H atoms are omitted. Left: Numbering of atoms refers to Table 71. Right: Numbering of skeletal atoms according to the systematic nomenclature

310 pm. The distances Si(2), Si(4) and Si(3), Si(5) as well as C(24), C(35) and C(123), C(145) fall within the expected range (311.5(2) pm, and 299.5(9) pm respectively). In contrast, the distances Si(2), Si(3) and Si(4), Si(5) are 322.0(2) pm, due to the enlarging of the angles (Si(2)—C(123)—Si(3) and Si(4)—C(145)—Si(5)) to 117.4(1)°.

The four-membered ring formed by Si(2), Si(3), Si(5) and Si(4) is not totally planar; the silicon atoms deviate alternatively by about ±1.0 pm from the common plane. Furthermore, the diagonals of this four-membered ring have different lengths Si(2)—Si(5) = 466.2(3) pm, Si(3)—Si(4) = 449.7(3) pm. Both adamantane skeletons are slightly rotated with respect to each other. The molecule displays C_2 symmetry within the limits of experimental error, with slight deviation from the highest possible symmetry C_{2v}. The enlargement of the Si—CH—Si angles in this eight-membered ring, along with the contraction of the C(24,35)—Si—CH$_2$ angles to 105.8(5)° and the slight deviation from C_{2v} symmetry, causes an enlargement of the distance between the opposite methylene C atoms C(24) and C(35) to 350.8(19) pm. The C(i)—Si(i)—C(ij) angles, where i = 2, 3, 4, 5 and j = 6, 7, are narrowed to 108.3(8)°; the mean value for the other CH$_3$—Si—CH$_2$ angles is 111.8(9)°. Neighboring methyl C atoms C(2), C(3) and C(4), C(5) in the eight-membered ring also exhibit distances of 349.5(10) pm.

In the region of the bridge positions this molecule shows analogies to the octamethyl-tetrasilacyclooctane 404 [184], in which the Si—C eight-membered ring also exists in the saddle conformation. The means of the atomic distances and bond angles are directly comparable with corresponding values in other Si—C six-membered ring systems (Table 69).

1.2.3 Octamethyl-octasilaheptacyclo-octadecane, Si$_8$C$_{18}$H$_{40}$, 69 [185]

Carbosilanes that have been clarified structurally thus far are of two classes: either carborundanes (Si—C six-membered rings existing in the chair conformation) or scaphanes (Si—C six-membered rings in the boat conformation). Compound 69 represents a further type of carbosilane, namely, one in which the six-membered rings exist in both the chair and boat conformations. The results of the X-ray crystal structure analysis are presented in Fig. 27 and Table 71.

Compound 69 (1,3,5,7,9,11,11,15-octamethyl-1,3,5,7,9,11,13,15-octasilaheptacyclo-[7.7.1.1^{3,15}.0^{4,13}.0^{5,10}.0^{13,17}]octadecane) contains m symmetry. The central polycyclic skeleton consists of trisilacyclohexane rings, some of which exist in the chair conformation and the rest in the boat conformation (Fig. 27). This skeleton can be compared to a section from the commonly occurring silicon carbide modification 4 H—III. It represents the smallest molecular unit which possesses all the essential and characteristic features of the 4 H—SiC structure, namely the adamantane and the complete wurtzitane skeleton [186, 187]. Atoms 1, 3, 13, 15 (systematic numbering) all participate in the construction of the tetrasilaadamantane skeleton. The wurtzitane skeleton itself emerges on the one hand from condensation of a hexasilaiceane (Si1,3,5,7,9,13) and from the complementary tetrasilabarrelane (Si5,9,11,13). Separate structural investigations of Si-adamantanes and Si-barrelanes [182, 172] have already been undertaken. The iceane system [189] was originally structurally characterized as a hexasila skeleton. Combining the wurtzitane skeleton, or at least parts of it, with the adamantane skeleton containing six-membered rings in the chair conformation

Table 71. Bond lengths (pm) and angles (°) of the molecule shown in Fig. 27. Standard deviations are given in brackets. (2×) stands for the multiplicity

Bond lengths

Si(1) - C(10)	184.7(15)	Si(3) - C(20)	190.8(17)	Si(4) - C(30)	189.0(8)	Si(5) - C(40)	192.6(9)
- C(50)	188.4(12)	- C(30)	187.5(8) (2x)	- C(40)	184.7(9)	- C(50)	190.5(7)
- C(11)	187.6(10) (2x)	- C(31)	186.3(12)	- C(70)	188.7(6)	- C(60)	189.0(7)
Si(2) - C(10)	190.7(15)	Si(6) - C(60)	186.7(9) (2x)	- C(41)	189.1(8)	- C(51)	187.5(9)
- C(20)	182.0(17)	- C(70)	189.6(12)				
- C(40)	189.3(6) (2x)	- C(61)	187.3(15)				

Angles

C(10) - Si(1) - C(50)	107.7(6)	C(20) - Si(3) - C(30)	106.4(4) (2x)	Si(1) - C(10) - Si(2)	113.5(7)
C(10) - Si(1) - C(11)	110.5(4) (2x)	C(20) - Si(3) - C(31)	111.9(6)	Si(2) - C(20) - Si(3)	113.9(6)
C(50) - Si(1) - C(11)	111.9(4) (2x)	C(30) - Si(3) - C(31)	112.9(3) (2x)	Si(3) - C(30) - Si(4)	113.9(4)
C(11) - Si(1) - C(11')	104.3(6)	C(30) - Si(3) - C(30')	105.8(5)	Si(2) - C(40) - Si(4)	111.8(4)
C(10) - Si(2) - C(20)	111.5(6)	C(40) - Si(5) - C(50)	108.8(4)	Si(2) - C(40) - Si(5)	109.7(4)
C(10) - Si(2) - C(40)	110.2(4) (2x)	C(40) - Si(5) - C(60)	109.2(3)	Si(4) - C(40) - Si(5)	112.8(3)
C(20) - Si(2) - C(40)	110.0(4) (2x)	C(40) - Si(5) - C(51)	109.7(3)	Si(1) - C(50) - Si(5)	111.2(4) (2x)
C(40) - Si(2) - C(40')	104.9(3)	C(50) - Si(5) - C(60)	107.8(4)	Si(5) - C(50) - Si(5')	107.5(6)
C(30) - Si(4) - C(40)	109.2(3)	C(50) - Si(5) - C(51)	114.5(4)	Si(5) - C(60) - Si(6)	112.6(4)
C(30) - Si(4) - C(70)	110.5(4)	C(60) - Si(5) - C(51)	106.8(4)	Si(4) - C(70) - Si(6)	112.3(4) (2x)
C(30) - Si(4) - C(41)	105.7(4)	C(60) - Si(6) - C(60')	101.0(6)	Si(4) - C(70) - Si(4')	110.0(5)
C(40) - Si(4) - C(70)	105.7(4)	C(60) - Si(6) - C(70)	110.4(3) (2x)		
C(40) - Si(4) - C(41)	112.6(3)	C(60) - Si(6) - C(61)	112.7(4) (2x)		
C(70) - Si(4) - C(41)	113.2(3)	C(70) - Si(6) - C(61)	109.5(7)		

actually enforces an ideal conformation for the six-membered rings in the boat conformation. That means that torsion angles of $0°$ or $\pm 60°$ exist. The mean bond length d(Si—C) = 188.6 pm corresponds to the distances obtained for the SiC modifications and to results obtained in earlier investigations on carbosilanes (Table 69).

Certain differences exist between the bond distances in the adamantane moiety (187.8 pm) and to its substituents (187.9 pm), on the one hand, and the distances in the wurtzitane moiety (190.0 pm in barrelane; 189.1 pm in iceane) on the other hand, indicating a somewhat larger strain existing in the six-membered rings in the boat conformation (cf. IV.1.1.3).

1.3 Structures of Si-Scaphanes

1.3.1 Heptamethyl-tetrasila[2.2.2]barrelane $Si_4C_{11}H_{28}$, 191 [172]

The triscaphane 191 is a carbosilane with a barrelane skeleton. The molecular skeletons of 200, 51 and 58 are constructed out of several barrelane skeletons.

The results of the X-ray structure analysis on $Si_4C_{11}H_{28}$ 191 are presented in Fig. 28.

All six-membered rings exist in the skewed boat form, as can be recognized from the diagram. This conformation indeed was at first surprising, since the [2.2.2]barrelane system, composed with tetrahedral angles, is completely rigid and fixes the six-

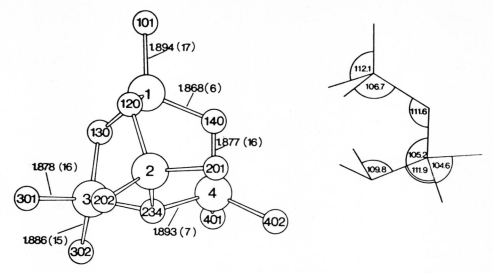

Fig. 28. A tetrasilabarrelane molecule. Si = large circles, C = small circles; the numbers designate particular atoms. The bond lengths and angles shown here are the averages of structurally equivalent values from the three different six-membered rings.

The labels of the atoms are chosen in such a way that the bonding of the C-atoms with the Si-atoms Si(1), Si(2), Si(3) and Si(4) is immediately recognized. C(234) is bound to three Si atoms, C(120) to two Si atoms and C(301) constitutes the methyl group (1) on Si(3)

membered rings in the normal boat form having torsion angles 0°, +60°, −60°, 0°, +60°, −60°, respectively. As in the Si-propellane system <u>405</u> [63, 173], it can be shown that this conformation leads to an intolerable H—H contact between CH_3 and CH_2 groups. Hence elimination of these disturbing interactions is also decisive in the barrelane system in which a skewed boat form prevails. Consequently, noticeable deviations of torsion angles must appear. The successive torsion angles stated here are completely analogous in all six-membered rings. The mean values found are −23.8°, +74.0°, −46.8°, −23.8°, +71.5°, −42.2°. The average deviation of the torsion angles from the normal boat conformation is 17.4°, and is practically completely analogous to that in the propellane system. This finding is an indication that carbosilane six-membered rings in the skewed boat form have assumed the optimal conformation to minimise unfavorable steric interactions between methyl substituents and ring methylene groups [190, 172, 173].

The average Si—C bond distance is 188.1(4) pm; this corresponds to the general trend. Within the carbosilane skeleton, a difference in the bond distances of Si—CH (189.3(7) pm) and Si—CH$_2$ (187.2(7) pm) exists. The bond distance to the methyl group of 188.6 pm can be compared with values found in silaadamantanes of 188.9 pm. However, the values of bond lengths observed in the unstrained silaadamantane system over the whole ring are distinctly shorter (186.6 pm) than in silabarrelane (187.9 pm) or in silapropellane (188.7 pm) [173]. A distinct tension exists, then, in the silabarrelane molecule, which obviously stems from the axially-oriented C atom (234) (mean Si—C (234) = 189.3 pm). These axially-oriented C atoms are bound firmly in proper tetrahedral angles by means of tension placed on the relevant six-membered rings in both the Si-barrelane and Si-propellane systems <u>405</u>, while peripheral CH_2 groups avoid this tension by a decisive angle enlargement (111.6° or 115.9°).

1.3.2 Octamethyl-hexasilahexascaphane $Si_6C_{15}H_{36}$, <u>200</u> [56]

The results of the crystal structure analysis are summarized in Table 72 and Fig. 29a, b. Compound <u>200</u> (1,3,3,5,5,7,9,12-octamethyl-1,3,5,7,9,12-hexasilatetracyclo-[5.3.3.0^{4,9}.0^{4,12}]tridecane) belongs to those carbosilanes whose polycyclic skeleton consists exclusively of six-membered rings existing in a skewed boat form. Compound <u>200</u> forms the first stage in the condensation of barrelane units <u>191</u> (triscaphane), which then proceeds further to the nonascaphane <u>57</u> and eventually to the dodeca-scaphane <u>58</u>. Similar to the other polyscaphanes, the hexascaphane in this case exists in a distorted boat conformation. The relevant successive torsion angles are 24.2°, 43.1° and −71.1°. As a result the conformation of all the six-membered rings lies between the ideal boat form (0°, 60°, −60°) and the ideal twisted boat form (33°, 33°, −71°).

Table 72. Bond lengths of the two crystallographically-independent molecules in pm (standard deviation). Below: the mean bond lengths (pm) of functionally equivalent bonds from both molecules, as well as their averaged bond angles (degrees).

Si1	- C1	191.1(4)	191.8(3)	Si2	- C2	186.7(5)	186.9(5)	Si3	- C1	191.2(4)	191.0(4)
	- C6	188.2(4)	187.9(4)		- C3	185.7(4)	186.8(4)		- C3	188.9(4)	188.7(4)
	- C7	189.7(4)	188.6(5)		- C7	187.4(4)	186.7(5)		- C4	189.5(5)	188.2(4)
	- C10	186.9(5)	187.8(5)		- C20	188.0(5)	187.4(6)		- C30	186.9(5)	188.3(5)
Si4	- C4	187.8(5)	187.0(5)	Si5	- C1	189.3(3)	189.4(4)	Si6	- C1	189.2(4)	188.6(4)
	- C5	187.5(5)	187.9(5)		- C5	188.4(4)	188.5(5)		- C2	188.6(4)	189.4(5)
	- C6	185.8(5)	186.6(4)		- C51	188.0(5)	186.8(5)		- C61	187.9(6)	187.5(6)
	- C40	187.5(5)	187.5(6)		- C52	187.7(5)	188.1(5)		- C62	188.1(5)	187.2(5)

Mean values

a = 191.3		f = 188.7		a - a = 104.0		d - e = 107.1	
b = 189.1		g = 187.5		a - b = 108.8		e - e = 105.8	
c = 188.7		h = 187.7		b - b = 117.0		e - f = 110.8	
d = 187.3		i = 187.7		b - c = 107.8		f - a = 108.5	
e = 186.7				c - d = 111.6		f - f = 110.3	

The torsion angles do not differ from those in barrelane <u>191</u>, but they change with respect to those of the ideal boat conformation (19°, 48°, −72°) in dodecascaphane <u>58</u>. Deviations from the ideal boat conformation are due to the steric hindrance between the CH_2 bridges [172, 190]. This is especially strong in the dodecascaphane and leads to considerable vibration. In the presently-investigated hexascaphane, further compensation can be made for the existing molecular strain through differentiation of bond lengths a and b from the central atom C1, which are longer than the average value quoted for endocyclic bonds (188.4 pm). In comparison to the Si2 and Si4 atoms situated opposite the barrelane part of the molecule, the bond distances d and e are distinctly smaller. The bond distances of type c and f correspond to the average value. The small differences of a and b from d and e result obviously from the actual functions of the bonds in the total skeleton. The exocyclic bond distance (d = 187.6 pm) corresponds exactly to the values determined in other structures. Also the angles between these bonds (Table 72) depend likewise on the function of the bond in the skeleton.

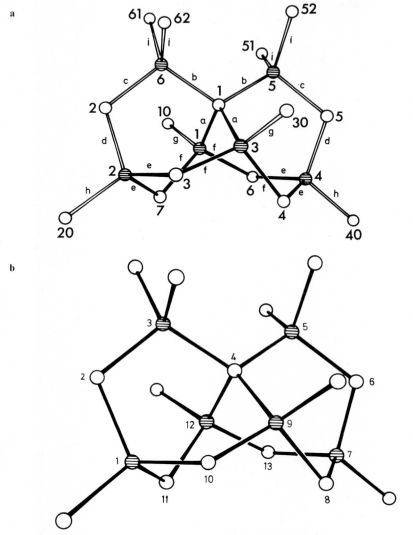

Fig. 29. a Hexascaphane molecule <u>200</u>, including crystallographic designation of atoms (Si-atoms hatched). Structurally-equivalent bonds are indicated by the same small letter. **b** The same molecule with the atoms numbered according to systematic nomenclature

1.3.3 Tetramethyl-octasiladodecascaphane $Si_8C_{17}H_{36}$, <u>58</u> [26, 190]

The results of the crystal structure analysis are shown in Fig. 30, with a view of the whole molecule showing bond lengths and angles as well as a picture of two six-membered rings in a skewed boat conformation.

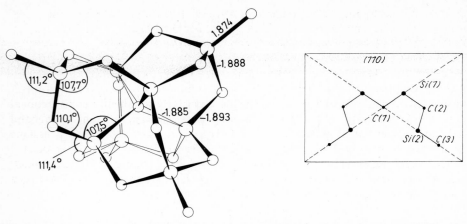

Fig. 30. Left: Octasila-dodecasphane <u>58</u> (3,7,11,15-tetramethyl-1,3,5,7,9,11,13,15-octasilaoctacyclo-[7.7.1.1^{1,7}.1^{3,13}.1^{5,11}.1^{9,15}.0^{5,17}]heneicosan) with bond lengths (Å) and bond angles. Right: Numbering of atoms and projection of the skewed boat conformation

From the Si—C bond lengths it is immediately clear that the bond distances Si—CH$_3$ = 187.4 pm are shorter than those in the scaphane skeleton (188.9 pm). Just as obvious is the fact that the bond lengths to the central carbon atom C(1) cannot be differentiated, within experimental error, from the bond distances to the CH$_2$ group C(2). The Si—C bond length of 188.9 pm corresponds to values obtained for other carbosilane ring systems containing a certain ring strain. The totally opposite situation occurs in the silaadamantane skeleton, with a bond length of 186.6(7) pm obviously resulting from a smaller ring strain. The distance to the CH$_3$ groups of 187.4(8) pm is considerably shortened. This value corresponds to similar observations made on other molecules, and emphasizes the particular situation of Si—CH$_3$ bond distances in silaadamantanes.

The bond angles for C(1) are the same as the tetrahedral angle due to symmetry requirements, while the C(2) bond angles (CH$_2$ groups) are only slightly larger. The bond angles on both Si atoms of 107.6° and 110.3° are somewhat smaller and larger than the tetrahedral angle but still average out with the tetrahedral value. From the model it can be recognized that these distortions depend on the torsion of the six-membered ring. This torsion is in fact the most noticeable structural characteristic in this molecule. The torsion angles (19°, 70°, −71°) show that the conformation of these six-membered rings deviates considerably from the boat form [190]. The reason for the twisting of the boat form to the skewed boat conformation is found in the repulsion of neighboring intramolecular methylene protons.

The conversion of all the six-membered rings to the skewed boat conformation is responsible for the formation of two chiral molecules, both of which are mirror images of each other. Both forms occur commonly in the overall structure. However, it is necessary that the two forms (A, B) process along (100) in the succession A, B, A, B ... , because otherwise the resulting intermolecular proton contacts would be too short (188 pm). Considering this sequence A, B ... as well as the F-translation of the unit cell, one recognizes the relationship to the NaCl type of structure. Every A molecule in the structure takes the position of Na while B substitutes for Cl, so that crystallographically the molecule contains 4 A and 4 B belonging to the same point position 8(a). The 8 molecules in 4 A and 4 B are distinguished by a rotation process, which has an opposite effect on the centres A or B.

The structures of the Si-scaphanes distinguish themselves quite decidedly from those of the Si-adamantanes. The Si-adamantanes present particular segments disected from the diamond type or the carborundum type structure, that is, a three-dimensional periodic structure emerges using exclusively tetrahedral constructional units. This is obviously not possible for the scaphanes. Three-dimensional tetrahedral skeletons only appear if a preponderate number of the six-membered rings assume the chair conformation. One can recognize from the wurtzite structure and associated structural relatives that six-membered rings only parallel to the trigonal (or hexagonal) bases exist in the boat conformation. We could regard the scaphane molecule as a seed from which spread out four boats in four directions. It would also be conceivable that constructional elements of the scaphane type contribute to the formation of polytypes of this class of substances, as during the growth stage of silicon carbide.

1.4 Dodecamethyl-heptasila[4.4.4]propellane, $Si_7C_{19}H_{48}$, <u>405</u> [63, 173]

The results of the crystal structure analysis are summarized in Fig. 31, which presents the molecule in two different ways. The sketches underneath present the mean bond lengths and bond angles of all three rings. For further information see [173]. The essential result concerning this structure determination is that the heptasilapropellane molecule consists of three practically identical six-membered rings in a distorted boat conformation. It can immediately be recognized from the dihedral and torsion angles that the conformation existing in the rings is not a transition of the boat to the chair conformation (half boat), rather a skewed boat conformation corresponding to:

Boat:	0°,	+60°,	−60°,	0°,	+60°,	−60°
<u>405</u>:	−20°,	+61°,	−35°,	−27°,	+67°,	−38°
Chair:	−60°,	+60°,	−60°,	+60°,	−60°,	+60°

The average deviation of all torsion angles from those of the normal boat conformation is 17°; from the chair conformation, it is however 63° and practically as large a deviation as that between a chair and boat conformation (60°). The deviation from a boat form consists essentially of a levelling of the six-membered ring, which is best recognized in the bond angles in the particular ring (mean: 111.8°).

From the NMR spectra of compound <u>405</u>, a highly symmetrical, fixed conformation of the chair form was originally presumed. Because all the rings in this molecule adhere to the twisted boat conformation, the rigidity of the molecule must be caused by the molecular arrangement in the overall system. All structural details show that the torsion on the rings, the ring flattening and the overall rigidity within the system all derive from the strong steric interactions between different CH_3 groups and between CH_3 and CH_2 groups. Especially noteworthy are the decreasing bond lengths starting from axial C(1). As Fig. 31 shows, the Si—C bond lengths in the ring can be reduced within the limits of normal error to three significantly different bond lengths: $R_1 = 191.6$ pm, $R_2 = 188.0$ pm, and $R_3 = 186.4$ pm. In all stages the Si—C bond

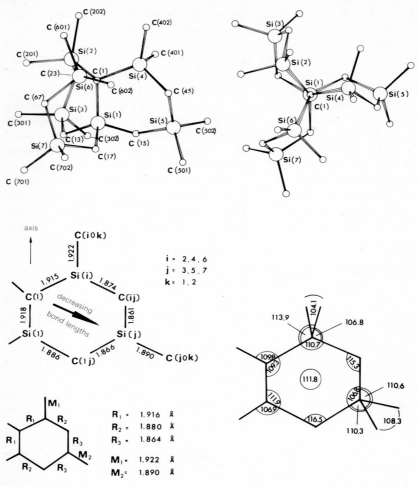

Fig. 31. Heptasilapropellane molecule, projected to the perpendicular molecular axis (above left) and along the molecular axis (above right). Sketches showing variations in the bond lengths (Å) and bond angles with mean values stated for all three six-membered rings appear below

distances fall as we proceed from C(1) to the periphery of the molecule. This also applies to bond distances to the methyl groups; in the region of C(1) they measure 192.2 pm, which is larger than at the periphery (189.0 pm). The mean value for the bond lengths Si—C = 189.3 pm, compared with the individual values, shows a very large standard deviation, indicating a systematic change in the bond lengths.

The increased Si—C bond lengths in the environment of C(1) illustrate the strong tension at least in this part of the molecule. This increase in bond distance is the result of the moving away from each other of atoms Si(2), Si(4) and Si(6) from 307 pm, as in carborundum SiC, to 313 pm. That is necessary in order to accommodate the axial CH_3 groups C(202), C(404) and C(601). It can easily be shown that a separation of at least 360 pm must exist between the C atoms of coaxially-situated CH_3 groups in order to guarantee H—H contacts of approximately 240–250 pm. This is achieved through not only the widening of the Si(i) basis but also through the outward rotation of the CH_3 groups from their coaxial position (C—C = 363–374 pm). Other qualities pertaining to this conformation can be clarified by similar considerations if one, for example, compares the CH_3—CH_2 and CH_2—CH_2 distances including H atom positions with the conformation of the pure boat form. On incorporating all H atoms into the discussion it can be recognized that the heptasilapropellane molecule has no remaining possibility to move, and hence is rigidly fixed.

Totally unexpected is the finding that the conformation of the six-membered rings in the hexasilapropellane $\underline{405}$ is practically the same as that in octasiladodecascaphane $\underline{58}$ and in tetrasilabarrelane $\underline{198}$. These are two basic examples for polycyclic six-membered ring systems in the boat conformation. The twist conformation is practically identical in all three molecules. This creates the impression that this particular conformation is typical for boat-shaped carbosilane six-membered rings.

1.5 trans-trans-1,3,3,5,7,7,9,11,11-Nonamethyl-1,3,5,7,9,11-hexasilatricyclo-[7.3.1.0^{5,13}]tridecane, $Si_6C_{16}H_{36}$, $\underline{209}$ [199]

It is particularly noteworthy that this molecule is formed like a segment of a sphere with a radius of 1300 pm. The repeatedly found flattening of the Si_3C_3 six-membered ring could be realized with angles of about the same size on Si and C atoms. However, the observed larger angles on carbon ($\sim 115°$) and the smaller angles on silicon ($\sim 106°$) cause a considerable arching of all Si_3C_3 chairs resulting in larger Si—Si then C—C distances in the six-membered rings. This makes readily understandable why skewed boat conformations do appear so often, for they represent the most appropriate means by which a strong levelling of the chair form is brought about through mutual substitution.

In compound $\underline{209}$, also named as hexasilaperhydrophenalene, all of the Si methyl groups are situated on top of the molecular "plane", while H atoms bonded to the skeletal carbons point below that "plane".

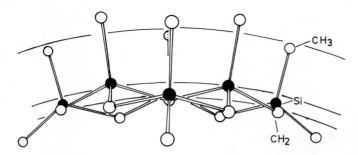

Fig. 32. Projection of trans-trans-hexasilatricyclo[7.3.1.0^{5,13}]tridecane showing particularly the arching of the connected six-membered rings. Si atoms are drawn in black, H atoms are omitted

1.6 Structures of Carbosilanes with Small Rings

1.6.1 Hexadecamethyl-octasiladispiro[5.1.5.1]tetradecane $Si_8C_{22}H_{56}$, <u>240</u> [78, 191]

The results of the crystal structure investigation are summarized in Fig. 33.

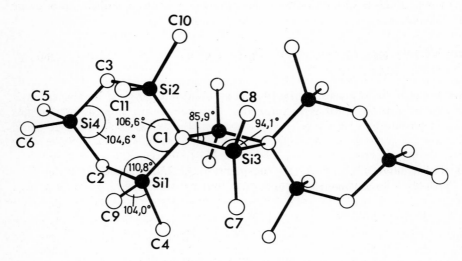

Fig. 33. Molecular structure of hexadecamethyl-octasiladispiro-tetradecane

The octasila-dispiro[5.1.5.1]tetradecane skeleton is formed by one central disila-cyclobutane ring bonded to two C-spiro bonded trisilacyclohexanes. Both six-membered rings exist in the chair conformation but are flattened due to the widened endocyclic bond angles in the six-membered ring: $\beta = 111.7°$ (chair conformation with a tetrahedral configuration: 109.5°; planar six-membered rings 120°). This serves as an indicator of strong levelling. The sequence of torsion angles ($-55.1°$; 54.3°; $-56.3°$; 49.8°; $-50.7°$; 58.2°) demonstrates the relationship to the chair conformation ($\pm 60°$) as well as the regularity in the flattening of the six-membered ring.

The planar C_2Si_2 four-membered ring is distorted to a rhombus, with angles of 85.9° on the C atoms and angles of 94.1° on the Si atoms. This conformation is directly comparable to the C_2Si_2 four-membered ring in octachloro-hexasilaasterane <u>335</u> (87° on C; 93° on Si). Noticeably large are the Si—C bond distances in the C_2Si_2 four-membered ring of <u>240</u>, with an average value of 192.2 pm. The four-membered rings of asteranes are distinctly shortened (189.2 pm) [67].

The Si—C spiro-bonds in the six-membered ring are likewise very long. The average value is 191.8 pm, including an angle of 106.6°. The opposite angle of Si4 is likewise very small (104.6°); however, bond lengths to neighboring C2 and C3 atoms are notice-ably short here (mean 186 pm). The most remarkable finding about the configuration of the six-membered ring is that the average bond angle on silicon within the ring is 108.9°, a value certainly within the experimental limits of a tetrahedral angle, while the corresponding C2 and C3 angles (mean value 118.5°) are considerably widened. It appears to be a typical quality of the carbosilanes that in the strained part of the molecule the C atoms usually yield to the strain exerted through changes in the tetrahedral configuration, while the Si atoms extensively maintain this angle. The spiro-dihedral angle is precisely 90.0°. The average exocyclic Si—C bond distance (187.6 pm) is rightly within the limits determined in previous investigations of cyclic carbosilanes.

1.6.2 cis-2,4-Dichloro-2,4-bis(trimethylsilyl)-1,1,3,3-tetramethyl-1,3-disilacyclobutane, $Si_4C_{12}H_{30}Cl_2$, 310 [129]

The results of the crystal structure analysis are presented in Fig. 34. The molecule almost complies with mm2-C_{2v} symmetry. The central disilacyclobutane skeleton is not planar, rather weakly folded (17.8° or 162.2°), and contains bond angles of 91.3° on the Si atoms and 87.2° on the C atoms. Compound 310 is a precursor in the reaction yielding eventually 2,2,4,4-tetramethyl-1,3-bis(trimethylsilyl)-2,4-disilabicyclo[1.1.0]butane 309 [127]. The reaction possibility of 310 is recognized by analysis of the configuration of both C atoms in the four-membered ring. While the methyl groups on the cyclic Si atoms stand practically in a mirror plane with the pertinent part of the four-membered ring (planes C2—Si1—C4 or C2—Si3—Si4; Fig. 35), the angle Si—C—Cl formed by the substituents SiMe₃ and Cl on the cyclic C atoms will be asymmetrically divided by the plane Si1—C2—Si3 (or Si1—C4—Si3). The rotation of these substituents results in a shorter distance of the Cl atoms.

Hence the bonding angle of the SiMe₃ groups to the bisecting axis diminishes (52° → 38°), while it increases considerably at the Cl substituents (52° → 66°). The configuration of the cyclic C atoms is leveled in relation to the three Si—C bonds, towards a planar sp^2 configuration. The Cl substituents are seen to reorient themselves

Table 73. Bond lengths in pm (above) and bond angles in degree (below). Standard deviations are given in brackets

Si(1) – C(2)	191.9(5)	Si(3) – C(2)	191.4(4)	Si(20) – C(2)	188.4(5)	Si(40) – C(4)	187.7(4)
– C(4)	192.3(5)	– C(4)	191.8(5)	– C(21)	185.2(7)	– C(41)	185.9(9)
– C(11)	186.7(6)	– C(31)	187.2(7)	– C(22)	186.0(7)	– C(42)	184.5(9)
– C(12)	186.0(6)	– C(32)	186.2(7)	– C(23)	187.2(7)	– C(43)	186.7(8)
Cl(2) – C(2)	184.9(5)	Cl(4) – C(4)	184.2(5)				

C(2) – Si(1) – C(4)	91.2(2)	C(2) – Si(20) – C(21)	109.0(3)	Si(1) – C(2) – Si(3)	87.3(2)
C(2) – Si(1) – C(11)	111.3(2)	C(2) – Si(20) – C(22)	112.9(2)	Si(1) – C(2) – Si(20)	124.8(2)
C(2) – Si(1) – C(12)	116.9(2)	C(2) – Si(20) – C(23)	109.1(3)	Si(1) – C(2) – Cl(2)	107.0(3)
C(4) – Si(1) – C(11)	111.7(2)	C(21) – Si(20) – C(22)	108.3(3)	Si(3) – C(2) – Si(20)	124.6(3)
C(4) – Si(1) – C(12)	115.5(3)	C(21) – Si(20) – C(23)	109.0(3)	Si(3) – C(2) – Cl(2)	107.9(2)
C(11) – Si(1) – C(12)	109.3(3)	C(22) – Si(20) – C(23)	108.4(3)	Cl(2) – C(2) – Si(20)	103.6(2)
C(2) – Si(3) – C(4)	91.5(2)	C(4) – Si(40) – C(41)	112.6(3)	Si(1) – C(4) – Si(3)	87.1(2)
C(2) – Si(3) – C(31)	110.7(3)	C(4) – Si(40) – C(42)	108.6(3)	Si(1) – C(4) – Si(40)	125.5(3)
C(2) – Si(3) – C(32)	117.0(3)	C(4) – Si(40) – C(43)	108.3(3)	Si(1) – C(4) – Cl(4)	106.3(2)
C(4) – Si(3) – C(31)	112.1(3)	C(41) – Si(40) – C(42)	108.1(4)	Si(3) – C(4) – Si(40)	124.6(3)
C(4) – Si(3) – C(32)	116.3(3)	C(41) – Si(40) – C(43)	109.0(4)	Si(3) – C(4) – Cl(4)	107.3(2)
C(31) – Si(3) – C(32)	108.5(4)	C(42) – Si(40) – C(43)	110.2(4)	Cl(4) – C(4) – Si(40)	104.0(2)

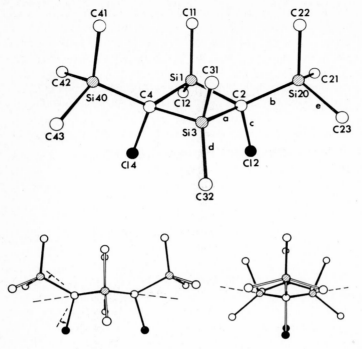

Fig. 34. Above: Molecule maintaining nomenclature of Tab. 73 (Si atoms hatched, Cl atoms black); small letters indicate characteristic bond types. Below: View along the Si—Si axis (left) or along the C—C axis (right) of the four-membered ring. The bonds on the central skeleton are prominent. Symmetry planes of the four-membered ring are shown by dashed lines. The folding of the four-membered ring and the swiveling of the substituents on the C atoms are recognized

into orthogonal positions. In connection with this the enlarged C—Cl bonds are noteworthy. They correspond to a value of 184.6 pm in comparison to the expected single bond length of 172 pm but having bond order according to Pauling [192] of only PBO = 0.66. The mean bond lengths (Fig. 34) amount to a = 191.9 pm, b = 188.0 pm, c = 184.6 pm, d = 186.5 pm and e = 186.9 pm. The associated bond orders are PBO = 0.91, 1.03, 0.66, 1.09 and 1.11. The sum of bond orders also shows very distinctly that the molecule, particularly in relation to C—Cl bonds, is in a transition state: Σ PBO: Si(ring) = 4.0; Si(exocyclic) = 4.4; C(ring) = 3.5; Cl = 0.7. The Cl substituents become closer to each other (from 487 pm to 370 pm) as a result of the bonding of the substituents, hence enabling a van der Waals contact. It is important here that the other side of the C—Cl bond is not affected by atoms of neighboring molecules.

1.6.3 2,2,4,4-Tetramethyl-1,3-bis(trimethylsilyl)-2,4-disilabicyclo[1.1.0]butane, Si$_4$C$_{12}$H$_{30}$, 309 [127]

Because the melting point of 309 is 28 °C, crystals were selected for analysis at temperatures <215 K. Intensity measurements were carried out using a Syntex-Temperature device. The results of the investigation are summarized in Fig. 35, which shows different views of the molecule and includes bond lengths, bond angles and the dihedral angles between the perpendicular lines of the planes of the 1,3-bridged cyclobutane. The relatively strong mutual angle incurred by both three-membered ring planes is a striking feature of this molecule ($\gamma = 58.2°$). Similar overall conformations have been observed in substituted heterocyclobutanes (also on diazadiphosphetinidine) with a dihedral angle existing of between 13° and 37° [193–196]. These conformations are suitably explained by an abatement of substituent interactions and by a partial adjustment of favorable bond angles of the ring atoms. The C—C bridge bond through the ring obviously causes the three-membered ring planes to incline as they do to the observed value.

The transannular C—C bond is a weak bond. It has a length of 178.1 pm which corresponds to a bond order n $\doteq$ 0.5. The lengths for Si—C bonds lie in the normal

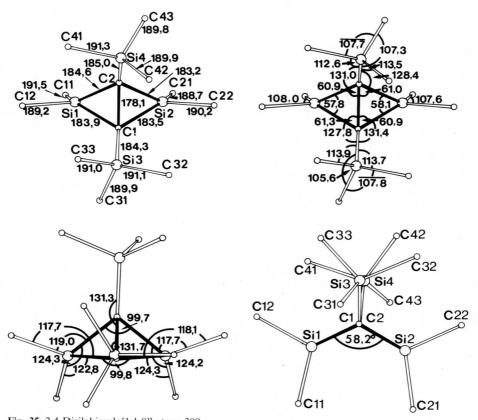

Fig. 35. 2,4-Disilabicyclo[1.1.0]butane 309.
Left above: Naming of respective atoms and bond lengths in pm. Right above, left below: Bond angles in degrees. Right below: Projection along the C1/C2 bond axis showing also the angle between perpendicular lines of the three-membered ring planes. The staggered conformation of the exocyclic methyl groups is easily recognized

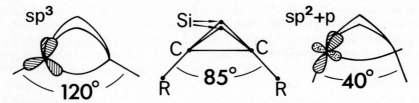

Fig. 36. Compound <u>309</u>: Scheme of the topological arrangement of the bonds for the canonical forms, sp³ (left) and sp² + σ$_p$ (right) together with those in the experimentally determined molecule (middle). The angles between the exocyclic SiMe₃ substituents are shown likewise and were derived from an analysis of framework molecular models

range for such bonds but show nevertheless a strong dependance on their individual functions [d(Si—C)$_{endo}$ = 183.8 pm and d(Si—C)$_{exo}$ = 190.2 pm]. The bond angle in the three-membered ring varies according to the varying positions of Si—C or C—C bonds (58–61°). The mean value for the bond angles including exo-bonds amounts to 108.3° on the Si atoms and 102.1° on the C atoms. While the mean bond angle here for Si corresponds to that in a tetrahedral configuration, this value diverges considerably from the ideal value for the C atoms. This indicates also the particular situation of the C atoms. The substituents are accommodated as regularly as possible to the three bonds of the ring C atoms, forming bond angles of approximately 130°. The deviation of the endocyclic C atoms out of the plane of the three directly-bound Si atoms amounts only to 9 pm, similar to the value already observed for ylides [132]. Fig. 35 shows (below right), that the methyl groups on the exocyclic Si atoms (3, 4) maintain a staggered conformation. Of interest in connection with these facts is the problem of the configuration of the C atoms in the central ring system. Because of the bent bonds in this system, the configuration cannot be elucidated immediately from the bond angles. On consideration of possible canonical forms both an sp³ hybridized configuration as well as an sp² hybridized configuration are possibilities. In the sp² case the extra C—C bond would arise as a σ$_p$ bond by using the remaining p orbitals. Figure 36 shows schematically a topological representation of these bonds from simple framework molecular models, with elastic bond contributions being used on the C atoms for both extreme configurations in comparison to the topological linkages obtained by actual experimental determination for the real molecule. One can recognize quite distinctly, that the configuration for the real experimentally determined molecule lies somewhere between the extreme models.

Further investigations in this area [132, 174] show that a number of factors contribute to bond lengthening or shortening in Si—X bonds. In some compounds it is the strongly electronegative substituents, in others it is the ability to utilize the d orbitals in silicon, hence achieving a higher bond order than 4. From a calculation utilizing values obtained earlier (d$_1^*$(Si—C) = 189 pm, d$_1^*$(C—C) = 154 pm) and from the Donnay-Allmann [175] relationship, the bond orders n(C—C)$_{endo}$ = 0.48, n(Si—C)$_{endo}$ = 1.15 and n(Si—C)$_{exo}$ = 0.98 are achieved. The sum of the bond orders for endocyclic atoms is Σ n(C)$_{endo}$ = 3.90 or Σ n(Si)$_{endo}$ = 4.25 and for exocyclic Si atoms Σ n(Si)$_{exo}$ = 4.0. The values show that a shift of the bond orders occurs in the ring from the C to the Si atom, which likewise confirms our estimations on the ability of Si to utilize d orbitals in forming bonds.

Robinson and Gillespie showed that simple framework molecular models enable surprisingly precise estimations to be made on the topological shortening of double bonds, triple bonds and even bonds existing in strained systems (bent bonds) [197]. This applies to the existing problem in two ways:

a) The endocyclic Si—C bonds are only shortened through bond bending; this bond contains $n(Si—C)_{endo} = 1.0$.

b) The transannular C—C bond must be derived from an "actual" bond length of 184 pm (not 178 pm), whereby the bond order reduces to $n(C—C)_{endo} = 0.41$.

However, this analysis does not change the overall effect of small summed bond orders on the endocyclic C atom.

1.6.4 Octachlorohexasilaasterane $Si_6C_6H_8Cl_8$, 335 [171, 144]

The most important results pertaining to the crystal structure analysis of 335 are summarized in Fig. 37.

Particularly conspicuous here is the deformation of the four-membered rings. The angles at the C atoms are $87.2°$ ($\pm 1.4°$), which is significantly smaller than the angles at the Si atoms ($92.7° \pm 0.6°$). The atoms belonging to the ring do not lie exactly in the plane, because C atoms C1 and C2 lie approximately 50 pm closer to the plane of the bridging atoms (Si3, Si3', C3, C3') than do Si atoms Si1 and Si2. The ring strain makes the relatively large Si—C bond lengths in the four-membered ring (mean value 188.7 pm) as well as the short bond length to the $SiCl_2$ bridge (Si—C = 181.9 pm) understandable. In addition, the observed Si—Cl bond lengths to the four-membered ring (Si—Cl = 202.5 pm) are shorter than the distance in the $SiCl_2$ bridge (205.7 pm). In total contrast to these bond lengths is the bond distance to the CH_2 bridge (189 pm), which is not shortened (Si1—C3 or Si2—C3). The Si atoms of the four-membered rings participate in both of these bonds. The deformation of the ring causes an unexpected approach of these two Si atoms to within 260 pm of each other. This distance is only about 10% larger than a normal Si—Si bond length. Even if one assumes that the bonding electrons are localized in the region of the Si—C bond axis, this approach of Si atoms makes it possible for an interaction to occur. Indeed, the d orbitals d_{z^2} and $d_{x^2-y^2}$ are oriented in such a direction to make contact feasible.

1.6.5 Tetrasilabicyclo[3.3.0]oct-1(5)ene $Si_4C_{12}H_{28}$, 364 [149, 150]

The results of the crystal structure analysis are summarized in Fig. 38.

The individual molecule of 364 possesses D_{2h} symmetry, while the closest packed arrangement within the crystal lattice leads to a lower symmetry. The shortest distance between individual molecules in the lattice lies at around 400 pm. The size of the thermal ellipsoids indicates that 364 undergoes considerable thermal vibrations.

The C1—C1' distance in the ring system is distinctly larger than that of a normal isolated double bond. The bond angles especially at the atoms adjacent to the double bond show also characteristic deviations from the ideal value of $108°$ for a regular five-membered ring. These bond distances and angles suggest that a considerable ring tension exists.

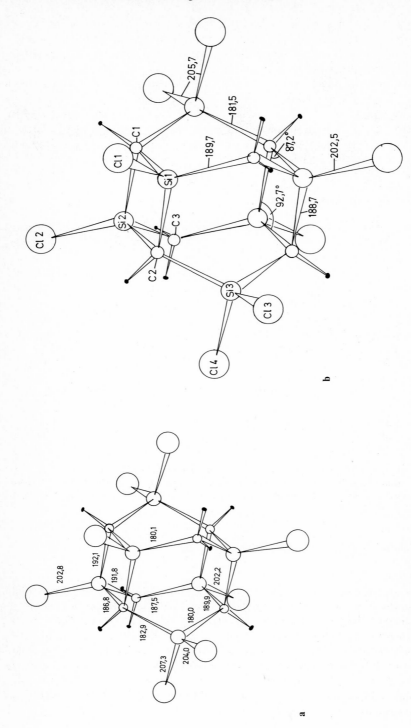

Fig. 37. Molecules of octachloro-hexasila-asterane 335 (1,3,3,5,7,9,9,11-octachloro-1,3,5,7,9,11-hexasilapentacyclo[6.4.0.0^{2,7}.0^{4,11}.0^{5,10}]dodecane). **a** shows bond lengths (pm) and angles

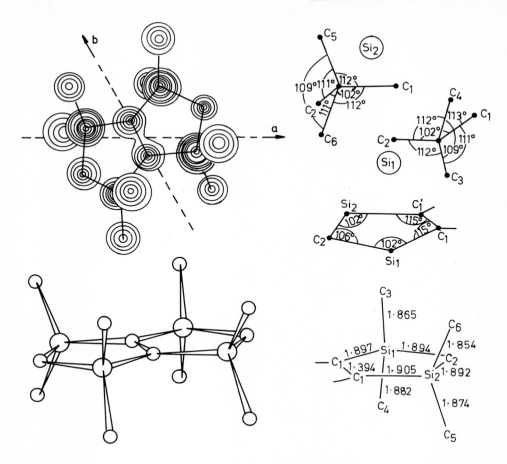

Fig. 38. Structure of 2,2,4,4,6,6,8,8-octamethyl-2,4,6,8-tetrasilabicyclo[3.3.0]oct-1(5)ene.
Above left: Section of an electron density map; below left: Perspective projection of the molecule,
Si = large circles, C = small circles; right: Bond lengths (Å) and angles of the five-membered ring
as well as with atoms Si(1) and Si(2)

The Si—C bond lengths amount on average to 188.3 pm, corresponding exactly
to the sum of the covalent radii (188 pm). Even taking into consideration the width of
experimental error, the bond lengths in the five-membered rings certainly appear to be
larger than those to the methyl groups, namely, $\overline{Si—C_R}$ = 190 pm, $\overline{Si—C_M}$ = 187 pm,
respectively. The bond angles on the Si atoms likewise show a characteristic deviation
from the tetrahedral angle, an effect easily explained by the actual position of the Si
atoms in the molecule. Only the $C_3—Si_1—C_4$ angle of 108.6° and the $C_5—Si_2—C_6$
angle of 108.9° compare exactly to the ideal values for those angles formed exclusively
by the C atoms and associated methyl groups.

1.7 Octamethyl-tetrasilacyclooctane, $Si_4C_{12}H_{32}$, <u>404</u> [149, 184]

The results of the investigation are summarized in Fig. 39.

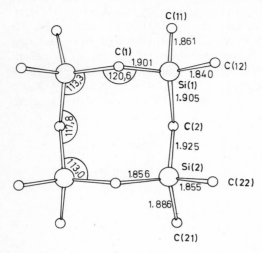

Fig. 39. 1,1,3,3,5,5,7,7-Octamethyl-1,3,5,7-tetrasilacyclooctane; bond lengths (Å) and bond angles

The molecular skeleton exists in a saddle conformation. The 4 Si atoms form an almost perfect planar square. However, they deviate by approximately 4 pm from the averaged plane, whereby the symmetry of the molecule is decreased from mm to 2. Correspondingly, the C atoms do not lie in a plane above and below the Si atoms. The plane of the C bridges is closer to the average plane of the Si atoms than that of the external methyl groups, which causes a widening of the angles at the CH_2 groups. This effect is explained by a flattening of the molecule.

The Si—C bond lengths vary between 184.0 pm and 192.5 pm. The standard deviation in the difference is $\sigma = 2$ pm, which is relatively large. Averaged over all bond distances, a value is obtained of Si—C = 187.9 ± 1.1 pm, which corresponds well with the summation of the covalent radii and also with values determined in other molecules measured. Despite the fact that the single values for the Si—CH_2 and Si—CH_3 bond distances overlap, the following applies: $\bar{d}(Si—CH_2)$ = 189.7 pm > $\bar{d}(Si—CH_3)$ = 186.1 pm. The bond angles show significant deviations from the tetrahedral angle, with the angle in the ring increasing considerably. This amounted to 119.2 ± 1.4° at the CH_2 carbons and to 113.2 ± 0.2° at silicon. The angles between free methyl groups bonded to an Si atom are smaller than the tetrahedral angle; they amount to 106.4 ± 0.5°. All other angles at silicon amount on average to 109.3 ± 1.7°, corresponding to the tetrahedral angle.

The decreased symmetry and the slight bond extension of Si—C bonds in the ring system are probably caused by repulsions of endocyclic methylene H atoms and the resulting (Pitzer) strain of the ring. The flattening of the molecular skeleton could be a result of interactions between methyl and methylene groups in the same plane of the molecule.

2. Electron Diffraction Studies

The electron diffraction investigations on 1,3-disilapropanes $(X_3Si)_2CR_2$ (X = Cl, F; R = H, Cl) can only be reviewed very briefly here. These studies were carried out to establish the influence of different substituents on the silicon and carbon atoms on bond lengths and angles. Such changes have to be attributed primarily to the different atomic radii and different electronegativities of the substituted groups. A comprehensive discussion including experimental method, data refinement, and detailed results is given in Refs. [208, 209]. Table 74 presents a selection of basic parameters and a schematic view of the molecule indicating the additional geometrical information.

Table 74. Structural parameters of 1,3-disilapropanes $(X_3Si)_2CR_2$ (C_2 model)[a, d]

	$(Cl_3Si)_2CH_2$	$(Cl_3Si)_2CCl_2$	$(F_3Si)_2CH_2$	$(F_3Si)_2CCl_2$
Independent distances (pm)				
Si—X	202.6(4)	201.8(4)	156.7(3)	156.0(3)
Si—C	186.4(6)	190.0(9)	182.7(4)	187.1(8)
C—R	109.8(24)	179.4(9)	114.2(9)	179.8(7)
Independent angles (°)				
X—Si—X	107.9(1)	109.5(1)	111.0(1)	109.6(1)
Si—C—Si	118.3(7)	120.6(9)	117.7(4)	119.4(6)
R—C—R	109.5 (fixed)	110.9(16)	109.5 (fixed)	112.6(6)
$\emptyset$[b]	11.0(5)	8.4(8)	10.3(25)	8.9(6)
τ[c]	3.9(4)	0.9(6)	1.8(4)	0.3(4)

[a] For error estimates see [208, 209]
[b] $\emptyset$ is the rotation angle of the SiX_3 groups; $\emptyset$ = 0 for the staggered C_{2v} model
[c] τ is the angle between the Si—C bond and the C_3 axis of the corresponding SiX_3 group. τ is positive when the angle of the two C_3 axes is larger than the Si—C—Si angle
[d] $(Me_3Si)_2CH_2$: r(Si—C) = 189.1(4) pm; Si—C—Si = 123.2(9)° [210]
  $(H_3Si)_2CH_2$: r(Si—C) = 187.4(12) pm; Si—C—Si = 114.1(6)° [211]

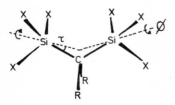

V. References

1   Fritz, G.: Angew. Chem. *70* (1958) 701
2   Voronkov, M. G.: Pure Appl. Chem. *19* (1969) 399
    Voronkov, M. G., Zelchan, G. I., Lukevits, E.: Silicium und Leben, Akademie-Verlag Berlin 1975
    Wannagat, U.: Jahrbuch der Akademie der Wissenschaften in Göttingen 1976
3   Kummer, D., Gaisser, K. E., Seifert, J., Wagner, R.: Z. anorg. allg. Chem. *459* (1979) 145
    Corriu, R. J. P., Guérin, Ch., Moreau, J. J. E.: Topics in Stereochemistry *15* (1984) 43
4   Brook, A. G., Abdesaken, F., Gutekunst, B., Gutekunst, G., Kallury, R. K.: J. Chem. Soc. Chem. Commun. 1981, 191
    Brook, A. G., Nyburg, S. C., Abdesaken, F., Gutekunst, B., Gutekunst, G., Kallury, R. K. M. R., Poon, Y. C., Chan, G. M., Wong-Ng, W.: J. Am. Chem. Soc. *104* (1982) 5667
5   West, R., Fink, M. J., Michl, J.: Science *214* (1981) 1344
6   Conlin, R. T., Gill, R. S.: J. Am. Chem. Soc. *105* (1983) 618
    Drahnak, T. J., Michl, J., West, R.: J. Am. Chem. Soc. *103* (1981) 1845
7   Hogness, T. R., Wilson, T. L., Johnson, W. C.: J. Am. Chem. Soc. *58* (1936) 108
    Stockland, K.: Trans. Faraday Soc. *44* (1948) 545
    Emeléus, H. J., Ried, C.: J. Chem. Soc. [London] 1939, 1021
    Fritz, G.: Z. Naturforsch. *7b* (1952) 507
8   Schwarz, R., Heinrich, F.: Z. anorg. allg. Chem. *221* (1935) 277
9   Kautsky, H.: Z. anorg. allg. Chem. *117* (1921) 209
    Kautsky, H., Herzberg, G.: Ber. Dt. Chem. Ges. *52* (1924) 1665
    Kautsky, H.: Z. Naturforsch. *7b* (1952) 174
10 Fritz, G.: Z. Naturforsch. *7b* (1952) 207
11 Fritz, G.: Z. Naturforsch. *5b* (1950) 444
12 Fritz, G.: Z. Naturforsch. *7b* (1952) 379
13 Wood, R. W.: Philos. Mag. *42* (1921) 729; ibid. *44* (1922) 538
    Bonhoeffer, K. F., Harteck, P.: Z. phys. Chem. A *139* (1928) 65
    Cario, G., Franck, J.: Z. Phys. *11* (1922) 161
    Emeléus, H. J., Stewart, K.: Trans. Faraday Soc. *32* (1936) 1578
14 Fritz, G.: Z. Naturforsch. *8b* (1953) 776;
    Fritz, G.: Z. anorg. allg. Chem. *280* (1955) 332
15 Fritz, G.: Z. anorg. allg. Chem. *273* (1953) 273
16 Fritz, G.: Z. Naturforsch. *12b* (1957) 66; ibid. *12b* (1957) 123
17 Fritz, G., Raabe, B.: Z. Naturforsch. *11b* (1956) 57; Z. anorg. allg. Chem. *286* (1956) 149; ibid. *299* (1959) 232
18 Fritz, G., Grobe, J.: Z. anorg. allg. Chem. *299* (1959) 302; ibid. *315* (1962) 157
19 Fritz, G.: Fortschr. Chem. Forsch. *4* (1963) 459
20 Fritz, G., Grobe, J., Kummer, D.: Adv. Inorg. Chem. Radiochem. *7* (1965) 349
21 Fritz, G., Wick, D.: Z. anorg. allg. Chem. *342* (1966) 130
22 Fritz, G., Götz, N.: Z. anorg. allg. Chem. *375* (1970) 171
23 Fritz, G., Köhler, H., Kummer, D.: Z. anorg. allg. Chem. *374* (1970) 54
24 Fritz, G., Diem, F., Köhler, H., Kummer, D., Scheer, H.: Angew. Chem. *82* (1970) 445
25 Fritz, G., Marquardt, G., Scheer, H.: Angew. Chem. *85* (1973) 587
26 Fritz, G., Marquardt, G.: Z. anorg. allg. Chem. *404* (1974) 1
27 Fritz, G., Dannappel, H. J., Marquardt, G.: Z. anorg. allg. Chem. *404* (1974) 51

254   References

28   Fritz, G., Habel, D., Kummer, D., Teichmann, G.: Z. anorg. allg. Chem. *302* (1959) 60
      Fritz, G., Teichmann, G.: Chem. Ber. *95* (1962) 2361
      Fritz, G., Ksinsik, D.: Z. anorg. allg. Chem. *322* (1963) 46
      Fritz, G., Buhl, H. J., Kummer, D.: Z. anorg. allg. Chem. *327* (1964) 165
      Fritz, G.: Angew. Chem. *79* (1967) 657
      Fritz, G., König, W., Scheer, H.: Z. anorg. allg. Chem. *377* (1970) 240
29   Fritz, G., Tögel, D.: unpublished
30   Fritz, G., Maas, J., Hornung, A.: Z. anorg. allg. Chem. *460* (1980) 115
31   Fritz, G., Maas, J.: Z. anorg. allg. Chem. *460* (1980) 144
32   Bürger, H., Kilian, W.: J. Organomet. Chem. *18* (1969) 299
33   Fritz, G.: Z. anorg. allg. Chem. *280* (1955) 134
34   Fritz, G., Blome, K.: unpublished
35   Galesworthy, R. C., Drake, J. E.: Proc. Roy. Soc. (London) *292* (1965) 489
36   Anderson, J. W., Drake, J. E.: J. Chem. Soc. A 1971, 1424
37   Fritz, G., Wörns, K.-P.: Z. anorg. allg. Chem. *512* (1984) 163
38   Cottrell, T. L.: The Strength of Chemical Bonds, Butterworths London 1958
39   Helm, D. F., Mack, E. J.: J. Am. Chem. Soc. *59* (1937) 60
40   Shiina, K., Kumada, M.: J. Org. Chem. *23* (1958) 139
      Davidson, I. M. T., Eaborn, C.: J. Chem. Soc., Faraday Trans. *70* (1974) 249
41   Fritz, G., Grunert, B.: Z. anorg. allg. Chem. *419* (1976) 249
      Davidson, I. M. T., Lawrence, F. T., Fritz, G., Matern, E.: Organometallics *1* (1982) 1453
42   Fritz, G., Grunert, B.: Z. anorg. allg. Chem. *473* (1981) 59
43   Fritz, G., Huber, R.: Z. anorg. allg. Chem. *421* (1976) 163
44   Fritz, G., Reuss, H., Wörns, K.-P., Wörsching, A.: Z. anorg. allg. Chem. *512* (1984) 93
45a   Rochow, E. G.: J. Am. Chem. Soc. *67* (1945) 963
45b   Müller, R.: review in: Wiss. Zeit. Univ. Dresden *12* No. 6 (1963)
45c   Patnode, W. I., Schiessler, R. W.: CA *39* (1945) 4889; A.P. 2 381 000, 2 381 001, 2 381 002 (1945)
46   Fritz, G., Thielking, H.: Z. anorg. allg. Chem. *306* (1960) 39
47   Fritz, G., Haase, R., Kummer, D.: Z. anorg. allg. Chem. *365* (1969) 1
48   Fritz, G., Wörsching, A.: Z. anorg. allg. Chem. *512* (1984) 131
49   Müller, R., Seitz, G.: Chem. Ber. *91* (1958) 22
50   Müller, R., Beyer, H.: Chem. Ber. *92* (1959) 1018
      Müller, R., Müller, W.: Chem. Ber. *96* (1963) 2894; ibid. *97* (1964) 1111
51   Fikhtengol'ts, V. S., Klebanskij, A. L.: J. Gen. Chem. USSR *27* (1957) 2535; ibid. *26* (1956) 2795
      Voorhoeve, R. J. H., Vlugter, J. A., Lipe, J. A.: J. Catalysis *3* (1964) 414; ibid. *4* (1965) 220
      Voorhoeve, R. J. H.: Organohalosilanes, Elsevier Amsterdam 1967
52   Kumada, M., Ishikawa, M.: J. Organomet. Chem. *1* (1964) 411
53   Kumada, M., Tamao, K.: Adv. Organomet. Chem. *4* (1968) 74
54   Frye, C. L., Weyenberg, D. R., Klosowski, J. M.: J. Am. Chem. Soc. *92* (1970) 6379; Intra
      Sci. Chem. Rep. *7* (1973) 4, 147
      Frye, C. L., Klosowski, J. M.: J. Am. Chem. Soc. *94* (1972) 7186
55   Fritz, G., Neutzner, J., Volk, H.: Z. anorg. allg. Chem. *497* (1983) 21
56   Fritz, G., Volk, H., Peters, K., Peters, E.-M., Schnering, H. G. v.: Z. anorg. allg. Chem. *497*
      (1983) 119
57   Fritz, G., Volk, H.: unpublished; Volk, H.: PhD thesis, Karlsruhe 1983
58   Fritz, G., Brauch, G.: Z. anorg. allg. Chem. *497* (1983) 134
59   Fritz, G., Brauch, G.: unpublished; Brauch, G.: PhD thesis, Karlsruhe 1984
60   Fritz, G.: Top. Curr. Chem. *50* (1974) 44
61   Fritz, G., Hähnke, M.: Z. anorg. allg. Chem. *390* (1972) 137
62   Fritz, G., Hähnke, M.: Z. anorg. allg. Chem. *390* (1972) 157
63   Fritz, G., Vogt, H. J.: Z. anorg. allg. Chem. *419* (1976) 1
64   Fritz, G., Arnason, I.: Z. anorg. allg. Chem. *419* (1976) 213
65   Fritz, G., Gompper, K., Wilhelm, H.: Z. anorg. allg. Chem. *478* (1981) 20
66   Walsh, R.: Acc. Chem. Res. *14* (1981) 246
67   Sawitzki, G., Schnering, H. G. v.: Z. anorg. allg. Chem. *399* (1973) 257
68   Fritz, G., Kummer, D.: Z. anorg. allg. Chem. *308* (1961) 105
69   Hengge, E., Kovar, F.: J. Organomet. Chem. *125* (1977) C 29

70   Fritz, G., Gompper, K.: Z. anorg. allg. Chem. *478* (1981) 94
71   Fritz, G., Hähnke, M.: Z. anorg. allg. Chem. *390* (1972) 104
72   Speier, J., Webster, J. A., Barnes, G. M.: J. Am. Chem. Soc. *79* (1975) 974
73   Benkeser, R. A., Hickner, R. A.: J. Am. Chem. Soc. *80* (1958) 5298
     Benkeser, R. A., Burrous, M. L., Nelson, L. E., Swisker, J. V.: J. Am. Chem. Soc. *83* (1961) 4385
74   Reikhsfeld, V. O., Vinogradov, V. N., Filipov, N. A.: Zh. Obshch. Khim. *43* (1973) 2216
75   Wiberg, N., Wagner, G.: Angew. Chem. *95* (1983) 1027
76   Fritz, G., Wilhelm, H., Oleownik, A.: Z. anorg. allg. Chem. *478* (1981) 97
77   Fritz, G., Finke, U., Speck, W.: Z. anorg. allg. Chem. *481* (1981) 51
78   Fritz, G., Speck, W.: Z. anorg. allg. Chem. *481* (1981) 60 .
79   Seyferth, D., Lefferts, J. L.: J. Organomet. Chem. *116* (1976) 257
80   Fritz, G., Kreilein, K.: Z. anorg. allg. Chem. *433* (1977) 61
81   Eaborn, C., Bott, R. W. in: A. G. MacDiarmid (ed.), Organometallic Compounds of the Group IV Elements, Vol. 1, Part I Marcel Dekker, New York 1968
82   Aylett, B. J., Campbell, J. M.: J. Chem. Soc., Chem. Commun. 1965, 217
83   Marcińca, B.: Hydrosililowanie, Warszawa-Poznań 1981 Państwowe Wydawnictwo Naukowe
84   Fritz, G.: Comments Inorg. Chem. *1* (1982) 329
85   Petrov, A. D., Nikishin, G. I.: Zh. Obshch. Khim. *26* (1956) 1233; CA *50* (1956) 14515
     Bailey, D. L., Pines, A. N.: Ind. Eng. Chem. *46* (1954) 2363
     Sommer, L. H., Bailey, D. L., Whitmore, F. C.: J. Am. Chem. Soc. *70* (1948) 2869
86   Fritz, G.: IUPAC Colloquium Münster 1954, p. 53, Verlag Chemie, Weinheim
87   Müller, R., Reichel, R., Dathe, C.: Chem. Ber. *97* (1964) 1673
88   Brauer, G.: Handbuch der Präp. Anorg. Chemie I, p. 251 Enke Verlag, Stuttgart 1975
89   Fritz, G., Berndt, M.: Angew. Chem. *83* (1971) 500
     Fritz, G., Berndt, M., Huber, R.: Z. anorg. allg. Chem. *391* (1972) 218
90   Fritz, G., Dresel, W. H.: Z. anorg. allg. Chem. *439* (1978) 116
91   Fritz, G., Habel, D., Teichmann, G.: Z. anorg. allg. Chem. *303* (1960) 85
92   Fritz, G., Fröhlich, H.: Z. anorg. allg. Chem. *382* (1971) 217
93   Fritz, G., Fröhlich, H.: Z. anorg. allg. Chem. *382* (1971) 9
94   Fritz, G., Schmid, K.-H.: Z. anorg. allg. Chem. *441* (1978) 125
95   Stock, A., Somieski, C.: Ber. Dt. Chem. Ges. *50* (1917) 1739
96   Castenburg, A.: Liebigs Ann. Chem. *164* (1872) 300,
     Reynolds, H. H., Bigelow, C. A., Kraus, C. A.: J. Am. Chem. Soc. *51* (1929) 3067,
     Nebergall, W. H.: J. Am. Chem. Soc. *72* (1950) 4702
97   Fritz, G., Burdt, H.: Z. anorg. allg. Chem. *317* (1962) 35
98   Fritz, G., Speck, W., Keuthen, M., Peters, K., Schnering, H. G. v.: unpublished
99   Fritz, G., Grobe, J., Ksinsik, D.: Z. anorg. allg. Chem. *302* (1959) 175
100 Fritz, G., Keuthen, M.: unpublished; Keuthen, M.: PhD thesis, Karlsruhe 1985
101 Fritz, G., Szczepanski, N.: Z. anorg. allg. Chem. *367* (1969) 44
102 Fritz, G., Finke, U.: Z. anorg. allg. Chem. *430* (1977) 121
     Fritz, G., Finke, U.: Z. anorg. allg. Chem. *424* (1976) 238
104 Eberhard, G. G., Butte, W. A.: J. Org. Chem. *29* (1964) 2928
105 Fritz, G., Fröhlich, H., Kummer, D.: Z. anorg. allg. Chem. *353* (1967) 34;
     Fritz, G., Teichmann, G., Thielking, H.: Angew. Chem. *72* (1960) 209
106 Whitmore, F. G., Pietrusza, E. W., Sommer, L. H.: J. Am. Chem. Soc. *69* (1947) 2108
107 Dolgov, B. N., Borisov, S. N., Voronkov, M. G.: J. Gen. Chem. Moskow, *27* (1957) [716] 789
108 Kerr, J. A., Smith, B. A., Trotman-Dichenson, A. F., Young, J. G.: J. Chem. Soc., Chem. Comm. 1966, 157
109 Gilman, H., Melvin, H. W.: J. Am. Chem. Soc. *71* (1949) 4050
     West, R., Rochow, E. G.: J. Org. Chem. *18* (1953) 302
110 Gilman, H., Zuech, E. A.: J. Am. Chem. Soc. *79* (1957) 4570; ibid. *81* (1959) 5925
     Corriu, R. J., Massé, J. P. R., Meunier, U. B.: J. Organomet. Chem. *55* (1973) 73
111 Fritz, G., Grobe, J.: Z. anorg. allg. Chem. *309* (1961) 77
112 Fritz, G., Riekens, R., Günther, T., Berndt, M.: Z. Naturforsch. *26b* (1971) 480
113 Fritz, G., Himmel, W.: Z. anorg. allg. Chem. *448* (1979) 40
114 Fritz, G., Himmel, W.: Z. anorg. allg. Chem. *448* (1979) 55

115   Fritz, G., Braunagel, N.: Z. anorg. allg. Chem. *399* (1973) 280
116   Rice, M. J., Andrellos, P. J.: Naval Research, Contract ONR-494(04) (1956)
       Wiberg, E., Bauer, R.: Chem. Ber. *85* (1952) 593
       Wiberg, E., Strebel, P.: Liebigs Ann. Chem. *607* (1957) 9
       Dymova, T. N., Elisseeva, N. G.: Russ. J. Inorg. Chem. *8* (1963) 820
117   Schlosser, M.: Struktur und Reaktionsfähigkeit polarer Organometalle, Springer Verlag
       Heidelberg 1973
118   Huheey, J. E.: J. Phys. Chem. *69* (1965) 3284
119   Seyferth, D., Hanson, E. M., Armbrecht, F. M.: J. Organomet. Chem. *23* (1970) 361
       Bamford, W., Pant, B. C.: J. Chem. Soc. C 1967, 1470
120   Krieble, R. H., Elliot, J. R.: J. Am. Chem. Soc. *67* (1945) 1810
121   Fritz, G., Bosch, E.: Z. Naturforsch. *25b* (1970) 1313; Z. anorg. allg. Chem. *404* (1974) 103
122   Fritz, G., Chang, J. W., Braunagel, N.: Z. anorg. allg. Chem. *416* (1975) 211
123   Fritz, G., Böttinger, P.: Z. anorg. allg. Chem. *395* (1973) 159
124   Fritz, G., Portner, M.: Z. Naturforsch. *30b* (1975) 965
125   Fritz, G., Bublinski, G.: unpublished
126   Fritz, G., Wartanessian, S.: J. Organomet. Chem. *178* (1979) 11
127   Fritz, G., Wartanessian, S., Matern, E., Hönle, W., Schnering, H. G. v.: Z. anorg. allg. Chem.
       475 (1981) 87
128   Seyferth, D., Annarelli, D. C.: J. Am. Chem. Soc. *97* (1975) 7162
       Lambert jr., R. L., Seyferth, D.: J. Am. Chem. Soc. *94* (1972) 9248
129   Fritz, G., Thomas, J., Peters, K., Peters, E.-M., Schnering, H. G. v.: Z. anorg. allg. Chem.,
       514 (1984) 61
130   Fritz, G., Thomas, J.: J. Organomet. Chem., *271* (1984) 107
131   Fritz, G., Bauer, H.: Angew. Chem. *95* (1983) 740
132   Fritz, G., Braun, U., Schick, W., Hönle, W., Schnering, H. G. v.: Z. anorg. allg. Chem. *472*
       (1981) 45
133   Fritz, G., Schick, W., Hönle, W., Schnering, H. G. v.: Z. anorg. allg. Chem. *511* (1984) 95
134   Oberhammer, H., Becker, G., Gresser, G.: J. Mol. Struct. *75* (1981) 283
135   Fritz, G., Schick, W.: Z. anorg. allg. Chem. *511* (1984) 108
136   Fritz, G., Kemmerling, W., Sonntag, G., Becher, H. J., Ebsworth, E. A. V., Grobe, J.: Z. anorg.
       allg. Chem. *321* (1963) 10
137   Devine, A. M., Griffin, P. A., Haszeldine, R. N., Newlands, M. J., Tipping, A. E.: J. Chem.
       Soc., Dalton Trans. 1975, 1434
138   Fritz, G., Dannappel, H. J.: unpublished
       Dannappel, H. J.: PhD thesis, Karlsruhe 1972
139   Fritz, G., Matern, E.: Z. anorg. allg. Chem. *426* (1976) 28
140   Nametkin, N. S., Vdovin, V. M., Zavyalov, V., Grinberg, P. L.: Izv. Akad. Nauk, USSR,
       Ser. Khim. *929* [902] (1965)
141   Müller, R., Köhne, R., Beyer, H.: Chem. Ber. *95* (1962) 3030
       Kriner, W. A.: J. Org. Chem. *29* (1964) 1601
142   Devine, A. M., Griffin, P. A., Haszeldine, R. N., Newlands, M. J., Tipping, A. E.: J. Chem.
       Soc., Dalton Trans. 1975, 1822
143   Auner, N., Grobe, J.: Z. anorg. allg. Chem. *483* (1982) 53
144   Fritz, G., Dannappel, H. J., Matern, E.: Z. anorg. allg. Chem. *399* (1973) 263
145   Fritz, G., Kummer, D., Sonntag, G.: Z. anorg. allg. Chem. *342* (1966) 121
146   Fritz, G., Schober, P.: Z. anorg. allg. Chem. *374* (1970) 229
147   Fritz, G., Schober, P.: Z. anorg. allg. Chem. *372* (1970) 21
148   Fritz, G., Mittag, J.: Z. anorg. allg. Chem. *458* (1979) 37
149   Fritz, G., Haase, R., Kummer, D.: Z. anorg. allg. Chem. *365* (1969) 1
150   Schnering, H. G. v., Krahé, E., Fritz, G.: Z. anorg. allg. Chem. *365* (1969) 113
151   Fritz, G., Matern, E., Bock, H., Brähler, G.: Z. anorg. allg. Chem. *439* (1978) 173
       Bock, H., Brähler, G., Fritz, G., Matern, E.: Angew. Chem. *88* (1976) 765
152   Krieble, R. H., Elliott, J. R.: J. Am. Chem. Soc. *68* (1946) 2291
153   Chvalovsky, V., Bažant, V.: Coll. Czech. Chem. Commun. *21* (1956) 93
154   Ishikawa, M., Kumada, M., Sakurai, H.: J. Organomet. Chem. *23* (1970) 63
155   Sakurai, H., Watanabe, T., Kumada, M.: J. Organomet. Chem. *7* (1967) P 15

156   Beck, K. R., Benkeser, R. A.: J. Organomet. Chem. *21* (1970) P 35
157   Hengge, E., Peter, W.: J. Organomet. Chem. *148* (1978) C 22
      Hengge, E., Schuster, H. G., Peter, W.: J. Organomet. Chem. *186* (1980) C 45
158   Bordeau, M., Djamei, S. M., Dunogues, J., Calas, R.: Bull. Soc. Chim. France *1982*, II-159
159   Fritz, G., Brandwirth, H., Honold, J.: unpublished
160   Fritz, G., Honold, J.: unpublished
161   Chalk, A. J., Harrod, F. J.: J. Am. Chem. Soc. *87* (1965) 16
162   Aylett, B. J.: Organometallic Compounds, Fourth Edition, Vol. 1, Part 2, Chapman and Hall, London 1979
163   Fritz, G., Hohenberger, K.: Z. anorg. allg. Chem. *464* (1980) 107
164   Malisch, W., Kuhn, M.: Chem. Ber. *107* (1974) 979
165   Nicholson, B. K., Simpson, J.: J. Organomet. Chem. *155* (1978) 237
166   Abraham, K. M., Urry, G.: Inorg. Chem. *12* (1973) 2850
167   Baay, Y. L., MacDiarmid, A. G.: Inorg. Chem. *8* (1969) 986
168   Fritz, G., Burkhardt, F.: unpublished
      Burkhardt, F.: PhD thesis, Karlsruhe 1982
169   Fritz, G., Matern, E., Hönle, W., Schnering, H. G. v.: to be published
170   Peters, K., Schnering, H. G. v.: Z. anorg. allg. Chem. *502.*(1983) 55
171   Chalk, H. J., Harrod, J. F.: J. Am. Chem. Soc. *87* (1965) 1133
172   Lipka, A., Schnering, H. G. v.: Z. anorg. allg. Chem. *419* (1976) 20
173   Lipka, A., Schnering, H. G. v.: Z. anorg. allg. Chem. *419* (1976) 9
174   Hönle, W., Schnering, H. G. v.: Z. anorg. allg. Chem. *464* (1980) 139
175   Allmann, R.: Monatsh. Chem. *106* (1975) 779
176   Schomaker, V., Stevenson, D. P.: J. Am. Chem. Soc. *63* (1941) 37
177   Schmidbaur, H.: Acc. Chem. Res. *8* (1974) 62
178   Karsch, H. H., Zimmer-Gasser, B., Neugebauer, D., Schubert, U.: Angew. Chem. *91* (1979) 519
179   Fritz, G., Lauble, S.: unpublished
180   Bartell, L. S., Brockway, L. O.: J. Chem. Phys. *32* (1960) 512
181   Mootz, D., Zinnius, A., Böttcher, B.: Angew. Chem. *81* (1969) 398
182   Krahé, E. W., Mattes, R., Tebbe, K.-F., Schnering, H. G. v., Fritz, G.: Z. anorg. allg. Chem. *393* (1972) 74
183   Sawitzki, G., Schnering, H. G. v.: Z. anorg. allg. Chem. *425* (1976) 1
184   Schnering, H. G. v., Lipka, A., Krahé, E. W.: Z. anorg. allg. Chem. *419* (1976) 27
185   Fritz, G., Wörns, K.-P., Peters, K., Peters, E.-M., Schnering, H. G. v.: Z. anorg. allg. Chem. *512* (1984) 126
186   Klaus, R. O., Tobler, H., Ganter, C.: Helv. Chim. Acta *57* (1974) 2517
187   Compound *69* is named "wurtzitane", because "iceane" and "barrelane" together form a complete wurtzite skeleton.
188   Fritz, G., Hähnke, M.: Z. anorg. allg. Chem. *390* (1972) 191
189   Fieser, L. F.: J. Chem. Educ. *42* (1965) 408
190   Schnering, H. G. v., Sawitzki, G., Peters, K., Tebbe, K. F.: Z. anorg. allg. Chem. *404* (1974) 38
191   Peters, K., Peters, E.-M., Schnering, H. G. v.: Z. anorg. allg. Chem. *502* (1983) 61
192   Pauling, L.: Die Natur der chemischen Bindung, 2. ed. Verlag Chemie Weinheim 1964, p. 227
193   Vilkov, L. V., Kusakov, M. M., Nametkin, N. S., Oppengeim, K. D.: Dokl. Akad. Nauk. USSR *183* (1968) 830
194   Parkanyi, L., Sasvari, K., Barta, I.: Acta Chryst. *B34* (1978) 883
195   Hurt, C. J., Calabrese, J. C., West, R.: J. Organomet. Chem. *91* (1975) 273
196   Pohl, S.: Z. Naturforsch. *34b* (1979) 256
197   Robinson, E. A., Gillespie, R. J.: J. Chem. Educ. *57* (1980) 329
198   Baldwin, A. C., Davidson, I. M. T., Reed, M. D.: J. Chem. Soc., Faraday Trans. *74* (1978) 2171
199   Fritz, G., Volk, H., Peters, K., Peters, E.-M., Schnering, H. G. v.: in print
200   Brown, I. D.: Bidies 1969—77, Inst. for Materials Research, Mc Master Univ. Hamilton, Ontario, Canada
201   Schulz, H., Thiemann, K. H.: Solid State Comm. *32* (1979) 783
202   Jansen, R., Oskam, A., Olie, K.: Cryst. Struct. Comm. *4*(1975) 667

258   References

203   Smith, R. A., Bennett, M. J.: Acta Cryst. *B33* (1977) 1118
204   Manojlovic-Muir, L., Muir, K. W., Ibers, J. A.: Inorg. Chem. *9* (1970) 447
205   Schubert, U., Rengstil, A.: J. Organomet. Chem. *166* (1979) 323
206   Vancea, L., Bennett, M. J., Jones, C. E., Smith, R. A., Graham, W. A. G.: Inorg. Chem. *16* (1977) 897
207   Schnering, H. G. v., Lipka, A.: unpublished
        Lipka, A.: Diploma Thesis, Münster 1972
208   Vajda, E., Kolonits, M., Rozsondai, B., Fritz, G., Matern, E.: J. Mol. Struct. *95* (1982) 197
209   Vajda, E., Kolonits, M., Fritz, G., Thomas, J., Sattler, E.: J. Mol. Struct. *117* (1984) 329
210   Glidewell, C., Liles, D. C.: J. Organomet. Chem. *234* (1982) 15
211   Shiki, Y., Kuginuki, Y., Hasegawa, A., Hayaski, M.: J. Mol. Spectrosc. *73* (1978) 9
        Almenningen, A., Seip, H. M., Seip, R.: Acta Chem. Scand. *24* (1970) 1697
212   Schmidbaur, H., Zimmer-Gasser, B.: Z. Naturforsch. *32b* (1977) 603